Bioengineering: A Conceptual Approach

Bioengineering: A Conceptual Approach

Edited by Billy Malcolm

SYRAWOOD
PUBLISHING HOUSE

New York

Published by Syrawood Publishing House,
750 Third Avenue, 9th Floor,
New York, NY 10017, USA
www.syrawoodpublishinghouse.com

Bioengineering: A Conceptual Approach
Edited by Billy Malcolm

International Standard Book Number: 978-1-64740-086-6 (Hardback)

Cataloging-in-Publication Data

Bioengineering : a conceptual approach / edited by Billy Malcolm.
    p. cm.
Includes bibliographical references and index.
ISBN 978-1-64740-086-6
1. Bioengineering. 2. Synthetic biology. 3. Engineering. 4. Biology. I. Malcolm, Billy.
TA164 .B56 2022
660.6--dc23

TABLE OF CONTENTS

PREFACE

Bioengineering or biological engineering is the application of the principles of engineering and biology to create tangible, usable and economically viable products. Bioengineering can be applied to systems at molecular, cellular, tissue-based, organismal and entire ecosystem levels. The design of bioengineering products is often inspired from biological systems or based on their modification and control. It is vital for the development of medical diagnostic equipment and medical imaging technology, prosthetics, renewable bioenergy, biocompatible materials, ecological engineering, etc. It can also be used in different areas of engineering, biotechnology, biocatalysis and bioprocess engineering. Some of the major branches of bioengineering are biomedical engineering, biological systems engineering, environmental health engineering, biorobotics, biomimetics, etc. The aim of this book is to present researches that have transformed the discipline of bioengineering and aided its advancement. From theories to research to practical applications, case studies related to all contemporary topics of relevance to this field have been included herein. Through this book, we attempt to further enlighten the readers about the new concepts in this field.

This book unites the global concepts and researches in an organized manner for a comprehensive understanding of the subject. It is a ripe text for all researchers, students, scientists or anyone else who is interested in acquiring a better knowledge of this dynamic field.

I extend my sincere thanks to the contributors for such eloquent research chapters. Finally, I thank my family for being a source of support and help.

Editor

Disilicate Dental Ceramic Surface Preparation by 1070 nm Fiber Laser: Thermal and Ultrastructural Analysis

Carlo Fornaini [1,2,*] , Federica Poli [1] , Elisabetta Merigo [2] , Nathalie Brulat-Bouchard [2], Ahmed El Gamal [2], Jean-Paul Rocca [2], Stefano Selleri [1] and Annamaria Cucinotta [1]

[1] Department of Engineering and Architecture, University of Parma, Parco Area delle Scienze 181/A, 43124 Parma, Italy; federica.poli@unipr.it (F.P.); stefano.selleri@unipr.it (S.S.); annamaria.cucinotta@unipr.it (A.C.)
[2] Micoralis Laboratory, Faculty of Dentistry, University of Cote d'Azur, 24 Avenue des Diables Bleus, 06357 Nice, France; elisabetta.merigo@gmail.com (E.M.); Nathalie.BRULAT@unice.fr (N.B.-B.); Ahmed.ELGAMAL@unice.fr (A.E.G.); Jean-Paul.ROCCA@unice.fr (J.-P.R.)
* Correspondence: carlo@fornainident.it

Abstract: Lithium disilicate dental ceramic bonding, realized by using different resins, is strictly dependent on micro-mechanical retention and chemical adhesion. The aim of this in vitro study was to investigate the capability of a 1070 nm fiber laser for their surface treatment. Samples were irradiated by a pulsed fiber laser at 1070 nm with different parameters (peak power of 5, 7.5 and 10 kW, repetition rate (RR) 20 kHz, speed of 10 and 50 mm/s, and total energy density from 1.3 to 27 kW/cm^2) and the thermal elevation during the experiment was recorded by a fiber Bragg grating (FBG) temperature sensor. Subsequently, the surface modifications were analyzed by optical microscope, scanning electron microscope (SEM), and energy dispersive X-ray spectroscopy (EDS). With a peak power of 5 kW, RR of 20 kHz, and speed of 50 mm/s, the microscopic observation of the irradiated surface showed increased roughness with small areas of melting and carbonization. EDS analysis revealed that, with these parameters, there are no evident differences between laser-processed samples and controls. Thermal elevation during laser irradiation ranged between 5 °C and 9 °C. A 1070 nm fiber laser can be considered as a good device to increase the adhesion of lithium disilicate ceramics when optimum parameters are considered.

Keywords: disilicate ceramics; fiber lasers; fiber Bragg grating; energy dispersive X-ray spectroscopy

1. Introduction

The demand for ceramic prosthetic restorations has become increasingly common in daily dentistry. Moreover, the continuous need for an increased level of precision, particularly in cosmetic dentistry, where new materials—such as feldspathic ceramics—play an important role in prosthetic rehabilitations, is considered crucially important. Unfortunately, failure resulting from porcelain fracture has been reported as ranging from 2.3 to 8% [1]. Nevertheless, it seems to be a function of a multi-factorial reason [2–4], with the key cause attributed to the composite resin adhesion with porcelain. Therefore, it is necessary to condition the ceramic surface, which is considered very interesting [5,6].

The inside surface of the ceramic prosthetics must be conditioned for optimized micro-mechanical retention by the resin penetration into the ceramic micro-roughness; this treatment enhances the mechanical retention of cement by enlarging the surface in contact with the tooth structure through the creation of micro-porosities [7,8]. For producing surface roughness and for promoting micro-mechanical retention, different treatment methods, such as diamond roughening, air-particle abrasion with aluminium oxide, and acid etching have been proposed in the literature [7,8]. All these techniques have been investigated under in vitro conditions [9–11].

The use of laser technology for surface treatment has already been successfully applied in many industrial applications by the utilization of high power sources. Today, this technology represents a controllable and flexible technique for the modification of surface properties for different various materials [12,13], since laser parameters have the capability to influence and alter the surface microstructure [14]. The in vitro study reported here has the aim to verify the possibility of performing the surface treatment of Lithium disilicate ceramic specimens by the irradiation of a 1070 nm pulsed fiber laser.

Fiber lasers act as sources where the active medium is an optical fiber with a core doped with active ions, such as Nd (neodymium), Yb (ytterbium), Er (erbium), or Tm (thulium) [15]. Fiber lasers differ from traditional solid-state lasers mainly by the form of the gain medium: in fact, bulk crystal lasers are typically based on conventional rod or slab geometries while, in the case of fiber lasers, active ions are added into the core of an optical fiber, often with a length of many meters [16]. These lasers operate in continuous wave (CW) or pulsed mode and emit in a wide range of wavelengths, which is a function of the dopants and host materials. CW output powers of several kW [17] and pulse energies up to around 30 mJ [18,19] can be currently obtained with Yb-doped fiber lasers.

2. Materials and Methods

The circular faces of twelve cylinders of lithium disilicate ceramics (e.max Press, Ivoclar, Bolzano, Italy) with a 10 mm diameter and an 8 mm length were processed into three 3×3 mm square zones by using a 1070 nm pulsed fiber laser (AREX 20, Datalogic, Bologna, Italy). This source has a maximum average output power of 20 W and a fixed pulse duration of 100 ns, thus providing a maximum peak power of 10 kW for a repetition rate of 20 kHz.

The chemical composition of the ceramic used in the study, as described by the company, consists of SiO_2 as the primary component (57–80%) and Li_2O (11–19%), K_2O (0–13%), P_2O_5 (0–11%), ZrO_2 (0–8%), ZnO (0–8%) as additional components.

After a preliminary pilot study, performed by optical microscope to evaluate the effects of different parameters, we used a repetition rate (RR) of 20 kHz; output powers of 5, 7.5 and 10 kW; and speeds of 10 and 50 mm/s. Thus, the cylinders were divided in two groups: six samples were irradiated at a speed of 10 mm/s (group A) and six at 50 mm/s (group B). On each sample of the two groups three zones were prepared at 5, 7.5 and 10 kW, respectively (Figure 1). As control group, a zone of the cylinders not irradiated by laser device was analyzed.

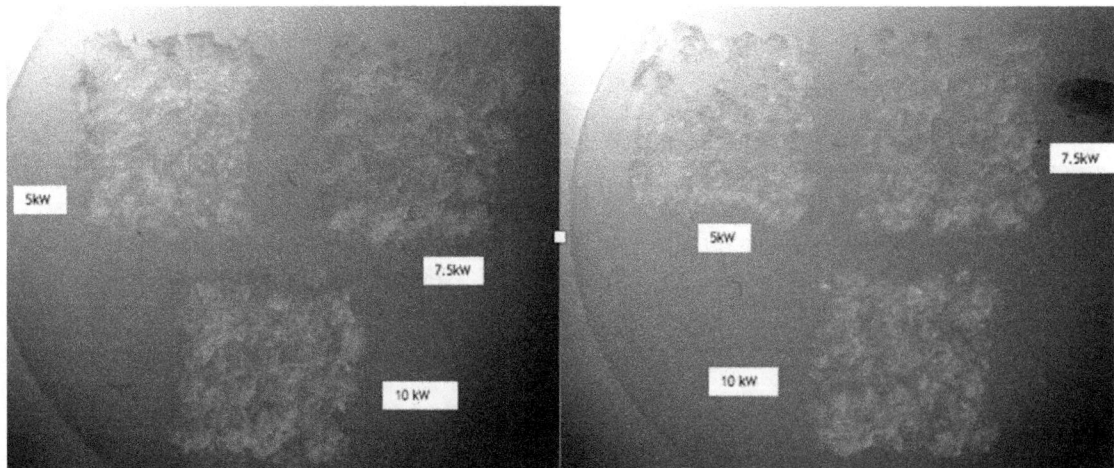

Figure 1. (**Left**) A Group A sample irradiated at 10 mm/s; (**Right**) A Group B sample irradiated at 50 mm/s.

The lens used with the AREX 20 laser has a focal length of 160 mm. In this configuration, the laser beam has a spot-size of 80 μm. Each square zone on the sample surface has been processed using a meshed filling pattern with a distance between lines of 0.03 mm. The laser beam focalization was checked by a metal cylinder of the same dimension of the samples. The power per unit area deposited on the material ranged between 1.3 and 27 kW/cm^2. The specimens were subsequently observed by an optical microscope (Olympus MTV-3, Tokyo, Japan), then metallized (ion sputter, Jeol JFC 1100E, Peabody, MA, USA) and analyzed by a Scanning Electron Microscope (SEM) (JSM-35CF, Jeol Ltd., Tokyo, Japan) and energy-dispersive X-ray spectroscopy (EDS) system (JED 2300F, Jeol Europe, Croissy-sur-Seine, France)

During the irradiation of the sample with the best laser parameters, the thermal elevation was recorded by a fiber Bragg grating (FBG)-based temperature sensor connected to an interrogator. The fiber sensor was positioned into the groove in the middle of the sample. A sm130-500 dynamic optical sensing interrogator (Micron Optics Inc., Atlanta, GA, USA) was used to measure the FBG wavelength shift induced by the temperature increase). This device is also considered as a compact, industrial-grade, dynamic optical sensor interrogation module, field-proven for robust, reliable, and long-term operation. The software included with the sensing interrogator system provides a single suite of tools for data acquisition, computation, and analysis of optical sensor networks. A 25 mm-long FBG with a center wavelength of 1550 nm, a reflectivity of 96%, and acrylate coating, imprinted in a standard SMF (AOS GmbH, Dresden, Germany), has been connected to the interrogator for performing the temperature change measurement. A temperature-induced wavelength shift of about 13 pm/°C has been considered for the FBG at 1550 nm.

The temperature increase during the laser irradiation has been measured only when the source operates with the best parameters, as per the observation of SEM and EDS analysis. The aim of this measurement was to provide the maximum value of the temperature rise, induced by the laser, that the disilicate ceramic material can withstand without being damaged. Higher-energy laser treatments provide a more significant temperature change, which is associated with the detrimental surface modifications, as shown by SEM and EDS analysis. With the FBG sensor it was also possible to see the time of the thermal elevation, also after the laser irradiation, until it decreased completely. The thermal elevation of the sample during the irradiation with the laser operating at a peak power of 5 kW, repetition rate of 20 kHz, and speed of 50 mm/s, has been recorded with a FBG connected to an interrogator.

3. Results

3.1. SEM Observation

By comparing the control group (non-irradiated samples) to the cylinders processed by the fiber laser at higher magnification, a greater difference can be noticed (Figure 2).

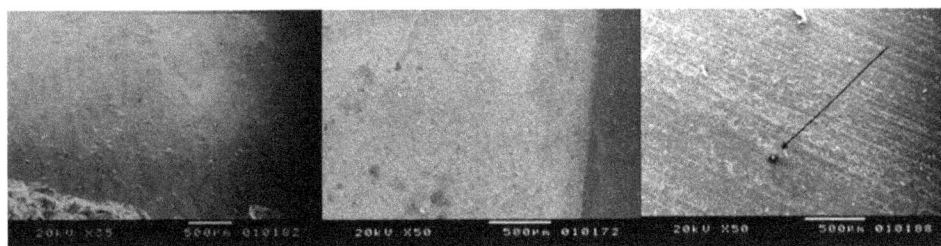

Figure 2. (**Left**) Non-irradiated sample; (**centre**) Peak power of 7.5 kW and 50 mm/s speed; (**right**) Peak power of 7.5 kW and 10 mm/s speed with a carbonization spot ((**left**) X35; (**centre,right**): X50).

In fact, all the treated surfaces show a rough surface with many holes and irregularities. It is evident that the samples irradiated at different lasing parameters experienced some areas of melting and burning when the highest energy level was used, due to the cumulative effect of the laser energy. The presence of some cracks with variable intensities are also found due to the thermal effects of laser irradiation (Figures 3–6). Only the samples of Group B irradiated at 5 kW did not show any thermal damages (Figure 7).

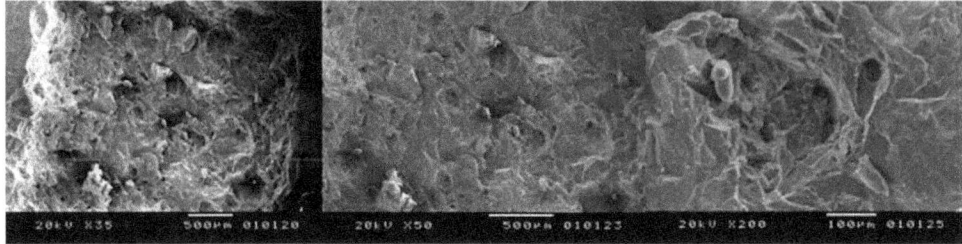

Figure 3. Peak power of 10 kW, speed of 10 mm/s: many zones with melting and carbonization are shown (**left**) X35; (**centre**) X200; (**right**) X500.

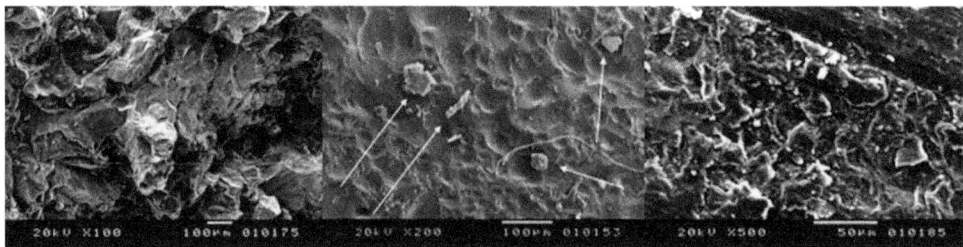

Figure 4. Peak power of 10 kW, speed of 50 mm/s: some points with melting are shown (**left**) X100; (**centre**) X200; (**right**) X500.

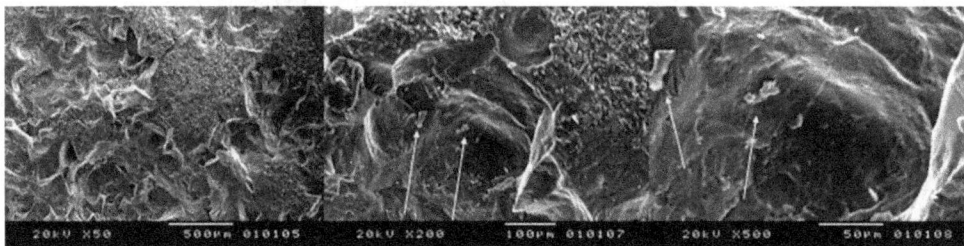

Figure 5. Peak power of 7.5 kW, speed of 50 mm/s: presence of melting and carbonization in some areas of the sample (**left**) X35; (**centre**) X50; (**right**) X200.

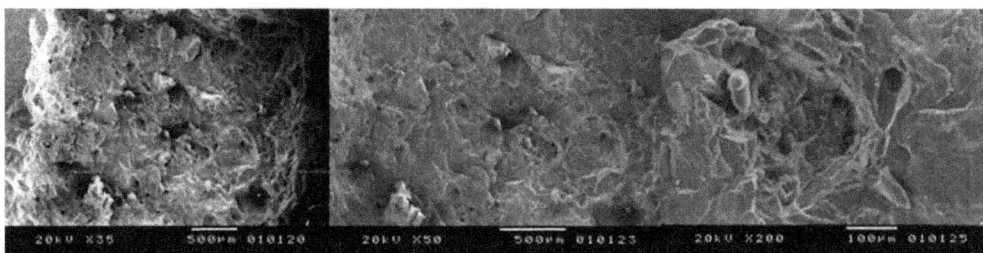

Figure 6. Peak power of 5 kW, speed of 10 mm/s: evidence of some zones with melting (**left**) X50; (**centre**) X100; (**right**) X500.

Figure 7. Peak power of 5 kW, speed of 50 mm/s: no evidence of carbonization and melting zones (**left**) X75; (**centre**) X100; (**right**) X200.

3.2. EDS Analysis

The EDS analysis consists of the percentage recording of chemical elements in the point where the probe is placed. Analyzed samples showed, in general, slight differences in the chemical composition between control groups and irradiated samples, even smaller variations by changing the lasing parameters were detected, thus confirming the information given by the SEM observation.

The differences of elemental composition between the non-irradiated areas in the different samples may be explained by the structure of the ceramic, which is not homogeneous, thus resulting in structural variations of the tested zones (Figure 1, left). The samples treated with the laser operating at a peak power of 10 kW, repetition rate of 20 kHz, and speed of 50 mm/s experienced some zones (red spots) of lower percentage of C when compared to the control group. On the other hand, O and Al elements were slightly higher in the affected zones (Figure 8).

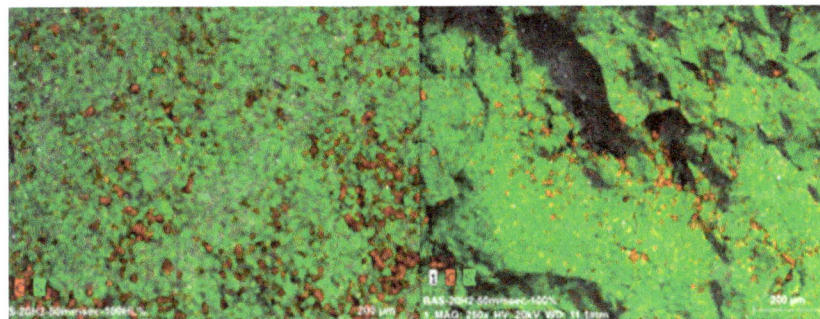

	C	O	Si	K	Al	Na
Laser	7.4	44.4	23.7	8.0	6.7	5.5
Non laser	16.7	41.4	25.7	6.1	4.3	3.4

Figure 8. (**Left**) Control group and (**right**) samples irradiated with peak power of 10 kW, and a speed of 50 mm/s: the C concentration is shown in red.

The samples irradiated with a peak power of 7.5 kW, repetition rate of 20 kHz, and speed of 50 mm/s showed that only the carbon concentration was higher in the control group (13.6%), while all the other elements—such as O, Si, K, Al, and Na—presented higher concentration values on the treated surfaces (Figure 9).

These data demonstrated a poor modification of the ceramic chemical structure caused by the laser operating with the optimum parameters (Figure 10).

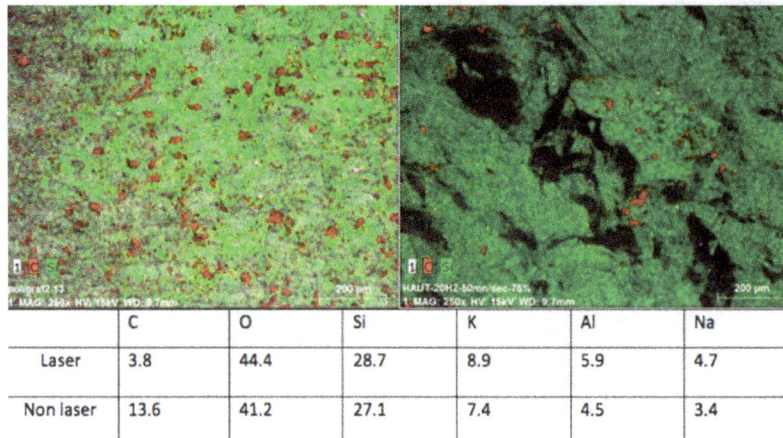

	C	O	Si	K	Al	Na
Laser	3.8	44.4	28.7	8.9	5.9	4.7
Non laser	13.6	41.2	27.1	7.4	4.5	3.4

Figure 9. (**Left**) Control group and (**right**) samples irradiated with peak power of 7.5 kW and speed of 50 mm/s: the C concentration is shown in red.

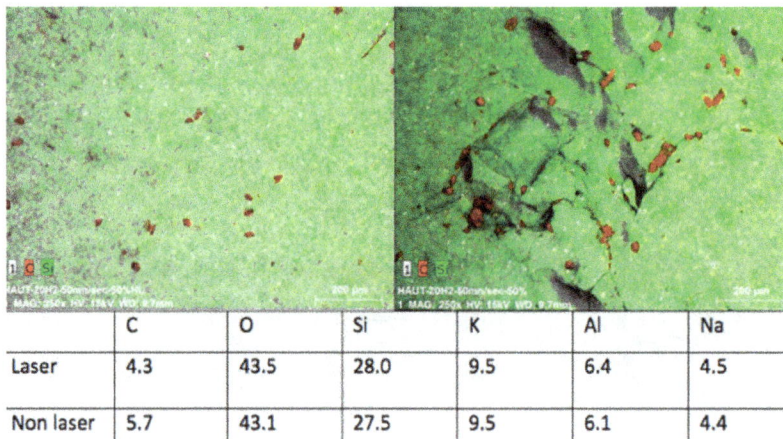

	C	O	Si	K	Al	Na
Laser	4.3	43.5	28.0	9.5	6.4	4.5
Non laser	5.7	43.1	27.5	9.5	6.1	4.4

Figure 10. (**Left**) Control group and (**right**) samples irradiated with peak power of 5 kW and a speed of 50 mm/s: the C concentration is shown in red.

3.3. Thermal Analysis

The FBG wavelength shift obtained in a time interval of 120 s, during the laser processing, is reported in Figure 11. The temperature measurement has been repeated three times by processing three square regions on the sample surface. The fiber sensor was placed in the center of the sample, approximately at the same distance from all the areas irradiated by the laser. Notice that the wavelength shift measured by the interrogator is between 65 pm and 115 pm, respectively, in the first and the third test. Consequently, the temperature rise due to the laser processing is between 5 °C and 9 °C. The slight growth of the temperature value measured in the second and the third test may be due to the gradual heating of the sample, originating from the previous laser processing. Moreover, slight differences in the distance of the three zones irradiated by the laser with respect to the sensitive part of the fiber sensor must be taken into consideration.

The measure of the temperature rise during laser irradiation may shed some light on the explanation behind the crack formations. After laser irradiation, which could be explained through the high thermal effects of laser processing, along with the consequence of an extreme physical stress in the re-hardening ceramic surface. The importance of the very short pulse duration given by the fiber laser used in this study (100 ns) must also be underlined, which may explain the greater difference between the fluences of these tests, compared to those given in the cited works where irradiation had been performed in CW or in μs.

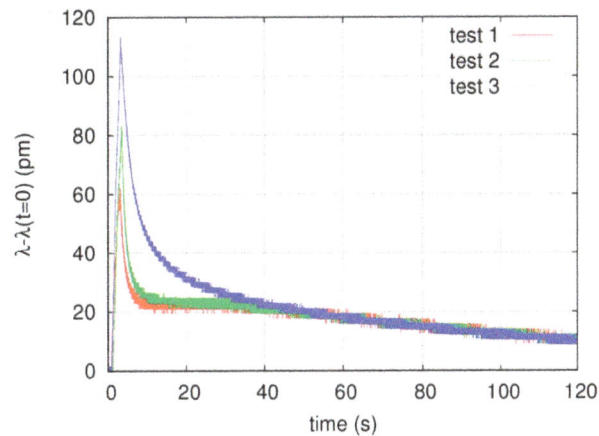

Figure 11. FBG sensor wavelength shift induced by temperature variations during and after the laser irradiation with the best parameters (peak power of 5 kW, repetition rate of 20 kHz, and speed of 50 mm/s).

4. Discussion

The current adhesion strategies using resin bonding for ceramics involve micromechanical interlocking and chemical adhesive bonding and, thus, the ceramic surface requires surface roughening for mechanical bonding and surface activation for chemical adhesion. The search for non-destructive methods to treat inert ceramics for modifying their surface texture and chemical properties help to produce an activated/functionalized surface [20]. Hydrofluoric acid etching changes the surface of the glass ceramic by dissolving its glassy phase [21], and this process creates irregularities on the surface and increases the contact area between the adhesive system and the glass ceramic.

The adhesion between dental ceramics and resin-based composites is the result of a physicochemical interaction across the interface between the resin (adhesive) and the ceramic (substrate). The physical contribution to the adhesion process is dependent on the surface treatment and topography of the substrate and can be characterized by its surface energy. Alteration of the surface topography by etching or airborne particle abrasion results in changes on the surface area and on the wettability of the substrate, which are related to the surface energy and the adhesive potential [22]. A rough surface increases the mechanical retention by enabling the adhesive to interlock with the surface irregularities created by the different conditioning methods [23]. Unfortunately, several studies demonstrated the possibility, by hydrofluoric acid, to damage the ceramic structure with the result increasing the fracture of the restorations [24].

Even the utilization of laser technology, recently introduced for this kind of treatment, is not free of problems. Particularly, some tests conducted on lithium disilicate [25] and CAD-CAM ceramics [26] with a CW CO_2 laser at 10.6 µm confirmed the presence of micro-cracks and melting textures, due to the thermal effect of the laser irradiation at output powers higher than 10 W CW (3184.7 W/cm^2). Moreover, the observation of the ceramics structure irradiated by a 10 W (14,185 W/cm^2) pulsed Nd:YAP laser at 1340 nm exhibited the presence of holes, micro-cracks, and melted grains [25,26]. This is probably caused by the effect of a high quantity of radiation energy given in a well-defined portion of the ceramic surface over a short period, thus leading to a very high energy density accumulation. Micro-crack formation on ceramics after CO_2 and Nd:YAP laser irradiations may be related to the high thermal effects of laser processing which leads to an extreme physical stress in the re-hardening ceramic surface [27,28]. Additionally, an Er:YAG laser was used for surface treatment of feldspathic porcelain, however, its effect resulted in a significantly weaker surface than that of the HF-treated surface. The probable assumption is that the laser energy from an Er:YAG laser is not well absorbed by the porcelain and, therefore, not sufficient to create a micro-mechanical retention pattern for more favorable bonding [29]. In agreement with this study, some authors affirmed that, even at a very high energy

(500 mJ), an Er:YAG laser is not able to cause on the porcelain surface to roughen sufficiently to promote reliable adhesion to the resin composite [30]. Recently, the so-called "ultra-short pulses" lit up a greater interest in the field of the mean roughness value [31]. However, due to the higher expense associated with this laser source, to date, it is still utilized in only a few laboratories.

This study, based on the utilization of a 1070 nm pulsed fiber laser, demonstrated the ability, with the described optimal parameters, to characterize the ceramic surface for enhancing its adhesion to the enamel and dentine avoiding, at the same time, the thermal damages in the structure evidenced with the utilization of the other different wavelengths before described.

The most common applications of fiber lasers regard the industrial field, where they are used mainly for material processing (i.e., for cutting and marking). The main utilizations of fiber lasers in medicine are related to lithotripsy [32], surgical treatment of vascular lesions [33], non-surgical skin aesthetic procedures [34,35], and eye surgery [36]. Recently, its use in the dental field also started to be considered, particularly in the surgery of soft oral tissues, where it demonstrated some advantages consisting of the scant overheating of the target and, consequently, scant tissue damages, probably also due to the shorter pulse duration (ns), compared to the emission normally used in dentistry (μs) [37].

This is also the reason for the great differences in the power densities utilized in this study (1.3 to 27 kW/cm^2), compared to those used in the similar cited works [25,26] performed with different wavelengths. The measure of temperature rise during laser irradiation may shed some light on the explanation behind the crack formation after laser irradiation, which could be explained through the high thermal effects of laser processing, along with the consequence of an extreme physical stress in the re-hardening ceramic surface.

In fact, main process parameters in the laser–material interaction involve laser pulse duration and, consequently, it significantly affects the quality of the produced micro-feature and the material removal rate.

Ablation of material can be facilitated by using short pulses as the laser energy is confined in a thin layer. For longer pulses, absorbed energy will be dissipated into the surrounding material by thermal processes. Absorption of long laser pulses also causes melting and substantial sputter evaporation of the material. These phenomena can contaminate the surrounding area, produce micro-cracks, and remove material over dimensions much larger than the laser spot. Other adverse effects include damage to adjacent structures, delamination, formation of recast material, and formation of a large heat affected zone (HAZ) [38]. This is also the reason for the low and short thermal elevation recorded by FBG during the irradiation of our samples, having a peak (5 to 9°) and coming back to the normality in 8 to 38 s as reported in Figure 10.

It must also be underlined once more that the importance of the very short pulse duration given by the fiber laser used in this study (100 ns), which may explain the greater difference between the fluences of these tests, compared to those given in the cited works where irradiation had been performed in CW or in μs.

In this research, the laser parameters which seem to be the most effective for surface conditioning of the materials without causing any damages are a peak power of 5 kW, a repetition rate of 20 kHz, and a speed of 50 mm/s. In fact, the samples irradiated with these parameters revealed a rough surface with holes, irregularities, cavities, and recesses, while the presence of thermal damaging effects—such as melting, burning, and cracks—was not evident.

EDS observations of the samples irradiated with these parameters confirmed the SEM observation. The analysis, in this case, was conducted in four different zones and the results showed slight differences for all the elements concentrated in each analyzed zone, except for the decreasing of the C concentration in the irradiated samples, which is a result present also in other similar studies performed by different kinds of treatment, both thermal and non-thermal [25,39].

This study demonstrated that, by the proper parameters (peak power of 5 kW, repetition rate of 20 kHz, and speed of 50 mm/s), a 1070 nm pulsed fiber laser may characterize the surface of lithium disilicate ceramics to enhance their resin adhesion, with a low and very short thermal increase and without damaging and modifying their chemical structure. The aim of this study was to investigate the possibility to condition the ceramic surfaces to improve their adhesion to the hard dental tissues. Despite further studies being necessary to fully confirm this hypothesis, both from the point of view of the mechanical properties of irradiated ceramic surfaces (micro-hardness, roughness) and the adhesion characteristics after ceramic sealing (wettability, shear bond strength, and micro-leakage), this preliminary study may be considered a demonstration of the hypothesis.

5. Conclusions

The use of a 1070 nm pulsed fiber laser represents a new interesting opportunity in the field of dentistry, thanks to the possibility of emission in ns pulse duration, limiting the collateral damage due to the overheating of the target, both tissues and materials. This preliminary study, performed on disilicate ceramic samples, showed the capacity of this device, with the proper parameters described before, to condition the dental porcelain, making it rougher without, by proper parameters, modifying its structure, and overheating and consequently damaging and destroying its surface layer.

Acknowledgments: The authors would like to thank the company Datalogic S.p.A. and, particularly, Lorenzo Bassi for providing the fiber laser source.

Author Contributions: S.S. and A.C. conceived and designed the experiments; E.M., C.F. and F.P. performed the experiments; J.-P.R. and N.B.-B. analyzed the data; A.E.G. contributed materials; and C.F. wrote the paper.

Conflicts of Interest: The authors declare no conflict of interest.

References

1. Cheung, G.S. A preliminary investigation into the longevity and causes of failure of single unit extracoronal restorations. *J. Dent.* **1991**, *19*, 160–163. [CrossRef]
2. Ulusoy, M.; Toksavul, S. Fracture resistance of five different metal framework designs for metal-ceramic restorations. *Int. J. Prosthodont.* **2002**, *15*, 571–574. [PubMed]
3. Kang, M.S.; Ercoli, C.; Galindo, D.F.; Graser, G.N.; Moss, M.E.; Tallents, R.H. Comparison of the load at failure of soldered and non-soldered porcelain-fused-to-metal crowns. *J. Prosthet. Dent.* **2003**, *90*, 235–240. [CrossRef]
4. Michalakis, K.X.; Stratos, A.; Hirayama, H.; Kang, K.; Touloumi, F.; Oishi, Y. Fracture resistance of metal ceramic restorations with two different margin designs after exposure to masticatory simulation. *J. Prosthet. Dent.* **2009**, *102*, 172–178. [CrossRef]
5. Appeldoom, R.E.; Wilwerding, T.M.; Barkmeier, W.W. Bond strength of composite resin to porcelain with newer generation porcelain repair systems. *J. Prosthet. Dent.* **1993**, *70*, 6–11. [CrossRef]
6. Demirel, F.; Muhtarogullari, M.; Yuksel, G.; Cekiç, C. Microleakage study of 3 porcelain repair materials by autoradiography. *Quintessence Int.* **2007**, *38*, 285–290.
7. Giordano, R.; McLaren, E.A. Ceramics overview: Classification by microstructure and processing methods. *Compend. Contin. Educ. Dent.* **2010**, *31*, 682–684. [PubMed]
8. Albakry, M.; Guazzato, M.; Swain, M.V. Fracture toughness and hardness evaluation of three pressable all-ceramic dental materials. *J. Dent.* **2003**, *31*, 181–188. [CrossRef]
9. Chung, K.H.; Hwang, Y.C. Bonding strengths of porcelain repair systems with various surface treatments. *J. Prosthet. Dent.* **1997**, *78*, 267–274. [CrossRef]
10. Suliman, A.H.; Swift, E.J., Jr.; Perdigao, J. Effects of surface treatment and bonding agents on bond strength of composite resin to porcelain. *J. Prosthet. Dent.* **1993**, *70*, 118–120. [CrossRef]
11. Diaz-Arnold, A.M.; Schneider, R.L.; Aquilino, S.A. Bond strengths of intraoral porcelain repair materials. *J. Prosthet. Dent.* **1989**, *61*, 305–309. [CrossRef]
12. Holand, W.; Schweiger, M.; Frank, M.; Rheinberger, V. A comparison of the microstructure and properties of the IPS empress 2 and the IPS empress glass ceramics. *J. Biomed. Mater. Res.* **2000**, *53*, 297–303. [CrossRef]
13. Della Bona, A.; Mecholsky, J.J., Jr.; Anusavice, K.J. Fracture behavior of lithia disilicate and leucite based ceramics. *Dent. Mater.* **2004**, *20*, 956–962. [CrossRef] [PubMed]

14. Piwowarczyk, A.; Ottl, P.; Lauer, H.C.; Kuretzky, T. A clinical report and overview of scientific studies and clinical procedures conducted on the 3 M ESPE lava all-ceramic system. *J. Prosthodont.* **2005**, *14*, 39–45. [CrossRef] [PubMed]

15. Tuchin, V. *Dictionary of Biomedical Optics and Biophotonics*; SPIE Press: Bellingham, WA, USA, 2012; pp. 394–397. ISBN 0819489735.

16. Pierce, M.; Jackson, S.; Golding, P.; Dickinson, B.; Dickinson, M.; King, T.; Sloan, P. Development and application of fiber lasers for medical applications. In Proceedings of the 2001 International Symposium on Biomedical Optics, San Jose, CA, USA, 20–26 January 2001; SPIE Press: Bellingham, WA, USA, 2001.

17. Shiner, B. The impact of fiber laser technology on the world wide material processing market. In Proceedings of the CLEO: Applications and Technology 2013, San Jose, CA, USA, 9–14 June 2013; Optical Society of America: Washington, DC, USA, 2013.

18. Zheng, C.; Zhang, H.; Yan, P.; Gong, M. Low repetition rate broadband high energy and peak power nanosecond pulsed Yb-doped fiber amplifier. *Opt. Laser Technol.* **2013**, *49*, 284–287. [CrossRef]

19. Stutzki, F.; Jansen, F.; Liem, A.; Jauregui, C.; Limpert, J.; Tünnermann, A. 26 mJ, 130 W Q-switched fiber-laser system with near-diffraction-limited beam quality. *Opt. Lett.* **2012**, *37*, 1073–1075. [CrossRef] [PubMed]

20. Thompson, J.Y.; Stoner, B.R.; Piascik, J.R.; Smith, R. Adhesion/cementation to zirconia and other non-silicate ceramics: Where are we now? *Dent. Mater.* **2011**, *27*, 71–82. [CrossRef] [PubMed]

21. Filho, A.M.; Vieira, L.C.; Araujo, E.; Monteiro Junior, S. Effect of different ceramic surface treatments on resin microtensile bond strength. *J. Prosthodont.* **2004**, *13*, 28–35. [CrossRef] [PubMed]

22. Della Bona, A.; Borba, M.; Benetti, P.; Pecho, O.E.; Alessandretti, R.; Mosele, J.C.; Mores, R.T. Adhesion to Dental Ceramics. *Curr. Oral Health Rep.* **2014**, *1*, 232–238. [CrossRef]

23. Saracoglu, A.; Cura, C.; Cotert, H.S. Effect of various surface treatment methods on the bond strength of the heat-pressed ceramic samples. *J. Oral Rehabil.* **2004**, *31*, 790–797. [CrossRef] [PubMed]

24. Blatz, M.B.; Sadan, A.; Kern, M. Resin-ceramic bonding-A review of the literature. *J. Prosthet. Dent.* **2003**, *89*, 268–274. [CrossRef] [PubMed]

25. Rocca, J.-P.; Fornaini, C.; Brulat-Bouchard, N.; Bassel Seif, S.; Darque-Ceretti, E. CO_2 and Nd:YAP laser interaction with lithium disilicate and Zirconia dental ceramics: A preliminary study. *Opt. Laser Technol.* **2014**, *57*, 216–223. [CrossRef]

26. El Gamal, A.; Fornaini, C.; Rocca, J.P.; Muhammad, O.H.; Medioni, E.; Cucinotta, A.; Brulat-Bouchard, N. The effect of CO_2 and Nd:YAP lasers on CAD/CAM Ceramics: SEM, EDS and thermal studies. *Laser Ther.* **2016**, *25*, 27–34. [CrossRef] [PubMed]

27. Liu, L.; Liu, S.; Song, X.; Zhu, Q.; Zhang, W. Effect of Nd: YAG laser irradiation on surface properties and bond strength of zirconia ceramics. *Lasers Med. Sci.* **2015**, *30*, 627–634. [CrossRef] [PubMed]

28. Ural, C.; KalyoncuoGlu, E.; Balkaya, V. The effect of different outputs of carbon dioxide laser on bonding between zirconia ceramic surface and resin cement. *Acta Odontol. Scand.* **2012**, *70*, 541–546. [CrossRef] [PubMed]

29. Sadeghi, M.; Davari, A.; Abolghasami Mahani, A.; Hakimi, H. Influence of different power outputs of Er:YAG laser on shear bond strength of a resin composite to feldspathic porcelain. *J. Dent.* **2015**, *16*, 30–36.

30. Shiu, P.; De Souza Zaroni, W.C.; Eduardo Cde, P.; Youssef, M.N. Effect of feldspathic ceramic surface treatments on bond strength to resin cement. *Photomed. Laser Surg.* **2007**, *25*, 291–296. [CrossRef] [PubMed]

31. Erdur, E.A.; Basciftci, F.A. Effect of Ti:Sapphire-femtosecond laser on the surface roughness of ceramics. *Lasers Surg. Med.* **2015**, *47*, 833–838. [CrossRef] [PubMed]

32. Li, R.; Ruckle, D.; Keheila, M.; Maldonado, J.; Lightfoot, M.; Alsyouf, M.; Yeo, A.; Abourbih, S.R.; Olgin, G.; Arenas, J.L.; et al. High-frequency dusting versus conventional holmium laser lithotripsy for intrarenal and ureteral calculi. *J. Endourol.* **2017**, *31*, 272–277. [CrossRef] [PubMed]

33. Park, J.A.; Park, S.W.; Chang, I.S.; Hwang, J.J.; Lee, S.A.; Kim, J.S.; Chee, H.K.; Yun, I.J. The 1470-nm bare-fiber diode laser ablation of the great saphenous vein and small saphenous vein at 1-year follow-up using 8–12 W and a mean linear endovenous energy density of 72 J/cm. *J. Vasc. Interv. Radiol.* **2014**, *25*, 1795–1800. [CrossRef] [PubMed]

34. Wu, D.C.; Friedmann, D.P.; Fabi, S.G.; Goldman, M.P.; Fitzpatrick, R.E. Comparison of intense pulsed light with 1927-nm fractionated thulium fiber laser for the rejuvenation of the chest. *Dermatol. Surg.* **2014**, *40*, 129–133. [CrossRef] [PubMed]

35. Wattanakrai, P.; Pootongkam, S.; Rojhirunsakool, S. Periorbital rejuvenation with fractional 1550-nm ytterbium/erbium fiber laser and variable square pulse 2940-nm erbium:YAG laser in Asians: A comparison study. *Dermatol. Surg.* **2012**, *38*, 610–622. [CrossRef] [PubMed]

36. Morin, F.; Druon, F.; Hanna, M.; Georges, P. MicroJoule femtosecond fiber laser at 1.6 μm for corneal surgery applications. *Opt. Lett.* **2009**, *34*, 1991–1993. [CrossRef] [PubMed]

37. Fornaini, C.; Merigo, E.; Poli, F.; Cavatorta, C.; Rocca, J.P.; Selleri, S.; Cucinotta, A. Use of 1070 nm fiber lasers in oral surgery: Preliminary ex vivo study with FBG temperature monitoring. *Laser Ther.* **2017**, *26*, 311–318. [CrossRef]

38. Rihakova, L.; Chmelickova, H. Laser Micromachining of Glass, Silicon, and Ceramics. *Adv. Mater. Sci. Eng.* **2015**, *2015*. [CrossRef]

39. Vechiato Filho, A.J.; dos Santos, D.M.; Goiato, M.C.; de Medeiros, R.A.; Moreno, A.; Bonatto Lda, R.; Rangel, E.C. Surface characterization of lithium disilicate ceramic after nonthermal plasma treatment. *J. Prosthet. Dent.* **2014**, *112*, 1156–1163. [CrossRef] [PubMed]

2

Guiding Device for the Patellar Cut in Total Knee Arthroplasty: Design and Validation

Erica L. Rex [1], Cinzia Gaudelli [2], Emmanuel M. Illical [3], John Person [4], Karen C. T. Arlt [1], Barry Wylant [5] and Carolyn Anglin [1,6,*]

[1] Biomedical Engineering and McCaig Institute for Bone and Joint Health, University of Calgary, Calgary, AB T2N 1N4, Canada; rexe4@hotmail.com (E.L.R.); karen.ct.arlt@gmail.com (K.C.T.A.)
[2] Section of Orthopaedic Surgery, Cumming School of Medicine, University of Calgary, Calgary, AB T2N 1N4, Canada; cinzia.gaudelli@gmail.com
[3] Department of Orthopaedic Surgery, SUNY Downstate Medical Center, Brooklyn, NY 11203, USA; eillical@gmail.com
[4] Box 13 Engineering, Calgary, AB T3L 2P5, Canada; john@box13eng.com
[5] Q Industrial Design Corporation, Calgary, AB T2T 0E7, Canada; bwylant@ucalgary.ca
[6] Department of Civil Engineering, University of Calgary, Calgary, AB T2N 1N4, Canada
* Correspondence: carolyn.anglin@gmail.com

Abstract: An incorrect cut of the patella (kneecap) during total knee arthroplasty, affects the thickness in different quadrants of the patella, leading to pain and poor function. Because of the disadvantages of existing devices, many surgeons choose to perform the cut freehand. Given this mistrust of existing devices, a quick, but accurate, method is needed that guides the cut, without constraining the surgeon. A novel device is described that allows the surgeon to mark a line at the desired cutting plane parallel to the front (anterior) surface using a cautery tool, remove the device, and then align the saw guide, reamer, or freehand saw with the marked line to cut the patella. The device was tested on 36 artificial patellae, custom-molded from two shapes considered easier and harder to resect accurately, and eight paired cadaveric specimens, each in comparison to the conventional saw guide technique. The mediolateral angle, superoinferior angle, difference from intended thickness, and time were comparable or better for the new device. Addressing the remaining outliers should be possible through additional design changes. Use of this guidance device has the potential to improve patellar resection accuracy, as well as provide training to residents and a double-check and feedback tool for expert surgeons.

Keywords: patella; resection; device; total knee arthroplasty; validation

1. Introduction

In total knee arthroplasty (TKA), worn and damaged surfaces of the knee are resected and replaced with artificial components. An incorrect cut of the patella, whether tilted, too thick, or too thin, occurs in at least 10% of cases, even amongst expert surgeons, leading to numerous clinical complications, in particular, a reduced range of motion, anterior knee pain, and patellofemoral impingement [1–7]. When rising from a low chair, only 15% of patients with a symmetric patellar cut experienced anterior knee pain compared with 44% of those with an asymmetric profile [4].

A symmetric cut is considered to be parallel to the front (anterior) surface of the patella [3,4,8]; however, existing devices either have no connection with the anterior surface or the contact points can cause the device to tilt because of high or low points on the surface [8–10]. A patellar computer-assisted surgery system addressed many of these factors, but the screw attachment of the bone reference frame was considered invasive; a better system to define the anterior surface was desired; and the greater simplicity of a mechanical interface was attractive [10,11].

Three main techniques are currently used to resect the patella: freehand with a saw, using a saw guide, and using a reamer [10]. None of these techniques has gained widespread acceptance because each has its own disadvantages. The freehand technique relies entirely on the experience of the surgeon. Achieving good accuracy is difficult because of the lack of distinctive landmarks, the small size of the patella, and the hard bone. The saw guide also depends on the subjective experience of the surgeon; it can be difficult to clamp securely around the patella; and it may be difficult to position at the correct height without further dissection of the soft tissues. The reamer is the easiest to apply, but can be inadvertently tilted, without the user realizing it, and can give the incorrect depth due to variability in the depth of the spikes [8,10].

On the basis of feedback from surgeons and residents, we developed a device with the objectives of guiding the cut rather than constraining it, so that surgeons can continue to use their judgment; be fast and simple to use; have minimal impact on the surgeon's current surgical procedure; provide a training tool for residents and a double-check and feedback tool for expert surgeons; and be able to be used either with everting the patella (as in conventional surgery) or without everting the patella (as in minimally-invasive surgery).

The purpose of this study was to evaluate the device using artificial bone models and cadaveric specimens, to determine the accuracy, the ease of use, and potential design improvements. Accuracy was judged by the mediolateral (ML) resection angle, superoinferior (SI) resection angle, and the difference from the intended thickness. The hypothesis was that the new device is more accurate than existing techniques, while taking comparable or less time.

2. Materials and Methods

2.1. Device Design

The central concept of the device is to mark a line on the patella parallel to the anterior surface (i.e., around the circumference), using a cautery tool or marker pen, and then remove the device, leaving the marked line. The surgeon then uses their technique of choice (freehand, saw guide, or reamer) to align the saw or reamer with the marked line. In this way, the surgeon can continue to use the device they are most comfortable with, but with greater confidence and accuracy. By not providing a saw slot, the device is lighter, smaller and non-invasive, and leaves the control in the surgeon's hands.

The surgeon or resident can compare the drawn line to what they would have done, learning in the process, and providing a second-thought evaluation of the patellar cut. This is similar to the common practice of drawing several guidelines on the femur (transepicondylar axis, posterior condylar axis, and Whiteside's line) to see how they compare. Residents may have the opportunity to do the patellar cut earlier in their residency because the surgeon, after checking the drawn line, will feel more confident in the resulting resection. The device is called TellaMark, to describe Marking the paTella (Figures 1 and 2).

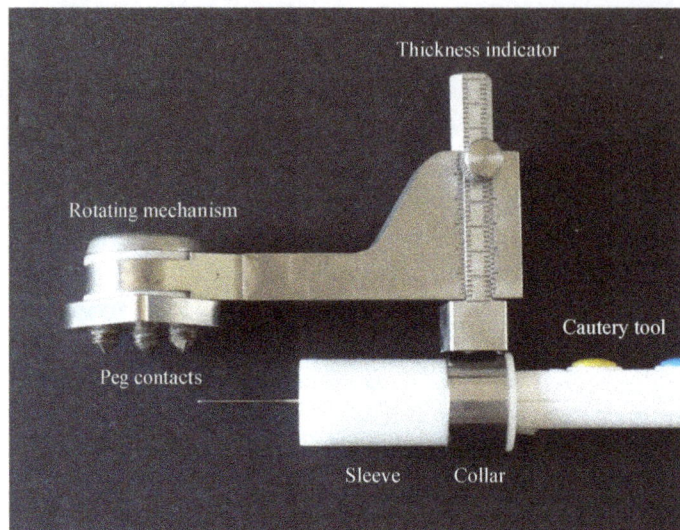

Figure 1. TellaMark device with cautery tool, showing the peg contacts, central rotating mechanism, and dovetail thickness indicator, suitable for all patellar shapes and sizes. The cautery tool collar is connected to the dovetail via a rare earth magnet to allow for free rotation. The cautery tool sleeve can translate freely in and out. Together, this makes it possible for the user to mark over 180° of the patella.

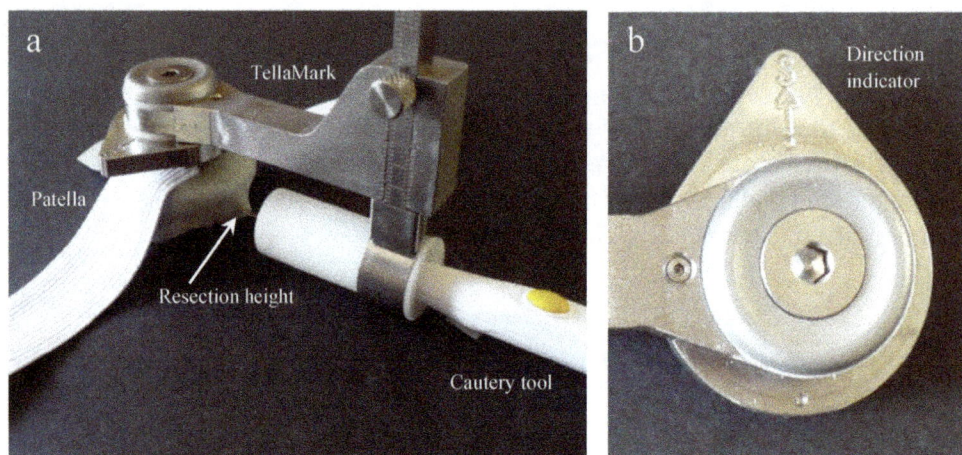

Figure 2. TellaMark device (**a**) placed on a patella, (**b**) with the arrow pointing superiorly. The device is placed centrally on the patella, with the pegs forming a 16 mm equilateral triangle, with two pegs positioned superiorly and one inferiorly. This configuration was determined after detailed analysis and surgeon input on 18 patellar models. The line marked by the user is then parallel to the anterior surface at the defined thickness.

Essential to the device is accurately defining the anterior surface to achieve the desired resection. The contact configuration was determined by first having four surgeons draw lines on 18 axial and sagittal X-ray images created from computed tomography (CT) images of patellae, marking the estimated anterior surface and desired resection line, and then testing different contact configurations virtually on the 3D CT images to achieve the desired resection plane. The best configuration was determined to be a 16 mm equilateral triangle with two points superiorly and one inferiorly, centred on the patella (Figure 1). The teardrop shape, with 'S' marked for superior, can be easily rotated to suit a right or left patella (Figure 2). The prototype design used for testing had cone-point set screws as the contact points to allow their depth to be adjusted during the initial stages of testing. The length was chosen to be short enough to promote stability while being long enough to allow visibility when applying the device to the anterior surface. The size and sharpness were tested to grab onto the bone without digging in excessively.

Rotating the cautery tool around the patella is achieved using a swing arm, a strong rare earth magnetic coupling, and a custom Delrin sleeve that fits around the cautery tool, sliding in and out of the metal collar (Figure 1). This sleeve could have a different inner profile for cautery tools with a different shape. The device could also be used with a marker pen, but the surgeons and residents preferred the cautery tool as it is more reliable and leaves a finer line. By pushing the cautery tool in and out of the metal collar while rotating it around the patella, the line can be drawn more than 180° around the patella, posterior to the tendon attachments, providing guidance in both the ML and SI directions.

The desired depth is set on the sliding dovetail mechanism (Figures 1 and 2), allowing the surgeon continuous depth adjustment. It was originally intended to be set exactly at the desired depth, but through this testing we discovered that it is advantageous to set the depth slightly thinner, resecting posterior to the line instead of on it, so that alignment with the marked line can be checked following resection. This could be part of the instruction procedure or could be incorporated directly into the device.

The device is held onto the patella with the thumb and forefingers, to avoid using an invasive bone screw or bulky clamping device, and to provide haptic feedback to the surgeon when applying the device to help avoid tilting the contacts off the anterior surface. Using the thumb and fingers works because the device is only used to mark the line rather than to create the saw cut, and is only held on for a short duration of time. The resulting profile provides good visibility of the patella while marking the line. The device is suitable for all patellar shapes and sizes, medial or lateral approaches, with right- or left-handed surgeons.

To our knowledge, no other device exists in which the desired cut line is drawn on the bone surface, for this or any other joint.

2.2. Artificial Bone Testing

To mimic the surgical setup in the artificial bone testing, and to perform pilot testing for design and use iterations before testing on valuable and limited cadaveric specimens, medium-sized right and left legs (Sawbones, Pacific Research Laboratories Inc., Vashon, WA, USA), without patellae, were set up at full extension and anchored onto a table. Custom patellae (see below) were attached to the femur and tibia models using materials simulating the tendons and lateral retinaculum, and covered with material representing skin. A standard incision represented the visibility and access during surgery.

Two custom-molded patellar geometries were created and used in the testing phase, an approach that could be useful to other researchers, as they were more realistic than previously-used Sawbones patellae and were derived from CT scans of cadaveric specimens with which we had done earlier resection analyses and could in turn be used to compare the resulting resections in the present experiments. Geometry 1 was a left patella, smaller, regularly-shaped, and considered the 'easier' geometry, based on the most consistent estimated resection lines drawn by surgeons on pseudo-X-rays generated from the CT scans. Geometry 2 was a right patella, larger, irregularly-shaped, and considered the 'harder' geometry based on the least consistent estimated resection lines. Patellar bone models were generated from the CT scans and rapid prototyped. A mold made from the rapid-prototyped model was used to generate the patellar bone models (Foam-it 15; SmoothOn Inc., Easton, PA, USA, for which the density is 15 pounds per cubic foot). The anterior surface was covered with a thin layer of Thera-band to provide compliance and to partially obscure the anterior surface. Since the foam is insulating and the cautery tool requires a conduction path, the experimenters instead dipped the cautery tool in calligraphy ink, leaving an ink line on the patella (Figure 3). Normal use of the cautery tool was verified during the cadaveric testing.

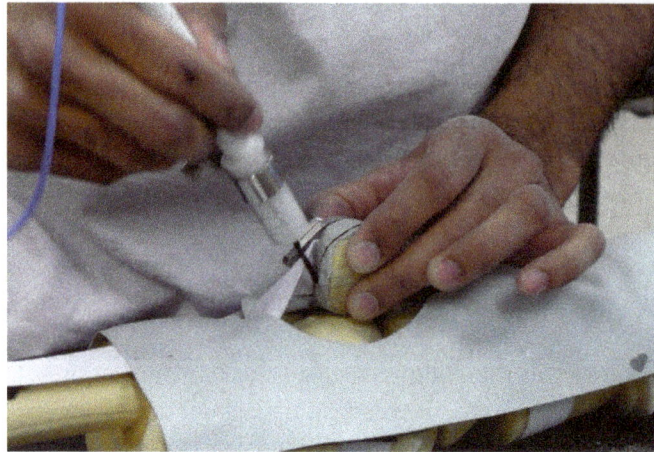

Figure 3. Using the TellaMark device to draw the desired resection line on artificial bones.

Two orthopaedic surgery residents (4th and 5th year) performed resections using three techniques: (1) using the conventional technique with a standard surgical saw guide (Zimmer; Warsaw, IN, USA); (2) TellaMark with the saw guide; and (3) TellaMark using a freehand technique, in each case with a surgical oscillating saw (Figure 4).

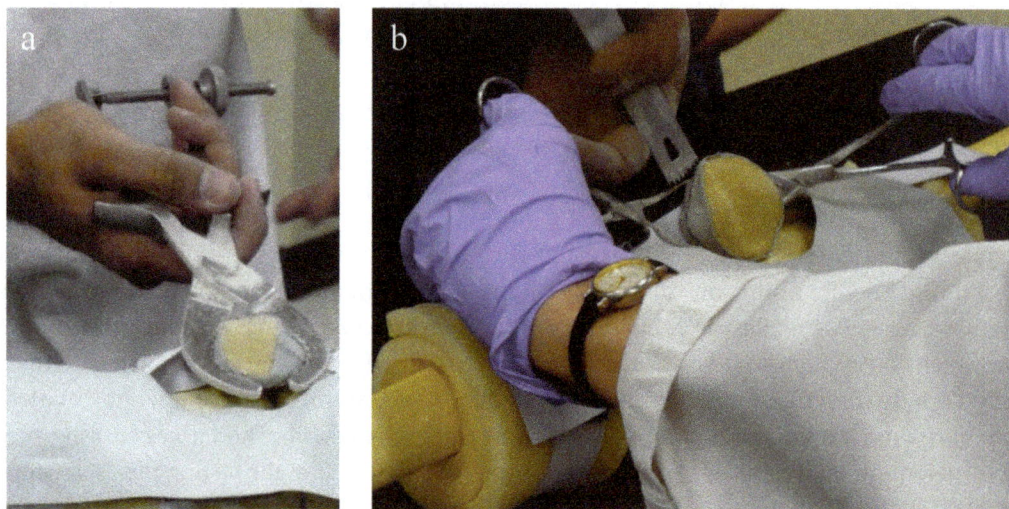

Figure 4. (**a**) Applying the saw guide to the patella and (**b**) securing the patella with towel clips for freehand resection. It worked best to mark the line slightly posterior to the desired thickness so that the line remained on the patellar bone remnant after resection to confirm the final cut, taking this into consideration in the thickness setting.

For the TellaMark resections, the initial resection was left as it was, whereas for the conventional resections, the experimenter measured the thickness and symmetry with calipers and had the option of revising the cut until satisfied. After initial practice with the instruments and experimental setup, each experimenter performed three repetitions of each of the three techniques on the two different geometries, for a total of 18 tests each. Tests were performed in a randomized order. The procedure time was recorded, including a breakdown of the steps.

The TellaMark procedure began by locating the center of the patella; this was done by feeling the height and width with the fingers, and marking the resulting central point with a marker pen or cautery tool (in the future, a dedicated device can be developed). The desired remaining thickness, determined from the patellar thickness minus the prosthesis height, was set on the depth gauge of the device and the device applied to the center of the patella with the arrow pointing superiorly. The line was then drawn with the cautery tool more than 180° around the patella, allowing both the ML and SI planes to be guided. The device was removed and the experimenter either aligned the saw or saw guide with the line to complete the cut.

The patellae were CT scanned before and after resection (0.6 mm slice thickness), followed by segmentation of the patellar bone (Amira Version 5.3.1; Visage Imaging, Andover, MA, USA). The resected patellae were aligned to the original surface models using the AlignSurface function in Amira, plus manual fine tuning, and then brought into AutoCAD (Version 2010, AutoDesk, San Rafael, CA, USA). In AutoCAD, an average plane was fitted visually to the resected surface of the patellar model, and then the average resection plane, determined previously from the four surgeons' input on pseudo X-rays, was applied to the model. This was the particular advantage of making custom molds of the previously-analyzed patellae. The ML and SI angles were measured between the resultant plane and the average surgeon-identified resection plane. The center of the patella was determined from the medial, lateral, superior, and inferior extents of the model, i.e., by drawing a box around the patella. The thickness from the anterior surface to the resected surface was measured at this central point and then compared to the intended remaining thickness specified in the testing process (13 mm for the left, 12 mm for the right).

The angle, thickness and time data were analyzed using ANOVA, followed by Student's t-tests when significant, using PASW Statistics 17.0 analysis software (Statistical Package for Social Sciences (SPSS) Inc., Chicago, IL, USA). Shapiro-Wilk tests confirmed the normality of the data. Angles within $\pm 7°$ were considered symmetric based on previous studies that showed greater anterior knee pain beyond this limit and represented a normal range of results [8,10].

2.3. Cadaveric Testing

Eight pairs of fresh-frozen cadaveric knee specimens (six female, two male; mean age 82, range 67 to 90 years) were used for testing, following ethics approval. They were CT scanned prior to testing and then prepared with a midline incision followed by a standard parapatellar capsulotomy: medial in 14 cases, lateral in two, providing the opportunity to test both approaches. The soft tissues were released to allow for eversion of the patella, and cleared around the circumference to allow for the application of the saw guide, as done clinically. The specimens varied from no arthritis to severe arthritis, with the majority having moderate arthritis (grades 2–3). The arthritic state did not affect the experiment since the device relies only on the anterior surface, not the articulating surface, one of the advantages of the device.

For each specimen pair, a TellaMark with saw-guide resection was performed on one side (Figure 5) and a conventional saw-guide resection was performed on the other (Figure 6), in a randomized order. The same two senior residents who performed the artificial bone testing performed the cadaveric testing. The cautery tool produced a clear, precise line, about 1 mm in thickness. As with the artificial bones, in the TellaMark case, the first cut was taken as the final cut; small corrections to the resection were allowed, such as removing a ridge, but the resection plane itself was not allowed to be recut or otherwise modified. In the conventional case, the experimenter could correct the cut until satisfied; cuts after the initial saw-guide cut were usually done freehand, with the patella being secured with towel clips (Figure 4). Experimenter 1 set the TellaMark device in such a way that the line would be cut off with the saw; Experimenter 2 set it in such a way that the saw cut just above the line, leaving the line visible afterward. This latter technique had the advantage of confirming that the cut made corresponded to the cut recommended by the device, and is now the recommended technique. The desired thickness was determined from caliper measurements, with the prosthesis thickness being subtracted from the total thickness.

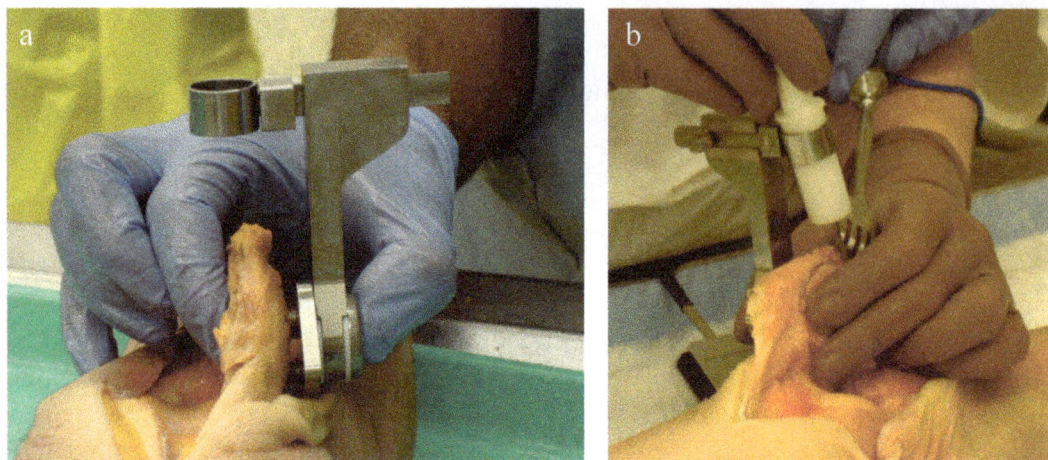

Figure 5. (**a**) Applying TellaMark to a cadaveric specimen; and (**b**) making a cautery line around the patellar circumference. Use of the thumb and forefingers provides haptic feedback on the contact of the device, leading to quick installation and removal. A single-pointed clamp (so as to keep the peg contacts on the surface) similar to towel-clamps could also be used to make the process hands-free.

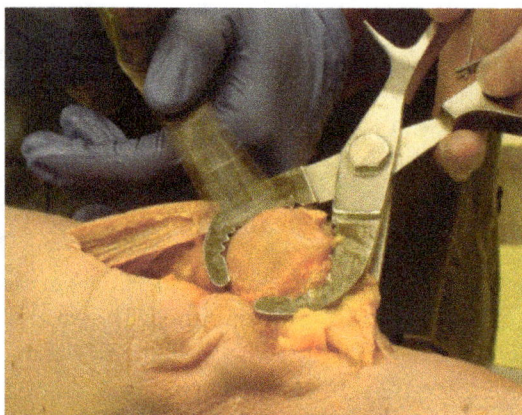

Figure 6. Using the conventional saw guide on a cadaveric specimen. Saw-guide use depends highly on the experience of the surgeon and the shape of the patella.

Once the resections were complete, CT images were acquired and used to calculate the desired resection plane as well as the achieved resection plane for each patella, by importing the segmented surfaces into AutoCAD. From this the ML and SI angles as well as the remaining bone thickness were measured, using the same method as for the artificial bone models. The three-peg model of the device was applied to the surface to determine the expected TellaMark resection angle. ANOVA tests of the ML angle, SI angle, bone remnant thickness, and time results were performed with $p < 0.05$ being considered significant. Normality was confirmed.

3. Results

The TellaMark device produced comparable or better results on average when compared to the conventional saw-guide technique, with an average reduction in time. In most cases, the results were not statistically different. Almost all of the ML angles were within the symmetry limit, i.e., acceptable. The cadaveric SI angles were mostly outside the symmetry limit for the conventional technique. Individual results are discussed below.

3.1. ML Resection Angle

There were no significant differences in ML resection accuracy between techniques in either the artificial or cadaveric testing (Figure 7). All but two of the conventional and one of the TellaMark angles were within the symmetry thresholds (Figure 7, red dashed lines). On average, the ML angles when using TellaMark were slightly closer to the desired angle than with the conventional technique.

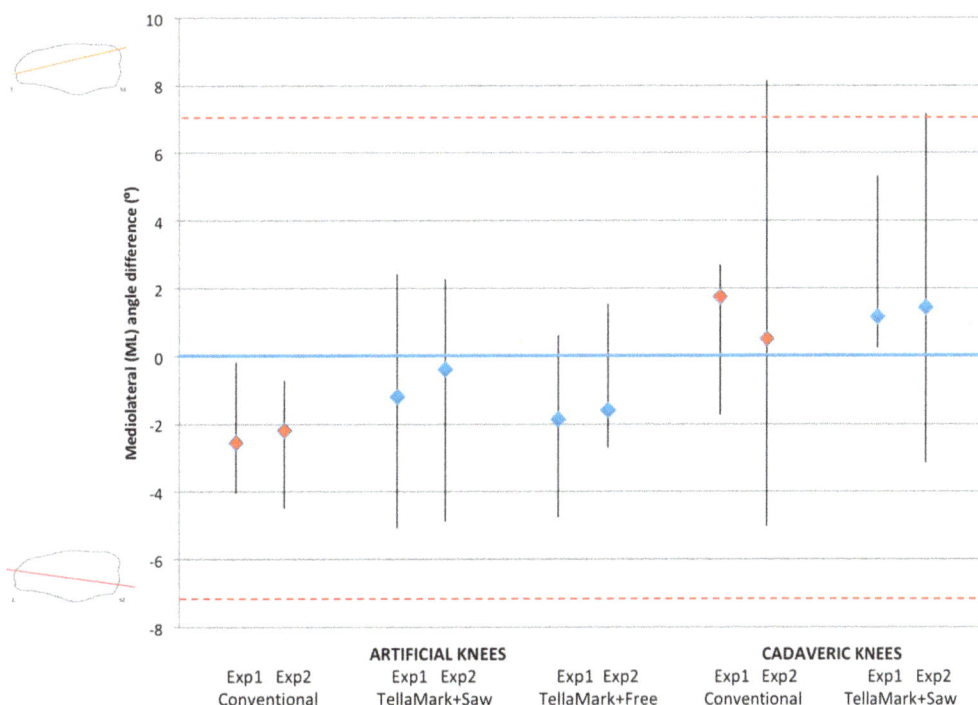

Figure 7. Mediolateral (ML) angle difference from the desired plane, showing the median, minimum and maximum, for the artificial bone models and cadaveric specimens, using the conventional saw guide (red markers), TellaMark + saw guide and TellaMark + freehand (blue markers). The dashed red lines indicate the symmetry goal ($\pm 7°$). The one TellaMark outlier had a lateral approach and capsulotomy, which the experimenter found confusing because of the truncated specimen.

3.2. SI Resection Angle

For the artificial results, all but one conventional resection and all of the TellaMark results were within the symmetry limit (Figure 8, red dashed lines). By contrast, for the cadaveric results, all but two of the conventional results were outside of the symmetry limit.

There were no significant differences in SI resection accuracy between techniques in either the artificial or cadaveric testing (Figure 8). On average, the SI angles when using TellaMark were closer to the desired angle than with the conventional technique. The TellaMark results differed by experimenter: Experimenter 1, who cut off the guide line during the resection, had similar SI accuracy to the conventional technique (median, 9.2° in each case); Experimenter 2, who retained the guide line during the resection, had dramatically better results with TellaMark (median 0.9° with TellaMark compared to 10° for the conventional technique; $p = 0.02$).

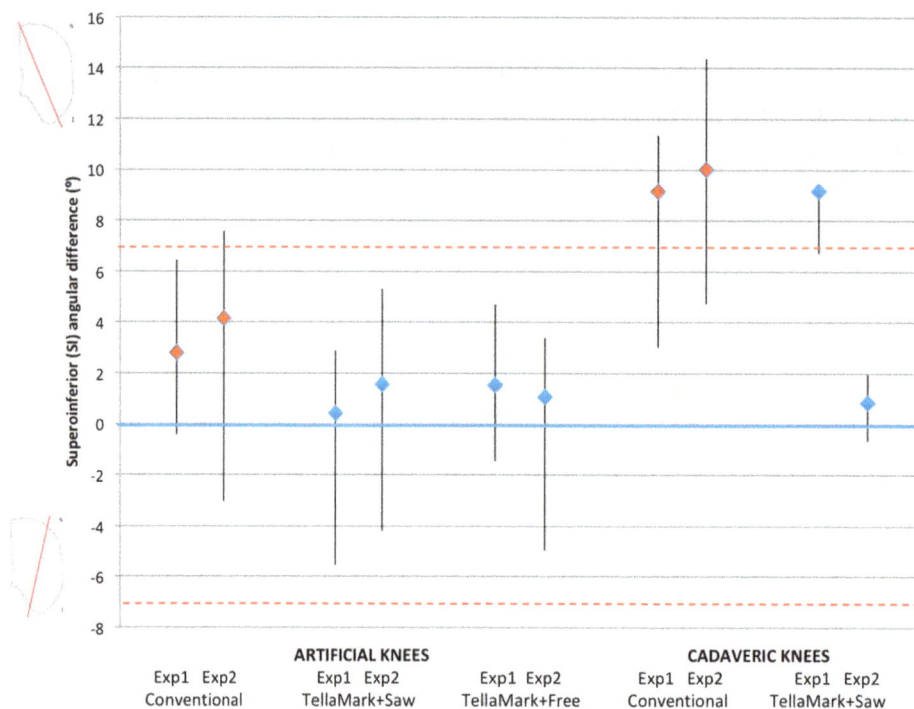

Figure 8. Superoinferior (SI) angle difference from the desired plane, showing the median, minimum and maximum, for the artificial bone models and cadaveric specimens, using the conventional saw guide (red markers), TellaMark + saw guide and TellaMark + freehand (blue markers). The dashed red lines indicate the symmetry goal ($\pm 7°$). The substantially more accurate TellaMark vs. conventional results for Experimenter 2 vs. Experimenter 1 may be due to leaving the guideline visible (which is now recommended) and carefully centering the device on the patellar surface (a centering device will be added in the future).

3.3. Bone Remnant Thickness

In the artificial bone testing, there were no significant differences in thickness between the three techniques (averaging 0.6 mm, 0.5 mm, and 0.7 mm thinner than intended, for the conventional technique, TellaMark + saw guide and TellaMark + freehand, respectively, with standard deviations of 0.6 mm, 1.0 mm, and 1.0 mm, respectively). In the cadaveric testing, the bone remnant tended to be thinner than intended with the conventional technique (-0.9 ± 0.5 mm) and thicker than intended with the TellaMark + saw guide (1.0 ± 1.0 mm), resulting in a significant difference between the two techniques ($p < 0.001$), although both averaged close to the intended thickness.

3.4. Procedure Time

In the artificial bone testing, there were no significant differences in time between techniques (3.8 ± 1.1 min for TellaMark + saw guide, 3.6 ± 0.7 min for TellaMark + freehand, 4.0 ± 0.7 min for the conventional technique), however the time breakdowns did differ, with more time required to perform recuts with the conventional saw guide whereas TellaMark required more initial time leading up to the resection. The main time consumption in both cases related to securing the patella with the saw guide or towel clips, and performing the cut. The time taken was significantly different between experimenters ($p = 0.002$), whereby Experimenter 2 took 50 s longer on average to complete a resection.

In the cadaveric testing, resection time was significantly shorter with TellaMark than the conventional technique (4.8 ± 1.8 min for TellaMark vs. 7.2 ± 1.2 min for the conventional technique; $p = 0.03$). As with the artificial bone testing, the TellaMark device took more time initially whereas the conventional device took more time later.

4. Discussion

In this study, a novel device for patellar resection was tested on artificial bone models and cadaveric specimens, demonstrating symmetries that were equivalent or better on average than the conventional technique in a similar amount of time, with a substantial improvement in SI symmetry (0.9° for TellaMark vs. 10° for the conventional technique) for the experimenter who retained the TellaMark guide line, even with multiple cuts permitted for the conventional technique, in contrast to the single cut permitted with TellaMark. Further improvements in the TellaMark design, based on this testing, together with more experience using the device, should lead to additional increases in accuracy and reductions in time.

Because of the advantages of keeping the guideline visible, we now recommend setting the depth greater by 1 mm, or building this into the device. The outlier ML cadaveric resection was likely caused by the lateral approach, which the experimenters said threw them off (the truncated knee specimen can easily be turned one way or the other, so the directions were less intuitive) as well as the hard, sclerotic bone, which the saw had trouble getting through.

For the SI resections, Experimenter 1 appears to have mounted the device higher than the patellar center (based on later analysis of the CT scans), which caused the device to have an angle relative to the anterior surface. In the future, we recommend incorporating a viewing hole that can be placed over the central mark to ensure that the desired centering is achieved, as well as possibly adding a device or procedure to aid with the centering. Experimenter 2 took on average 30 s longer than Experimenter 1 to ensure that the device was centred. A simple mechanical device could achieve this centering quickly and without subjectivity. The apparent value for SI symmetry is important since a previous study showed that SI symmetry was even more strongly correlated to anterior knee pain ($p = 0.001$) than ML symmetry ($p = 0.02$) [4].

Regarding thickness, TellaMark can be used to resect at the final depth directly, or can be used first with a conservative thickness to guide the angle, followed by a final cut after checking the thickness. This could be particularly valuable if the patella is not everted. If there is a consistent bias, e.g., to a thicker patellar as in this study, this can be incorporated into the design and instructions of the device.

A related device that uses the same three-peg contact on the anterior surface, designed to verify the patellar cut after it is made, demonstrated good accuracy [12], suggesting that deviations in this study related mainly to the execution of the cut and that the estimation of the anterior surface is robust.

The main limitation of this study was the small number of experimenters and specimens, reflecting the availability of specimens and the time required to perform the testing. Also, while the artificial bone testing provided useful feedback on the design and use of the device as well as practice for the experimenters, it did not fully mimic the cadaveric testing. A larger cadaveric study should be conducted once the design is updated based on the results of this testing; nonetheless, the present study was able to reveal the fundamental similarities and differences between the TellaMark and conventional results. One other consideration is that conventional techniques are relatively successful in a lab setting (i.e., within the symmetry limit), despite wide ranges clinically, making it difficult to show a clear advantage of a new device, especially given the multiple cuts allowed for the conventional technique but not for the new device. After further testing to ensure the best usability and accuracy of TellaMark, our goal is to implement the device clinically, allowing expert surgeons to compare their intended resection with the proposed resection and then examine the final resection result to determine which would have been the better choice.

For clinical use, there should be fewer parts, which can be addressed through higher-volume manufacturing methods, and the plastic bearing should be replaced by metal for ease of sterilization. While the thumb and forefingers provide a good way of holding the device on the patella, it is also possible to hold it in this position mechanically: a pointed connection or small contact point similar to a towel clip can help hold it in place to release the fingers for applying the cautery mark; a clamp with a large contact area should be avoided as this tends to tilt the contact points off the anterior surface or tilt the patella [8]. To our knowledge, the concept of marking bone to indicate the desired resection plane has not been used previously and could benefit other joints as well.

The experimenters considered the device a useful learning tool to create better resections, especially for less experienced surgeons. They appreciated the guidance that it provided, so that they did not need to struggle to decide where to put the saw guide. They also appreciated the fact that the resident or surgeon could confirm that they are satisfied with the line before proceeding with the cut. The expert surgeons we interviewed like that it provides guidance without constraining them to a particular resection line, and allows them to continue using the freehand or saw guide technique that they are already familiar with.

5. Conclusions

With minor design and use changes to address the current outliers, TellaMark offers the potential to substantially improve patellar resection thickness and symmetry, particularly given the substantially better SI results for Experimenter 2 compared to the conventional technique. It also offers the opportunity to improve training and confidence. On the basis of previous studies, better patellar resection accuracy will translate to reduced pain and increased function after knee replacement surgery, leading to more satisfied patients.

Author Contributions: E.L.R., C.A., J.P., B.W., and K.C.T.A. designed the device (patent application US20150005772); E.L.R. and C.A. conceived and designed the experiments; C.G. and E.M.I. performed the experiments and provided feedback on the device design and use; E.L.R. analyzed the data; E.L.R. and C.A. wrote the paper; all authors reviewed the manuscript.

Funding: This research was funded by Natural Sciences and Engineering Research Council of Canada (NSERC) and Alberta Innovates.

Acknowledgments: We wish to thank Tak Fung for his statistical advice, and John Kornelson and Calvin Cockerline for their help in the Anatomy Lab.

Conflicts of Interest: The authors declare no conflict of interest. The funding sponsors had no role in the design of the study; in the collection, analyses, or interpretation of data; in the writing of the manuscript, and in the decision to publish the results.

References

1.   Shervin, D.; Pratt, K.; Healey, T.; Nguyen, S.; Mihalko, W.M.; El-Othmani, M.M.; Saleh, K.J. Anterior knee pain following primary total knee arthroplasty. *World J. Orthop.* **2015**, *6*, 795–803. [CrossRef] [PubMed]

2.   Feczko, P.Z.; Jutten, L.M.; van Steyn, M.J.; Deckers, P.; Emans, P.J.; Arts, J.J. Comparison of fixed and mobile-bearing total knee arthroplasty in terms of patellofemoral pain and function: A prospective, randomised, controlled trial. *BMC Musculoskelet. Disord.* **2017**, *18*, 279. [CrossRef] [PubMed]

3.   Baldini, A.; Anderson, J.A.; Cerulli-Mariani, P.; Kalyvas, J.; Pavlov, H.; Sculco, T.P. Patellofemoral evaluation after total knee arthroplasty: Validation of a new weight-bearing axial radiographic view. *J. Bone Jt. Surg. Am.* **2007**, *89*, 1810–1817. [CrossRef] [PubMed]

4.   Baldini, A.; Anderson, J.A.; Zampetti, P.; Pavlov, H.; Sculco, T.P. A new patellofemoral scoring system for total knee arthroplasty. *Clin. Orthop. Relat. Res.* **2006**, *452*, 150–154. [CrossRef] [PubMed]

5.   Pagnano, M.W.; Trousdale, R.T. Asymmetric patella resurfacing in total knee arthroplasty. *Am. J. Knee Surg.* **2000**, *13*, 228–233. [PubMed]

6.   Bengs, B.C.; Scott, R.D. The effect of patellar thickness on intraoperative knee flexion and patellar tracking in total knee arthroplasty. *J. Arthroplast.* **2006**, *21*, 650–655. [CrossRef] [PubMed]

7.   Mihalko, W.; Fishkin, Z.; Krackow, K. Patellofemoral overstuff and its relationship to flexion after total knee arthroplasty. *Clin. Orthop. Relat. Res.* **2006**, *449*, 283–287. [CrossRef] [PubMed]

8.   Anglin, C.; Fu, C.; Hodgson, A.J.; Helmy, N.; Greidanus, N.V.; Masri, B.A. Finding and defining the ideal patellar resection plane in total knee arthroplasty. *J. Biomech.* **2009**, *42*, 2307–2312. [CrossRef] [PubMed]

9.   Ledger, M.; Shakespeare, D.; Scaddan, M. Accuracy of patellar resection in total knee replacement: A study using the medial pivot knee. *Knee* **2005**, *12*, 13–19. [CrossRef] [PubMed]

10. Fu, C.; Wai, J.; Lee, E.; Hutchison, C.; Myden, C.; Batuyong, E.; Anglin, C. Computer-assisted patellar resection system: Development and insights. *J. Orthop. Res.* **2011**, *30*, 535–540. [CrossRef] [PubMed]
11. Fu, C.; Wai, J.; Lee, E.; Myden, C.; Batuyong, E.; Hutchison, C.R.; Anglin, C. Computer-assisted patellar resection for total knee arthroplasty. *Comput. Aided Surg.* **2012**, *17*, 21–28. [CrossRef] [PubMed]
12. Rex, E.L.; Illical, E.M.; Gaudelli, C.; Wylant, B.; Ho, K.C.; Person, J.G.; Anglin, C. Device for verifying the patellar cut during knee replacement surgery. *J. Med. Devices* **2016**, *10*, 024502. [CrossRef]

3

Effect of Pyruvate Decarboxylase Knockout on Product Distribution using *Pichia pastoris* (*Komagataella phaffii*) Engineered for Lactic Acid Production

Nadiele T. M. Melo [1,2,†], Kelly C. L. Mulder [3,†], André Moraes Nicola [4] 🆔, Lucas S. Carvalho [1,3], Gisele S. Menino [3], Eduardo Mulinari [3] and Nádia S. Parachin [1,*] 🆔

[1] Grupo de Engenharia Metabólica Aplicada a Bioprocessos, Instituto de Ciências Biológicas, Universidade de Brasília, CEP 70.790-900 Brasília-DF, Brazil; nadytamires@gmail.com (N.T.M.M.); u.lucas@gmail.com (L.S.C.)

[2] Pós-Graduação em Ciências Genômicas e Biotecnologia, Universidade Católica de Brasília, CEP 70.790-900 Brasília-DF, Brazil

[3] Integra Bioprocessos e Análises, Campus Universitário Darcy Ribeiro, Edifício CDT, Sala AT-36/37, CEP 70.790-900 Brasília-DF, Brazil; kellylmulder@gmail.com (K.C.L.M.); gisele.sa27@gmail.com (G.S.M.); edumulinari@gmail.com (E.M.)

[4] Faculty of Medicine, University of Brasilia, Campus Universitário Darcy Ribeiro, Faculdade de Medicina, Sala BC-103, CEP 70.790-900 Brasília-DF, Brazil; andre.nicola@gmail.com

*   Correspondence: nadiasp@unb.br

†   These authors contributed equally to this work.

Abstract: Lactic acid is the monomer unit of the bioplastic poly-lactic acid (PLA). One candidate organism for lactic acid production is *Pichia pastoris*, a yeast widely used for heterologous protein production. Nevertheless, this yeast has a poor fermentative capability that can be modulated by controlling oxygen levels. In a previous study, lactate dehydrogenase (LDH) activity was introduced into *P. pastoris*, enabling this yeast to produce lactic acid. The present study aimed to increase the flow of pyruvate towards the production of lactic acid in *P. pastoris*. To this end, a strain designated GLp was constructed by inserting the bovine lactic acid dehydrogenase gene (LDHb) concomitantly with the interruption of the gene encoding pyruvate decarboxylase (PDC). Aerobic fermentation, followed by micro-aerophilic culture two-phase fermentations, showed that the GLp strain achieved a lactic acid yield of 0.65 g/g. The distribution of fermentation products demonstrated that the acetate titer was reduced by 20% in the GLp strain with a concomitant increase in arabitol production: arabitol increased from 0.025 g/g to 0.174 g/g when compared to the GS115 strain. Taken together, the results show a significant potential for *P. pastoris* in producing lactic acid. Moreover, for the first time, physiological data regarding co-product formation have indicated the redox balance limitations of this yeast.

Keywords: *Pichia pastoris*; pyruvate decarboxylase; lactic acid; homologous recombination; arabitol; redox metabolism

1. Introduction

Lactic acid has a high commercial value due to its broad application in several areas of industry, such as the automobile, food, pharmaceutical, and textile industries, in addition to the production of biodegradable polymers such as poly-lactic acid (PLA) [1]. The production of lactic acid by fermentation becomes economically feasible when compared to chemical synthesis, as microorganisms can be modified to produce a single isomer. The production of only one isomer facilitates the process of purification and polymerization into poly L-lactic acid (PLLA), which is mainly used in biomedical applications [2]. Moreover, the metabolic conversion of L-lactic acid in humans is much faster when compared to D-lactic acid, thus being preferentially employed in the food and medical sectors [3]. Approximately 82% of the world's lactic acid production is used by the food industry for microbial fermentation in sauerkraut, yogurts, and butter, among others. Moreover, lactic acid also functions as a pH reducer, solvent, antimicrobial agent, humectant, flavor adjuvant and emulsifier [4]. Due to its various applications, it is estimated that the lactic acid market will be valued at USD \$3.82 billion by 2020, which would represent an annual growth rate of 18.6% (https://www.marketsandmarkets.com/Market-Reports/polylacticacid).

Many yeast species have been genetically modified in order to produce lactic acid, including *Saccharomyces cerevisiae* [5–7], *Kluyveromyces lactis* [8,9], *Zygosaccharomyces bailii* [10], *Candida* sp. [11,12], and *Pichia* sp. [13]. The yeast *P. pastoris* has been reclassified into the new gender Komagataella, and sub-divided into the three species *K. pastoris*, *K. phaffii* and *K. pseudopastoris* [14]. It has, as its most notable physiological feature, the ability to grow in media containing only methanol as a carbon source [15]. Another advantage is that this yeast grows as fast on crude glycerol as on glucose [16], and can utilize crude glycerol without being inhibited by its impurities [17]. Crude glycerol is the main residue during biodiesel production. For example, Brazil, the second largest biodiesel producer worldwide, reported a production of approximately 4.3 million cubic meters of biodiesel in 2017, which resulted in an estimated 429,129.4 m^3 of crude glycerol (http://www.anp.gov.br/wwwanp/dados-estatisticos). Thus, the use of glycerol for lactic acid production is advantageous since it adds value to the biodiesel production chain.

The most frequently used and commercially available *K. phaffii* strains are GS115 and X-33, which are derived from the wildtype CBS7435 [18]. The genome of the latter is arranged in four chromosomes, with 5313 open reading frames identified [19]. The GS115 strain is known for its mutation in the enzyme *histidinol dehydrogenase* (HIS4), which makes it auxotrophic for histidine. Another strain derived from CBS7435 is the Ku70 mutant [20]. This harbors the deletion of the gene Ku70, resulting in the absence of a protein involved in the non-homologous end joining repair mechanism. Its deletion is reported to significantly increase the efficiency of homologous recombination and reduce false positives [20].

P. pastoris strain GS115 has an annotated gene in its genome that encodes for a putative lactate dehydrogenase (LDH) enzyme (EC 1.1.1.27). However, when growing it on glycerol as a substrate, lactate production is almost absent. In a recent study, this strain was engineered for lactic acid production through the insertion of a LDH activity [21]. Here, for the first time to the authors' knowledge, deletion of a pyruvate decarboxylase (PDC)-encoding gene has been performed in combination with LDH over-expression, with the aim of funneling further pyruvate to the lactic acid production pathway. Genetically modified *PDC*-knockout strains, as well as GS115, were used in oxygen limited cultivation in order to assess substrate consumption, lactic acid production, and by-product formation. In the *PDC*-deleted strains a 32% and 75% reduction of biomass and acetic acid production was observed, respectively, when compared to the wildtype strain GS115. However, a 2.6-fold increase in arabitol production was observed. In addition, arabitol production increased nearly seven-fold when LDH activity was associated with *PDC* disruption, demonstrating that this alcohol is now the main byproduct of recombinant lactic acid production in the strains of *P. pastoris* tested.

2. Materials and Methods

2.1. Strains and Plasmids

All of the plasmids and strains used in this study are listed in Table 1. The *Escherichia coli* strains utilized during the cloning steps were grown at 37 °C in Luria broth media (0.5% yeast extract, 1% peptone and 0.5% sodium chloride) and supplemented with ampicillin (100 $\mu g/mL^{-1}$). The *P. pastoris* strains were grown at 30 °C in YPD (1% yeast extract, 2% peptone and 2% dextrose) and supplemented with Geneticin (G418) (500 $\mu g/mL^{-1}$) and/or zeocin (100 $\mu g/mL^{-1}$) when necessary.

Table 1. Plasmids and strains used and developed in this work.

Plasmids	Relevant Genotype	Ref.
pUG6	loxP-PTEF-KanMX-TTEF-loxP	Life Technologies
pGAP-LDH	LDH+. *Bos taurus* gene encoding the LDH enzyme	[21]
pUG6-PDC	loxP-PTEF-KanMX-TTEF-loxP+PDC-	This work
Strains	**Relevant Genotype**	**Ref.**
DH5α™	F-Φ80*lacZ*ΔM15 Δ(*lacZYA-arg*F) U169 *rec*A1 *end*A1 *hsd*R17 (rK–, mK+) *phoA supE44 λ-thi1 gyr*A96 *rel*A1	Life Technologies
X-33	Wildtype	Invitrogen
CBS7435 ku70	Δ*ku70*	[20]
GS115	Δ*his4*	[18]
XL	X-33 + pGAP-LDHBos taurus	[21]
GLp	GS115: Δ*pdc* + pGAP-LDHBos taurus	This work
Gp	GS115:Δ*pdc*	This work

2.2. Identification of Putative Genes Encoding Pyruvate Decarboxylase

The NCBI platform was used to identify putative genes encoding the PDC enzyme in the genome of the yeast *Komagataella phaffii* GS115. The search for the *P. pastoris* pyruvate decarboxylase gene revealed an ORF annotated as coding for pyruvate decarboxylase XP_002492397, as well as two putative isozymes: XP_002492304 and XP_002492397. Therefore, the reference enzyme sequence chosen for this work was XP_002492397, which refers to the ORF annotated as pyruvate decarboxylase, and previously kinetically characterized in *S. cerevisiae* [22].

2.3. Construction of a PDC Knockout Cassette

Once the DNA sequence for the PDC enzyme had been obtained, PCR was performed to amplify the complete ORF. For this, the genomic DNA of *P. pastoris* GS115 was used as a template. The primers used were PDC5′F and PDC3′R (Table 2), and the fragment amplified was approximately 1.6 Kb, matching the expected size. The strategy used for construction of the PDC interruption cassette is summarized in Figure 1. For *PDC* gene interruption, the sequence was divided into two fragments: PDC5′ and PDC3′ (840 and 843 base pairs, respectively—see Figure 1). First, the pUG6 plasmid was treated with *Pvu*II and *Sal*I and the PDC5′ fragment was amplified using the primers PDC5′F and PDC5′R (Table 2). The PDC5′ fragment was then cloned into the vector at the upstream region from the kanamycin resistance marker. After confirmation by PCR, the plasmid was treated with the restriction nucleases *Sep*I and *Sac*II. The PDC3′ fragment, amplified by PCR using the primers PDC3′F and PDC3′R (Table 2), was then cloned downstream from the kanamycin resistance marker and its insertion was also confirmed by PCR (data not shown). PCR amplifications were performed in a 20 μL reaction mix containing 2.5 pmol of each primer, 2.5 units Taq DNA polymerase, 0.2 μM of each dNTP, 2.0 μL 10× reaction buffer (10 mM Tris-HCl pH 8.3, 50 mM KCl, 1.5 mM MgCl$_2$), and 50 ng of chromosomal DNA. Amplifications were performed with the following conditions: 95 °C for 30′, 95 °C for 15′, 58 °C for 30′, 72 °C for 3 min (30 cycles) and 72 °C for 10 min. This vector was named pUG6-PDCK, and it contained a cassette composed of PDC5′, followed by the kanamycin marker and the PDC3′, a fragment of 3.3 Kb in all.

Table 2. Primers used in this study. Sequences of restriction enzymes are highlighted in bold.

Primer	Sequence 5′-3′	Endonuclease
PDC5′F	**CAGCTG**ATGGCTGAAATAACACTAGGAACT	*Pvu*II
PDC5′R	**GTCGAC**ATCAGCCTTCTCCACGAACT	*Sal*I
PDC3′F	**ACTAGT**CTTGTCATCTCTGTTGGTGC	*Spe*I
PDC3′R	**CCGCGG**TTAAGCTGCGTTGGTCTTGG	*Sac*II
KanF	AGCTTGCCTCGTCCCC	
KanR	TCGACACTGGATGGCG	

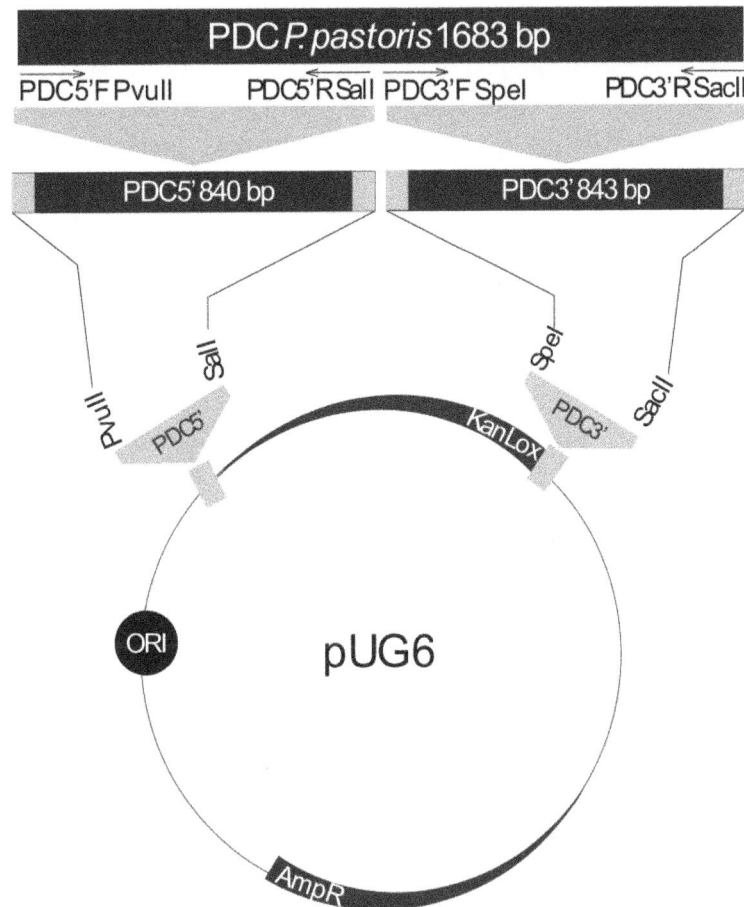

Figure 1. Construction of *PDC* knockout cassette. The entire PDC-encoding gene has 1683 bp. This was divided into two fragments of 840 and 843 bp. Each fragment was cloned into the pUG6 plasmid flanking the Kanamycin resistance cassette with the indicated restriction enzymes. Final cassette has a total of 3317 Kb.

2.4. PDC Knockout in P. pastoris Strains

The *P. pastoris* strains X-33, Ku70, XL and GS115 (Table 1) were transformed with the pUG6-PDCK cassette after linearization with PvuII and SacII. Transformation was carried out by electroporation, following the Easy Select *P. pastoris* Expression Kit (Invitrogen, EUA) protocol, with modifications. Briefly, a single colony was inoculated into 25 mL of YPD medium, and after 24 h at 28 °C and 250 rpm, 1 mL of cells was used to inoculate 100 mL of YPD medium. When the OD_{600nm} reached approximately 1.5 the cells were collected by centrifugation and re-suspended three times with 50 mL of cold and sterile water, followed by one step with 5 mL of cold 1 M sorbitol. All centrifugation steps were performed at $1500 \times g$ for 5 min at 4 °C. Afterward, the cells were re-suspended in 150 μL of cold 1 M sorbitol and 80 μL of these cells were homogenized with 5–10 μg DNA. This mix was then transferred to a 2 mm electroporation cuvette (Bio-Rad, Berkeley, CA, USA) and incubated on ice for 5 min. An electrical pulse was applied with the following conditions: 1500 V, 400 Ω, and 25 μF. Immediately after the pulse, 1 mL of cold 1 M sorbitol was added and the cells were incubated at 30 °C for 1 h. Finally, the cells were plated in YPD medium supplemented with geneticin (G418) (500 μg/mL^{-1}) and incubated at 30 °C for three days. The GS115 strain transformed with the cassette from pUG6-PDCK was then named Gp. The integrative plasmid pGAP-LDH from a previous study [21], harboring the codon-optimized *LDH* encoding the LDH enzyme from *Bos taurus*, was linearized using AvrII and inserted into the Gp strain, resulting in the GLp strain (Table 1).

2.5. Ploidy Determination in P. pastoris Strains

Inoculum of 5 mL YPD containing either X-33 or GS115 cells was grown at 30 °C and 200 rpm until the log phase. The cells were then centrifuged at 1500 × *g* for 5 min and maintained for 12 h at 4 °C in 10 mL of cold ethanol (70%). After centrifugation for 5 min at 2500 × *g*, the cells were washed with 1 mL of 50 mM sodium citrate (pH 7.5). The centrifugation was then repeated and the cells were resuspended in 1 mL of 50 mM sodium citrate (pH 7.5), containing 0.25 g/L RNAse (250 mg/mL). After 1 h at 55 °C, 50 μL of proteinase K (20 mg/mL) was added to the cell suspension. After a further hour at 55 °C, the permeabilized cells were washed, counted and resuspended at a concentration of 10^7 cells/mL in PBS supplemented with 50 μg/mL propidium iodide. Following 30 min of incubation, the cells were analyzed in a FACS Verse flow cytometer (BD Biosciences) equipped with a 488 nm laser. All samples were collected with the same cytometer settings, and propidium iodide fluorescence was set in linear mode. A forward angle versus side angle scatter area gate was used to remove debris and a forward scatter width versus forward scatter height gate was used for doublet discrimination. The experiment was repeated in triplicate.

2.6. LDH Enzyme Activity

Enzyme assays were carried out as described previously, with modifications [23]. Briefly, a primary inoculum culture was prepared in YPD medium, with zeocin (100 μg/mL), and maintained at 30 °C and 180 rpm overnight. Cells were harvested, re-inoculated in a new flask, and grown in a shaker at 30 °C until the exponential phase. After centrifugation, cells were resuspended in Yeast Protein Extraction Reagent (Y-Per, Thermo Scientific, Rockford, IL, USA) for 10 min. The reaction was assembled with 10 μL cellular extract, 8 μL NADH, 800 μL 50 mM phosphate buffer (pH 7), and ultra-pure water for a 1 mL final volume. After 150 s, 40 μL of pyruvate was added and the reaction was completed in 300 s. A unit of enzyme activity was defined as the amount of enzyme necessary to oxidize 1 μmol of NADH per minute. Protein concentration of cell extracts was determined using the Coomassie protein assay reagent (Pierce, Rockford, IL, USA) according to the manufacturer's instructions. BSA in known concentrations was used to construct the standard curve. Enzyme assays were carried out in three biological replicates.

2.7. Fermentation Parameters

For cultivations in the bioreactors, a defined medium was utilized as previously described [24], with modifications. The composition of the medium (per liter) was: 1.8 $C_6H_8O_7$, 0.02 g $CaCl_2 \cdot 2H_2O$, 12.6 g $(NH_4)2HPO_4$, 0.5 g $MgSO_4 \cdot 7H_2O$, 0.9 g KCl and 4.35 mL PTM1 trace salts stock solution. pH was adjusted to 5.0 with 25% HCl. PTM1 trace salts stock solution (per liter) was composed of: 6 g $CuSO_4$ $5H_2O$, 0.08 g NaI, 3 g $MnSO_4 \cdot H_2O$, 0.2 g Na_2MoO_4 $2H_2O$, 0.02 g H_3BO_3, 0.5 g $CoCl_2$, 20 g $ZnCl_2$, 14.3 g $FeSO_4$, 0.4 g biotin and 5 mL H_2SO_4 (95–98%). 0.04 g/L histidine was supplemented for the GS115 strain.

A 100 mL pre-culture was prepared with 20 g/L glycerol, and was grown for approximately 48 h at 30 °C and 200 rpm in a 1 L bioreactor (Infors HT., Bottmingen, Switzerland). Cultivations in the bioreactors were performed with 500 mL medium at an initial OD_{600nm} of 2, and with initial glycerol concentration of 80 g/L. The batch phase was performed under the following conditions: 30 °C, 500 rpm, dissolved oxygen at 30%, and pH 5.0 controlled with 5 M NH_4. Feeding started after glycerol depletion by supplementing the culture with 40 g/L glycerol in a single pulse. When the pH went above 5.0, dissolved oxygen was kept at 3%. Samples were collected every 90 min and centrifuged at 12,000 g for 2 min. The supernatant was stored at −20 °C for HPLC analysis.

2.8. Substrate Consumption and Cellular Products Quantification

Glycerol, lactic acid, acetic acid, ethanol and arabitol were quantified using High-performance Liquid Chromatograph (HPLC) (Shimadzu, Kyoto, Japan) equipped with UV (210-nm) and refractive index detectors as previously described [25]. A pre-column Guard Column SCR (H) (50 mm $\times$ 4 mm id) with stationary phase sulfonated styrene-divinylbenzene copolymer resin was used. The chromatography flow rate was 0.6 mL/min and an injection volume of 20 μL, using a Shim-pack SCR-101H (Shimadzu) (300 mm $\times$ 7.9 mm id) column equilibrated at 60 °C with 5 mM H_2SO_4 as the mobile phase. For biomass determination, samples collected for OD_{600nm} measurement were dried and then weighed for analysis of biomass dry cell weight (DCW). A calibration curve (1 unit of OD_{600nm} corresponded to 0.31 g DCW/L) was used to convert OD_{600nm} to DCW (g/L).

3. Results and Discussion

3.1. Construction of pUG6-PDCK and PDC Knockout P. pastoris Strains

Pyruvate Decarboxylase activity has been deleted in other yeast species to reduce by-product formation such as acetate and ethanol as well as concomitant increases in lactic acid production [5,6,26]. Therefore, in order to verify whether lactic acid production could also be increased in *P. pastoris*, the PDC gene was interrupted. To that end, a knockout cassette with a kanamycin resistance marker was constructed (Figure 1) and used for yeast transformation.

After transformation into the *P. pastoris* X-33 strain, 300 colonies were screened for the *PDC*-knockout cassette using PCR amplification, without success. The amplification of a fragment size of approximately 3.3 Kb was expected, indicating the insertion of the cassette. However, only the fragment of 1.6 Kb was detected, showing amplification of only the intact *PDC* gene (Figure 2 upper panel).

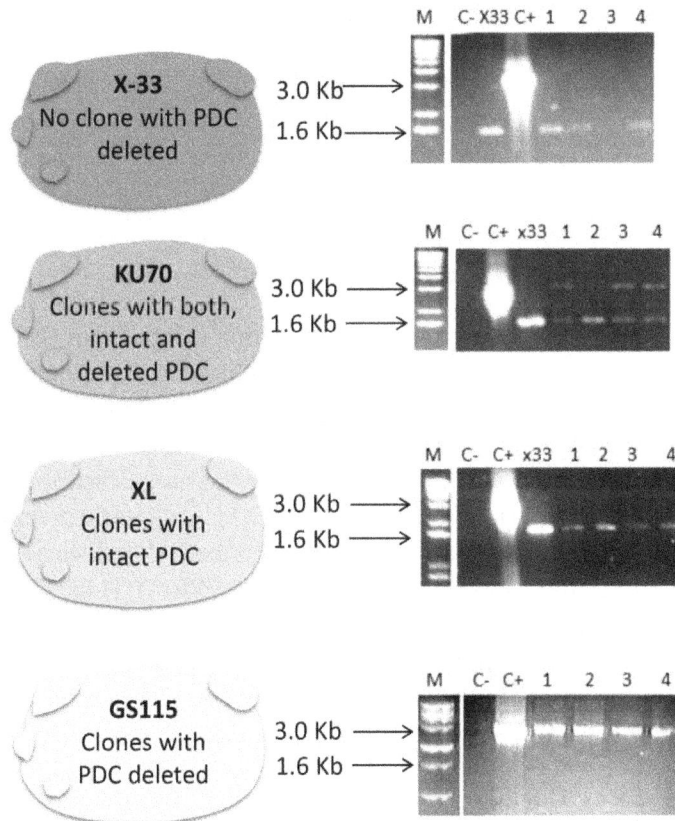

Figure 2. The four *P. pastoris* strains used for the insertion of the *PDC* knockout cassette with their respective results among all screened clones. The PCR results using the primers PDC5′F and PDC3′R are shown after electrophoresis on a 0.8% agarose gel. Kb: kilobases; M: molecular marker; C−: negative control (no DNA template in the PCR reaction); C+: positive control (Plasmid pUG6-PDCK as template in the PCR reaction), X-33-*Pichia pastoris* wildtype strain where *PDC* is not deleted. Numbers 1–4 are samples of different clones that were evaluated for deletion of *PDC*.

Therefore, to improve the efficiency of homologous recombination, the Ku70 strain was utilized. This strain has been previously described as having an increased frequency of homologous recombination [20]. After yeast transformation, all the selected clones presented amplification of both intact *PDC* (1.6 Kb) and *PDC*-knockout (3.3 Kb) (KU070 results in Figure 2). All clones were then re-plated on YPD supplemented with geneticin (G418) (500 $\mu g/mL^{-1}$), 5 times. After the final plating, isolated colonies had their genomic DNA extracted and used as template for PCR reactions. Of these, all the clones presented only the 1.6 Kb fragments, which indicated the presence of intact *PDC* (data not shown).

The deletion of the *PDC* gene has been shown to reduce or impair growth in other yeast species [8,27], which may explain why no single Ku70 colony with interrupted *PDC* could be isolated. Therefore, the *P. pastoris* XL strain that has lactate dehydrogenase activity [21] was tested. The hypothesis was that the presence of a novel route for NAD regeneration, which was not respiratory, could reestablish growth in PDC-defective strains. Nevertheless, after screening approximately 200 colonies that were resistant to Geneticin (G418), all clones had amplification profiles matching intact *PDC* (XL results in Figure 2). To understand why antibiotic resistant colonies did not have *PDC* knockout, PCR using the specific primers KanF and KanR (Table 2) as a Geneticin marker were utilized. In all tested colonies, the amplification of uninterrupted *PDC* genes and the Geneticin-resistant marker were observed (data not shown). Hence, the resistance marker was integrated into *P. pastoris* genome by non-homologous integration.

Lastly, transformation of the *P. pastoris* GS115 strain was performed. This strain has the HIS4 enzyme gene deleted, and is considered the most popular strain for production of heterologous proteins [28]. Since it has a well-established deletion in its genome, the knockout of *his4*, it was hypothesized that it could facilitate a further insertion. After GS115 transformation, colonies selected in the presence of geneticin (G418) had their genomic DNA extracted. As can be seen in the lower panel of Figure 2, evaluation of all clones resulted in the amplification of a 3.3 Kb fragment that indicated the insertion of the *PDC* knockout cassette into the yeast chromosome (Table 2).

3.2. Determination of Ploidy in Different P. pastoris Strains

In order to understand why *PDC* knockout was easily performed in GS115 and could not be achieved in a single colony for X-33, Ku70 and XL, flow cytometer experiments were performed to determine the total amount of DNA for each strain. This technique has been previously utilized for separating metabolically-defective cells during fed-batch cultures of recombinant X-33 producing both trypsinogen and horseradish peroxidase [29]. The X-33 strain has been described as wildtype and there are no reports stating whether it is haploid [30]. Ku70 and XL, being derivatives of X33, should have, in theory, the same DNA content. Finally, the GS115 strain has been described as haploid and has been generated by random mutagenesis from the parental strain NRRL Y-11430 [18,31]. Flow cytometer experiments showed that X33 and GS115 have the same DNA content, thus not explaining why integration occurred in the PDC fragment in GS115 and not in X-33 (Supplementary Material S1). As such, a possible explanation for obtaining amplification of the deletion cassette and intact PDC in Ku70 strains would be the integration of the entire deletion cassette in another region of the genome.

However, this does not explain the successful results obtained when *PDC* knockout was attempted in these strains (GS115). For the strains where *PDC* intact and knockout profile could be detected within the same colony, serial dilution was carried out unsuccessfully in order to obtain a pure profile (data not shown). Therefore, it has been hypothesized that as happens in other yeasts such as *S. cerevisiae* [5,6] and *K. lactis* [8,9], *PDC* knockout reduces biomass which makes its selection difficult when performing serial dilution. Later in this study, the fermentation results corroborated this hypothesis where a decrease of 35% of biomass could be seen in Gp when compared to GS115 (Table 3).

Table 3. Kinetic parameters during fermentation experiments at aerobic and oxygen limited phases. Y: yield, s: substrate, x: biomass, lac: lactate, ac: acetate, ara: arabitol, Y: g/g, q: g/g/h, r: g/L/h. Experiments were performed in biological triplicates. Yields during the oxygen limited phase were calculated upon 40% glycerol feeding at the end of the aerobic phase.

Strain	$Y_{x/s}$	$Y_{lac/s}$	$Y_{lac/x}$	$Y_{ac/s}$	$Y_{ara/s}$	μ	q_{lac}	q_{ac}	q_{ara}	
GS115	0.197 ± 0.025	0.015 ± 0.002	0.075 ± 0.002	0.062 ± 0.001	0.170 ± 0.064	0.023 ± 0.001	0.002 ± 0.001	0.007 ± 0.001	0.021 ± 0.011	Aerobic
Gp	0.204 ± 0.014	0.045 ± 0.001	0.219 ± 0.012	0.003 ± 0.001	0.247 ± 0.028	0.028 ± 0.005	0.006 ± 0.001	0.000 ± 0.000	0.033 ± 0.009	
GLp	0.138 ± 0.018	0.110 ± 0.003	0.807 ± 0.126	0.008 ± 0.001	0.308 ± 0.003	0.021 ± 0.003	0.017 ± 0.000	0.001 ± 0.000	0.047 ± 0.000	
GS115	0.319 ± 0.021	0.007 ± 0.001	0.022 ± 0.002	0.102 ± 0.000	0.025 ± 0.016	0.015 ± 0.001	0.000 ± 0.000	0.005 ± 0.001	0.002 ± 0.001	Oxygen Limited
Gp	0.217 ± 0.014	0.013 ± 0.002	0.058 ± 0.014	0.026 ± 0.011	0.067 ± 0.023	0.017 ± 0.001	0.001 ± 0.000	0.002 ± 0.001	0.006 ± 0.002	
GLp	0.194 ± 0.012	0.646 ± 0.054	3.34 ± 0.072	0.081 ± 0.015	0.174 ± 0.038	0.015 ± 0.003	0.050 ± 0.009	0.006 ± 0.002	0.014 ± 0.007	

3.3. Product Distribution in P. pastoris Strains during Bioreactor Cultivation

After confirming the deletion of *PDC* in GS115 generating the GP strain (Table 1), the *LDHb* gene was inserted into that strain to produce lactic acid, and this was designated the GLp strain (Table 1). Subsequently, LDH activities were measured in both the XL and GLp strains (Supplementary Information S2). As can be seen, GLp had about 40% lower LDH activity when compared to GL, most probably due to a difference in copy integration of the LDH-encoding gene. Nevertheless, since this was the highest activity achieved in a strain where PDC was interrupted, the experiments were conducted using this strain.

The impact of both *PDC* gene knockout and insertion of the bovine lactate dehydrogenase gene on the production of L-lactic acid in *P. pastoris*, as well as the evaluation of by-products such as ethanol, acetate and arabitol, were investigated (Figure 3 and Table 3). The cultivations in the bioreactors were divided into two phases (Figure 3). In the first, termed the aerobic phase, the dissolved oxygen was maintained at 30%, and 8% glycerol was fed to the strains. In the second, which was initiated by pH level alteration to >5.0, the dissolved oxygen was limited to 3%, and this was therefore named the restricted aerobic phase, where 40% glycerol was added in a single pulse.

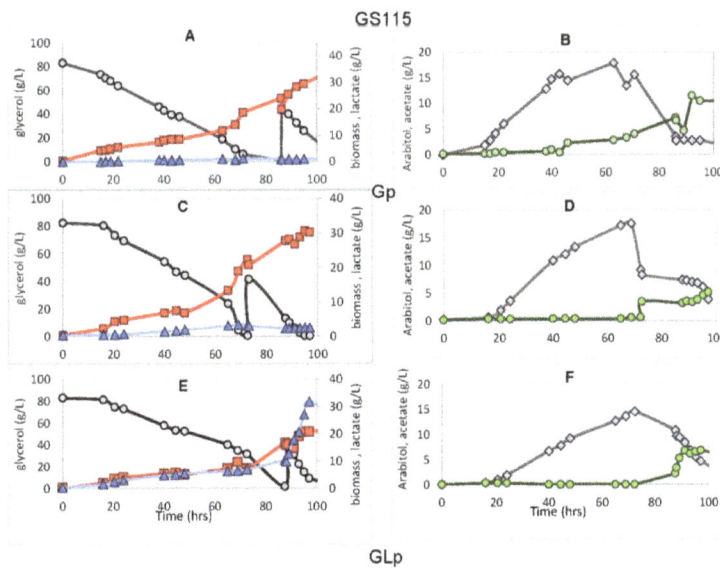

Figure 3. Fermentation profile of the strains GS115 (**A,B**), Gp (**C,D**), and GLp (**E,F**) showing consumption of glycerol (gray circle) and the production of biomass (red square), lactate (blue triangle), arabitol (gray diamond) and acetate (green circle). 4% glycerol was added at 70–80 fermentation hours when oxygen was limited. Experiments were performed in biological triplicates, and the figure shows a typical fermentation profile.

The Gp strain harboring *PDC* knockout reduced acetate formation by approximately 20-fold and 4-fold in the aerobic and restricted aerobic phases, respectively, when compared with GS115 (Table 3). On the other hand, it can be observed that the production of arabitol was approximately 1.5-fold and 3.0-fold higher than GS115 in the same phases, respectively (Figure 3 and Table 3).

Regarding biomass formation, PDC deletion did not reduce biomass formation aerobically. Nevertheless, for the GLp strain that had both *PDC* deleted and insertion of LDH activity, the biomass yield was reduced by approximately 30% and 10% in the aerobic and restricted aerobic phases, respectively, when compared to the Gp strain harboring only *PDC* knockout (Table 3). Nevertheless, when aerobic and oxygen-limited conditions are compared within the same strain, biomass yields are slightly higher for the latter, even though the specific growth rate is lower (Table 3). Since this is not a fermentative yeast even under oxygen-limiting conditions, the main yeast product is still biomass. Altogether these results show that under aerobic conditions the main yeast products are biomass and arabitol for all strains independently of the inserted genetic modification. Nevertheless, under oxygen-limited conditions, the lactate production was favored over biomass, where the lactic acid yield was approximately five times higher than at the aerobic phase.

However, different than what was observed from the comparison between GS115 and Gp, GLp led to an acetate yield increase of 3-fold for the restricted aerobic phase (Table 3). Although the production of arabitol followed the same pattern, the yield increase was lower from Gp to GLp (1.2-fold and 2.5-fold for aerobic and restricted aerobic phases, respectively), than from GS115 to Gp. Moreover, under oxygen-limited conditions, GLp presented an arabitol yield decrease of approximately 45%, reaching a specific productivity rate of approximately 3.5-fold lower than that observed in the first phase (Table 3). This suggests that during the aerobic phase, excess NADH may be oxidized via arabitol production (Figure 4). On the other hand, under oxygen-limited conditions this was minimized by the addition of the LDH activity, which is also a pathway that leads to NADH oxidation.

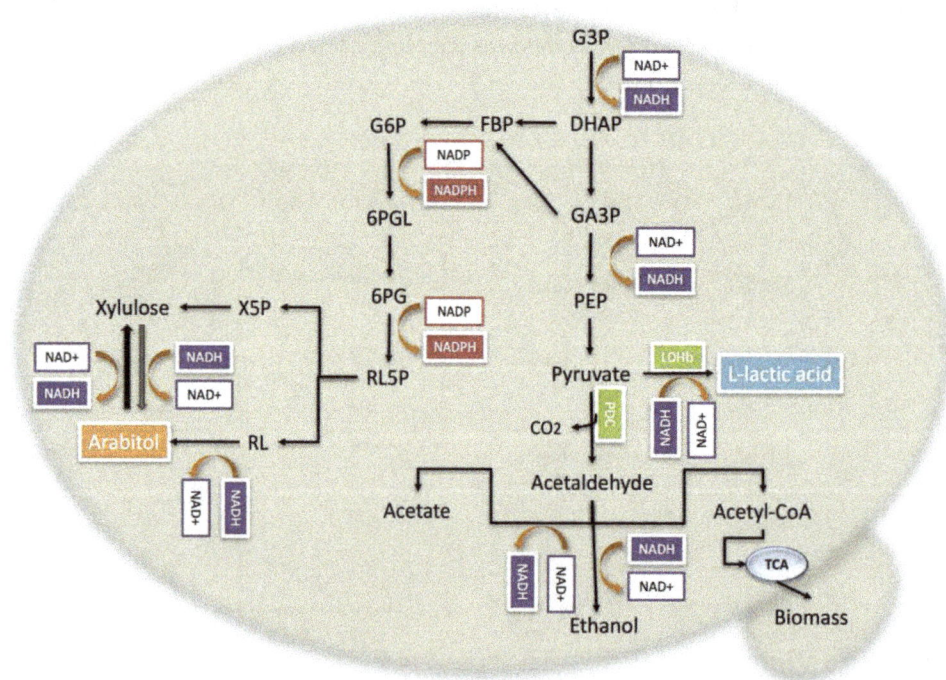

Figure 4. Glycerol metabolism of *P. pastoris* GLp strain showing the main pathways involved in NADH oxidation. G3P: Glycerol-3-phosphate; DHAP: Dihydroxyacetone; GA3P: glyceraldehyde-3-phosphate; PEP: phosphoenolpyruvate; LHHb: bovine L-lactic acid dehydrogenase; PDC: pyruvate decarboxylase; FBP: fructose-biphosphate; G6P: glucose-6-phosphate; 6PGL: phosphogluconolactone; 6PG: phosphogluconate; TCA: tricarboxylic acid; NAD: nicotinamide adenine dinucleotide; NADH: reduced NAD.

When compared with a previous study, where a putative lactate transporter from *P. pastoris* was inserted into the GS115 strain harboring the *LDHb* gene [21], the GLp strain proved to be a more efficient biocatalyst, converting substrate into L-lactic acid faster than GLS: qlac = 0.146 h^{-1}; GLp: qlac = 0.05 h^{-1}, while producing the same amount of biomass (GLS: qx = 0.014 g/g/h; GLp: qx = 0.015 g/g/h) (Table 3) [21]. In fact, the yields for GLS and GLp are 0.673 g/g and 0.646 g/g respectively. Although very similar, the GLS strain is much slower at reducing, requiring about 48 h more to achieve the same lactic acid yield.

3.4. Effect of PDC Interruption on L-Lactic Acid Production

P. pastoris strains engineered for production of lactic acid have not yet reached the maximum theoretical yield [21]. In fact, only a few yeast species have been modified to increase lactic acid yield, including *K. lactis* and *S. cerevisiae* [32,33]. *K. lactis* has only one *PDC* gene, which is non-essential for its survival when grown on glucose [26]. Suppression of this gene together with knockout of the pyruvate dehydrogenase E1 alpha sub-unit gene increases lactic acid production, reaching a yield of 0.85 g/g [8].

S. cerevisiae is the most widely studied yeast for production of lactic acid, mainly due to its fermentative metabolism [8,33–35]. *S. cerevisiae* has two active structural *PDC* genes, *PDC1* and *PDC5*, and a third inactive gene named *PDC6* [27]. It has been previously reported that double deletion of *PDC1* and *PDC5* led to efficient production of L-lactic acid (82 g/L), however, cell growth was also reduced in the knockout strain [7]. Among the *PDC* genes, *PDC1* has been shown to have the greatest effect on cell growth and suppression of ethanol production (yields from 0.35 g/g to 0.20 g/g), concomitant with an improvement in lactate yield from glucose (from 0.155 g/g to 0.20 g/g) [36]. Another study has shown that double disruption of the *PDC1* and alcohol dehydrogenase genes results in an almost two-fold increase in L-lactic acid production, reaching a yield of 0.75 g/g in glucose [5]. However, it has been recently reported that while *LDH A* from the fungus *Rhizopus oryzae* has been used instead of the bovine *LDH* gene, no gene deletions are required to reach yields of up to 0.69 g/g of lactic acid using *S. cerevisiae* when xylose is used instead of glucose [37]. This latter study showed that when culturing the yeast on xylose under oxygen-limited conditions, the major fermentation product was lactic acid, showing an increase of 3-fold while ethanol production decreased 30-fold [37].

PDC knockout was shown to eliminate acetate production and thus improve lactic acid formation in other yeasts [8,26]. In this study, higher lactic acid titers were also achieved using the strain with interrupted *PDC*. Nevertheless, although acetic acid was reduced, it can be seen that it was not completely eliminated. Examination of the *P. pastoris* genome annotation revealed that this yeast has one gene annotated as *PDC* (XP_002492397) and two other putative PDC enzymes. Since we deleted only XP_002492397, our hypothesis is that the other isoforms are still converting pyruvate to acetaldehyde.

3.5. The Redox Mechanism when PDC Is Interrupted

The insertion of the *PDC* knockout cassette into the GS115 strain led to a 1.6-fold increase in arabitol production (Table 3). Arabitol belongs to the pentitol family and is used in the food, health and chemical industries [38]. It has been demonstrated that several yeasts, mainly from the genera *Debaryomyces*, *Geotrichum*, *Metchnikowia*, *Candida* and *Dipodascus*, are able to produce significant amounts of arabitol from glycerol (up to 41.7 g/L) [39]. The *P. pastoris* GS115 strain has been previously shown to produce D-arabitol from glucose [40].

In yeast cells, arabitol synthesis from glycerol is expected to follow similar routes to those leading from glucose [38]. According to the genome annotation, after glycerol is assimilated by the cells, it follows two steps of phosphorylation and is converted into pyruvate, producing two NADH molecules (Figure 4). This latter compound may then follow one of two pathways, one being the conversion to glyceraldehyde-3–phosphate (GA3P) by the enzyme triose phosphate isomerase, with subsequent conversion to pyruvate, which is in turn converted into L-lactic acid by the NADH-dependent bovine lactate dehydrogenase, resulting in a NADH surplus per mol of consumed glycerol under oxygen-limited conditions.

In the GLp strain, since the *PDC* gene is interrupted, NADH cannot be regenerated under oxygen-limited conditions by the formation of ethanol. This unbalanced co-factor concentration would then drive DHAP into arabitol formation by reduction of either ribulose or xylulose through reoxidation of NADH, which then reestablishes the redox balance of the yeast (Figure 4). Indeed it has been previously reported for *P. pastoris* that the effect of reduced oxygen supply from 21% to 8% in the core metabolism led to an increase in arabitol yield of 220%, reflecting adaptation from a respiratory to a respiratory-fermentative metabolism [41].

4. Conclusions

In this study, the disruption of *P. pastoris* genes required the analysis of many clones, since homologous recombination was not highly efficient even when performed with more than 500 bp of homology. Although different strain backgrounds were tested for deletion of the *PDC* gene by homologous recombination, a positive result was only achieved for GS115. Fermentation experiments showed that during the oxygen-limited phase the GLp strain achieved a lactic acid yield of 0.65 g/g. In this strain, acetate formation was reduced by 20% but with a concomitant 7-fold increase in arabitol production when compared to wildtype strain GS115. This may be explained as a route for NADH regeneration under oxygen-limited conditions.

Acknowledgments: This work was supported by CNPQ and FAPDF.

Author Contributions: Nadiele T. M. Melo Strain construction and enzymatic assays, and helped write the manuscript; Kelly C. L. Mulder Wrote the manuscript and analyzed all fermentation data; André Moraes Nicola Performed the experiment with the flow cytometer; Lucas S. Carvalho Strain construction and helped analyze the data and built the figures for the molecular biology experiments; Gisele S. Menino Performed Fermentation experiments; Eduardo Mulinari Performed Fermentation experiments; Nádia S. Parachin Supervised the research, designed the experiments, analyzed the data.

Conflicts of Interest: The authors declare no conflict of interest.

References

1. Tsuji, H. Poly(lactic acid) stereocomplexes: A decade of progress. *Adv Drug Deliv Rev.* **2016**, *107*, 97–135. [CrossRef] [PubMed]

2. Lasprilla, A.J.R.; Martinez, G.A.R.; Lunelli, B.H.; Jardini, A.L.; Filho, R.M. Poly-lactic acid synthesis for application in biomedical devices—A review. *Biotechnol. Adv.* **2012**, *30*, 321–328. [CrossRef] [PubMed]

3. Okino, S.; Suda, M.; Fujikura, K.; Inui, M.; Yukawa, H. Production of D-lactic acid by *Corynebacterium glutamicum* under oxygen deprivation. *Appl. Microbiol. Biotechnol.* **2008**, *78*, 449–454. [CrossRef] [PubMed]

4. Juturu, V.; Wu, J.C. Microbial production of lactic acid: the latest development. *Crit. Rev. Biotechnol.* **2016**, *36*, 967–977. [CrossRef] [PubMed]

5. Tokuhiro, K.; Ishida, N.; Nagamori, E.; Saitoh, S.; Onishi, T.; Kondo, A.; Takahashi, H. Double mutation of the PDC1 and ADH1 genes improves lactate production in the yeast *Saccharomyces cerevisiae* expressing the bovine lactate dehydrogenase gene. *Appl. Microbiol. Biotechnol.* **2009**, *82*, 883–890. [CrossRef] [PubMed]

6. Ishida, N.; Saitoh, S.; Onishi, T.; Tokuhiro, K.; Nagamori, E.; Kitamoto, K.; Takahashi, H. The Effect of Pyruvate Decarboxylase Gene Knockout in *Saccharomyces cerevisiae* on L-Lactic Acid Production. *Biosci. Biotechnol. Biochem.* **2006**, *70*, 1148–1153. [CrossRef] [PubMed]

7. Ishida, N.; Saitoh, S.; Tokuhiro, K.; Nagamori, E.; Matsuyama, T.; Kitamoto, K.; Takahashi, H. Efficient production of L-lactic acid by metabolically engineered *Saccharomyces cerevisiae* with a genome-integrated L-lactate dehydrogenase gene. *Appl. Environ. Microbiol.* **2005**, *71*, 1964–1970. [CrossRef] [PubMed]

8. Bianchi, M.M.; Brambilla, L.; Protani, F.; Liu, C.L.; Lievense, J.; Porro, D. Efficient Homolactic Fermentation by *Kluyveromyces lactis* Strains Defective in Pyruvate Utilization and Transformed with the Heterologous LDH Gene. *Appl. Environ. Microbiol.* **2001**, *67*, 5621–5625. [CrossRef] [PubMed]

9. Porro, D.; Bianchi, M.M.; Brambilla, L.; Menghini, R.; Bolzani, D.; Carrera, V.; Lievense, J.; Liu, C.L.; Ranzi, B.M.; Frontali, L.; Alberghina, L. Replacement of a metabolic pathway for large-scale production of lactic acid from engineered yeasts. *Appl. Environ. Microbiol.* **1999**, *65*, 4211–4215. [PubMed]

10. Branduardi, P.; Valli, M.; Brambilla, L.; Sauer, M.; Alberghina, L.; Porro, D. The yeast *Zygosaccharomyces bailii*: A new host for heterologous protein production, secretion and for metabolic engineering applications. *FEMS Yeast Res.* **2004**, *4*, 493–504. [CrossRef]

11. Ilmén, M.; Koivuranta, K.; Ruohonen, L.; Rajgarhia, V.; Suominen, P.; Penttilä, M. Production of L-lactic acid by the yeast *Candida sonorensis* expressing heterologous bacterial and fungal lactate dehydrogenases. *Microb. Cell Factor.* **2013**, *12*, 1–15. [CrossRef]

12. Osawa, F.; Fujii, T.; Nishida, T.; Tada, N.; Ohnishi, T.; Kobayashi, O.; Komeda, T.; Yoshida, S. Efficient production of L-lactic acid by Crabtree-negative yeast *Candida boidinii. Yeast* **2009**, *26*, 485–496. [CrossRef] [PubMed]

13. Ilmén, M.; Koivuranta, K.; Ruohonen, L.; Suominen, P.; Penttilä, M. Efficient production of L-lactic acid from xylose by *Pichia stipitis. Appl. Environ. Microbiol.* **2007**, *73*, 117–123. [CrossRef] [PubMed]

14. Kurtzman, C.P. Biotechnological strains of *Komagataella (Pichia) pastoris* are Komagataella pha Y i as determined from multigene. *J. Ind. Microbiol. Biotechnol.* **2009**, *36*, 1435–1438. [CrossRef]

15. Kurtzman, C.P. Description of Komagataella phaffii sp . nov . and the transfer of *Pichia pseudopastoris* to the methylotrophic yeast genus Komagataella. *Int. J. Syst. Evolut. Microbiol.* **2005**, *55*, 973–976. [CrossRef]

16. Looser, V.; Brühlmann, B.; Bumbak, F.; Stenger, C.; Costa, M.; Camattari, A.; Fotiadis, D.; Kovar, K. Cultivation strategies to enhance productivity of *Pichia pastoris*: A review. *Biotechnol. Adv.* **2015**, *33*, 1177–1193. [CrossRef] [PubMed]

17. Anastácio, G.S.; Santos, K.O.; Suarez, P.A.Z.; Torres, F.A.G.; De Marco, J.L.; Parachin, N.S. Utilization of glycerin byproduct derived from soybean oil biodiesel as a carbon source for heterologous protein production in *Pichia pastoris. Bioresour. Technol.* **2014**, *152*, 505–510. [CrossRef] [PubMed]

18. Cregg, J.M.; Barringer, K.J.; Hessler, A.Y.; Madden, K.R. *Pichia pastoris* as a host system for transformations. *Mol. Cell. Biol.* **1985**, *5*, 3376–3385. [CrossRef] [PubMed]

19. De Schutter, K.; Lin, Y.; Tiels, P.; Van Hecke, A.; Glinka, S.; Weber-Lehmann, J.; Rouzé, P.; Van de Peer, Y.; Callewaert, N. Genome sequence of the recombinant protein production host *Pichia pastoris. Nat. Biotechnol.* **2009**, *27*, 561–566. [CrossRef] [PubMed]

20. Näätsaari, L.; Mistlberger, B.; Ruth, C.; Hajek, T.; Hartner, F.S.; Glieder, A. Deletion of the *Pichia pastoris* ku70 homologue facilitates platform strain generation for gene expression and synthetic biology. *PLoS ONE* **2012**, *7*, e.39720. [CrossRef] [PubMed]

21. De Lima, P.B.A.; Mulder, K.C.L.; Melo, N.T.M.; Carvalho, L.S.; Menino, G.S.; Mulinari, E.; de Castro, V.H.; dos Reis, T.F.; Goldman, G.H.; Magalhães, B.S.; Parachin, N.S. Novel homologous lactate transporter improves L-lactic acid production from glycerol in recombinant strains of *Pichia pastoris. Microb. Cell Fact.* **2016**, *15*, 158. [CrossRef] [PubMed]

22. Agarwal, P.K.; Uppada, V.; Noronha, S.B. Comparison of pyruvate decarboxylases from *Saccharomyces cerevisiae* and *Komagataella pastoris (Pichia pastoris). Appl. Microbiol. Biotechnol.* **2013**, *97*, 9439–9449. [CrossRef] [PubMed]

23. Tarmy, E.M.; Kaplan, N.O. Chemical characterization of D-lactate dehydrogenase from *Escherichia coli* B. *J. Biol. Chem.* **1968**, *243*, 2579–2586. [PubMed]

24. Maurer, M.; Kühleitner, M.; Gasser, B.; Mattanovich, D. Versatile modeling and optimization of fed batch processes for the production of secreted heterologous proteins with *Pichia pastoris. Microb. Cell Fact.* **2006**, *5*, 37. [CrossRef] [PubMed]

25. Doyon, G.; Gaudreau, G.; St-Gelais, D.; Beaulieu, Y.; Randall, C.J. Simultaneous HPLC Determination of Organic Acids, Sugars and Alcohols. *Can. Inst. Food Sci. Technol. J.* **1991**, *24*, 87–94. [CrossRef]

26. Bianchi, M.M.; Tizzani, L.; Destruelle, M.; Frontali, L.; Wésolowski-Louvel, M. The "petite-negative" yeast *Kluyveromyces lactis* has a single gene expressing pyruvate decarboxylase activity. *Mol. Microbiol.* **1996**, *19*, 27–36. [CrossRef] [PubMed]

27. Hohmann, S. Characterization of PDC6, a third structural gene for pyruvate decarboxylase in *Saccharomyces cerevisiae. J. Bacteriol.* **1991**, *173*, 7963–7969. [CrossRef] [PubMed]

28. Li, P.; Anumanthan, A.; Gao, X.G.; Ilangovan, K.; Suzara, V.V.; Duzgunes, N.; Renugopalakrishnan, V. Expression of recombinant proteins in Pichia pastoris. *Appl. Biochem. Biotechnol.* **2007**, *142*, 105–124. [CrossRef] [PubMed]

29. Hyka, P.; Züllig, T.; Ruth, C.; Looser, V.; Meier, C.; Klein, J.; Melzoch, K.; Glieder, A.; Kovar, K.; Zu, T.; Meyer, H. Combined Use of Fluorescent Dyes and Flow Cytometry To Quantify the Physiological State of Pichia pastoris during the Production of Heterologous Proteins in High-Cell-Density Fed-Batch Cultures. *Appl. Environ. Microbiol.* **2010**, *76*, 4486–4496. [CrossRef] [PubMed]

30. Küberl, A.; Schneider, J.; Thallinger, G.G.; Anderl, I.; Wibberg, D.; Hajek, T.; Jaenicke, S.; Brinkrolf, K.; Goesmann, A.; Szczepanowski, R.; et al. High-quality genome sequence of *Pichia pastoris* CBS7435. *J. Biotechnol.* **2011**, *154*, 312–320. [CrossRef] [PubMed]

31. Love, K.R.; Shah, K.A.; Whittaker, C.A.; Wu, J.; Bartlett, M.C.; Ma, D.; Leeson, R.L.; Priest, M.; Borowsky, J.; Young, S.K.; Love, J.C. Comparative genomics and transcriptomics of *Pichia pastoris*. *BMC Genomics* **2016**, *17*, 550. [CrossRef] [PubMed]

32. Porro, D.; Brambilla, L.; Ranzi, B.M.; Martegani, E.; Alberghina, L. Development of metabolically engineered *Saccharomyces cerevisiae* cells for the production of lactic acid. *Biotechnol. Prog.* **1995**, *11*, 294–298. [CrossRef] [PubMed]

33. Sauer, M.; Porro, D.; Mattanovich, D. 16 years research on lactic acid production with yeast—Ready for the market? *Biotechnol. Genet. Eng. Rev.* **2010**, *27*, 229–256. [CrossRef] [PubMed]

34. Lee, J.Y.; Kang, C.D.; Lee, S.H.; Park, Y.K.; Cho, K.M. Engineering cellular redox balance in *Saccharomyces cerevisiae* for improved production of L-lactic acid. *Biotechnol. Bioeng.* **2015**, *112*, 751–758. [CrossRef] [PubMed]

35. Porro, D.; Brambilla, L.; Ranzi, B.M.; Martegani, E.; Generali, B.; Comparata, S.B.; Milano, U. Development of Metabolically Engineered. *Society* **1995**, 294–298.

36. Adachi, E.; Torigoe, M.; Sugiyama, M.; Nikawa, J.I.; Shimizu, K. Modification of metabolic pathways of *Saccharomyces cerevisiae* by the expression of lactate dehydrogenase and deletion of pyruvate decarboxylase genes for the lactic acid fermentation at low pH value. *J. Ferment. Bioeng.* **1998**, *86*, 284–289. [CrossRef]

37. Turner, T.L.; Zhang, G.C.; Kim, S.R.; Subramaniam, V.; Steffen, D.; Skory, C.D.; Jang, J.Y.; Yu, B.J.; Jin, Y.S. Lactic acid production from xylose by engineered *Saccharomyces cerevisiae* without PDC or ADH deletion. *Appl. Microbiol. Biotechnol.* **2015**, *99*, 8023–8033. [CrossRef] [PubMed]

38. Kordowska-Wiater, M. Production of arabitol by yeasts: current status and future prospects. *J. Appl. Microbiol.* **2015**, *119*, 303–314. [CrossRef] [PubMed]

39. Koganti, S.; Kuo, T.M.; Kurtzman, C.P.; Smith, N.; Ju, L.K. Production of arabitol from glycerol: Strain screening and study of factors affecting production yield. *Appl. Microbiol. Biotechnol.* **2011**, *90*, 257–267. [CrossRef] [PubMed]

40. Cheng, H.; Lv, J.; Wang, H.; Wang, B.; Li, Z.; Deng, Z. Genetically engineered *Pichia pastoris* yeast for conversion of glucose to xylitol by a single-fermentation process. *Appl. Microbiol. Biotechnol.* **2014**, *98*, 3539–3552. [CrossRef] [PubMed]

41. Baumann, K.; Carnicer, M.; Dragosits, M.; Graf, A.B.; Stadlmann, J.; Jouhten, P.; Maaheimo, H.; Gasser, B.; Albiol, J.; Mattanovich, D.; Ferrer, P. A multi-level study of recombinant Pichia pastoris in different oxygen conditions. *BMC Syst. Biol.* **2010**, *4*, 141. [CrossRef] [PubMed]

4

Microbiological Sensing Technologies

Firouz Abbasian [ID]**, Ebrahim Ghafar-Zadeh * and Sebastian Magierowski**

Biologically Inspired Sensors and Actuators Laboratory, Department of EECS, Lassonde School of Engineering, York University, Toronto, ON M3J 1P3, Canada; fabbasian@cse.yorku.ca (F.A.); magiero@cse.yorku.ca (S.M.)
* Correspondence: egz@cse.yorku.ca

Abstract: Microorganisms have a significant influence on human activities and health, and consequently, there is high demand to develop automated, sensitive, and rapid methods for their detection. These methods might be applicable for clinical, industrial, and environmental applications. Although different techniques have been suggested and employed for the detection of microorganisms, and the majority of these methods are not cost effective and suffer from low sensitivity and low specificity, especially in mixed samples. This paper presents a comprehensive review of microbiological techniques and associated challenges for bioengineering researchers with an engineering background. Also, this paper reports on recent technological advances and their future prospects for a variety of microbiological applications.

Keywords: biosensors; bacteria; microbiology; Lab on chip (LoC)

1. Introduction

Microorganisms are present everywhere and are involved in many clinical, industrial, and environmental phenomena. While the outbreak and richness of infectious diseases and food poisoning that are caused by microorganisms has declined through the years, the total numbers of these diseases is still highly prevalent [1,2]. Furthermore, the contamination of food/drug and cosmetic products with microorganisms or their derivatives remains a big challenge for industries operating in these markets. Present in all of these issues is the need for timely, ideally on-line, analyses of dangers, and compromised materials. As a result, there is a growing emphasis on the realization and application of real-time detection devices with high sensitivity and specificity for identification of microorganisms present in clinical/environmental/industrial samples [3,4]. This research approach has attracted the attention of engineering researchers to contribute in this field by proposing new solutions for current microbiological challenges. To date, many papers have reported the design and implementation of engineered microbiological protocols, such as microorganism biosensors [5–9]. With anticipation that such technologies will be dominated by miniaturized systems in the future, in addition to introducing the current microbiological practices to engineering researchers, we also review the most recent technological advances that are associated with this field of research.

Laboratory-on-a-chip (LoC) devices, or micro total analysis systems (µTAS), offer great advantages in terms of their material (e.g., reaction substrates) and time (i.e., rapid loading and analysis overhead) requirements. As a result, they are one of the most promising systems for the development of high throughput and automated biosensors [6,7]. LoC systems featuring a number of microscale reaction chambers and channels are used to prepare samples and to deliver analytes (e.g., bacteria, DNA, etc.) toward miniaturized embedded sensing sites [8]. The core part of a LoC system designed for microbial detection is a biosensor, which itself consists of a recognition element and a readout system [8]. The recognition elements, e.g., antibodies, bacteriophages, antimicrobial peptides, and bacteriocins, are used to convert the biological phenomenon to a physical or chemical variation (Figure 1) [10,11]. Indeed, any microbial features, or the presence of any specific factor in a special microorganism, including genomic elements, antigenic properties, electromechanical properties, metabolic activities, and/or photographic indexes are potentially useful for the detection of microorganisms in a sample [10]. The readout system or physicochemical transducers are used to sense, amplify, and measure the signals (mechanical, optical, electrochemical, and acoustic changes) obtained from the recognition element [12]. A readout system can be implemented using microelectromechanical system (MEMS) technology or microelectronic technology [13].

Among various competing microelectronic technologies for reading out the biosensors, complementary metal–oxide–semiconductor (CMOS) offers the capability of hosting complex analog and digital integrated circuits for signal processing and sensors for transduction on the same chip to track and analyze millions of changes in the nano/micro scale levels in a time and cost effective manner [13]. The potential for the mass industrial production of such CMOS devices coupled with high-performance data processing and machine-learning back-ends can make inexpensive and autonomous devices available to personal health care systems [13]. To date, many papers have reported the advantages of CMOS for a multitude of biological applications, including DNA sequencing and bacterial growth analysis [14–16]. However, the majority of these devices suffer from the inability to differentiate between living and dead microorganisms and fail to separate and identify the etiologic microorganisms in a mixed microbial sample [14–16]. Consequently, the small size of bacterial cells (typically 0.5–5 μm), and the multi-epitope characterization of the bacterial cell surface, restricts the application of routine biosensors for the detection of whole intake cells of microorganisms [13,17]. As a result of the aforementioned biosensing limits, the majority of researchers in this field have focused on the detection of pure bacterial cultures using optical and electrical readout systems [13,17].

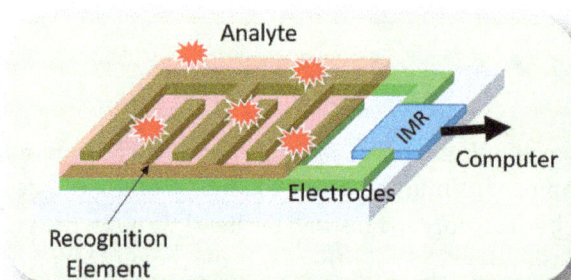

Figure 1. Illustration of a biosensor for the detection of microorganism including a recognition element which can be any specific agent, microelectrodes for impedance measurement, electronic impedance reader (IMR). The output of the IMR device is sent to computer.

LoC technologies can be split into culture-based and culture-free techniques. The culture-free systems rely on genomic and antigenic properties of microorganisms, and are limited to bench-top equipment that requires skilled technicians and are further hampered by the inability to differentiate between living and destroyed cells [18,19]. The principal for culture-based systems is, however, the measurement of changes in the electric and chemical properties of the microenvironment around the growing cells [20,21]. The main advantages of this system are direct detection of microorganisms without the complexity incumbent in techniques relying on fluorescent or genetic labels [20,21]. However, it must be taken into account that the selection of a method for the detection of microorganisms in a sample mostly depends on the nature of the microorganisms and their ability to respond to different chemical and physical treatments. It needs to be mentioned that there is the possibility of designing a detection device using a combination of these technologies to improve the quality and (maybe) the speed of microbial diagnosis.

The main focus of this paper is not placed on the recent advances of LoC, MEMS, or CMOS technologies for microbiological applications. Instead, the rest of this paper discusses the principles of the main microbiological methods. Among these methods, we will discuss the immobilization of bio-reporters on the surfaces for bio-sensing applications in Section 2. Also, the detection of intact microbial cells or a part of their cell contents, such as genomic contents, proteins, and fatty acids will be discussed in Sections 3–5. These sections will be followed up with a conclusion in Section 6. This section will also put forward the key challenges and critical future works in microbiological sensing technologies.

2. Immobilisation of Bio-Reporter on Functionalized Surfaces

Regardless of the technology used for the detection of microorganisms, the immobilisation of a biological particle on the surface of some measurement circuit is a real challenge. This step allows for the bio-reporter, which determines the specificity of a kit for a special target, to attach properly on the surface of the kit, and is therefore a crucial effort towards improved kit performance in terms of sensitivity, specificity and long-term stability [22]. Immobilization strategies differ mainly in the natures of the functionalized surface and the bio-reporter [22]. For different applications, the bio-reporter can be immobilized on different types of functionalized slide surfaces, including glass [23], silicon [24], PDMS (polydimethylsiloxane) [25], COC (cyclic olefin copolymer) [26], PS (polystyrene) [27], PMMA (polymethyl-methacrylate) [28], and metal films [29]. For the majority of these slides, the surface is first covered with a layer of a silanizing agent, such as siorganofunctional alkoxysilane molecules, which provide suitable functional residues for anchoring to functional groups (like -OH and $-NH_2$) of proteins [30]. Park et al., for instance, used sulfosuccinimidyl 6-[3-(2-pyridyldithio)propionamido]hexanoate (sulfo-LC-SPDP) as a heterobifunctional cross-linker that was able to make a N-S band with an antibody (Ab) at one side and thiol reaction with an aluminium layer on the surface of the slide [31]. The selection of synthetic layers mostly depends on their properties, such as softness, optical transparency, chemical resistance, as well as their final cost and facilitation of fabrication [32,33]. In addition to the superiority of these options, the slides made of PMMA, PS, and COC show lower auto-fluorescence and contain intrinsic functional groups for the attachment of proteins [34,35].

In addition to the composition of the functionalized surfaces, the application of three-dimensional surfaces can increase the levels of protein coverage on the surfaces, and therefore, can improve the sensitivity of the device [36,37]. The dimensional surfaces are created by the formation of microbeads [38], hydrogels (such as polyacrylamide gel and polyethylene glycol) [39], porous membranes [40], micropits [41], and microposts [42]. While the same silanization technology is used to build up the functional groups on the 3-D silicon/glass surfaces, the polymer-based 3-D surfaces, including polymer monoliths, hydrogels, and agarose beads can be adjusted for immobilization using oxidative activation of functional groups, graft polymerization, and copolymerization of protein [22].

Along with the challenges of analyte attachment on the electrodes, the deposition of undesired particles on the surfaces, referred to as fouling phenomenon, is another serious situation needed to be addressed for the design of an efficient functionalized surface. This problem is normally resolved with the application of various physical and chemical methods. PDMS, for instance, is commonly used for fabrication of microfluidic devices used in biological and chemical analysis. However, this material shows very high hydrophobicity, which causes a biofouling property and low wettability [43]. The physical methods used to resolve this failure, such as interaction with electrostatic or hydrophobic interactions and surface activation by ultraviolet (UV) light, ozone, and oxygen plasma, are very temporary and susceptible to a return of the hydrophobicity property [44]. However, chemical modification, in which multiple reagents are used for the preparation of the slides, improves the immobilization due to the creation of covalent cross-bonds between the solid layer and the bio-reporters [44].

Several factors, including the properties of the bio-reporters, the nature of the functional surface, the chemical activities of additives, such as buffers and co-factors, and also, the sensitivity and specificity of kits affect the bio-reporter immobilisation strategy most suitable for a specific purpose [45]. A bio-reporter is immobilised on the surface using covalent bonding (with the sulfhydryl, amine or carboxylated groups of aminoacids), bio-affinity interactions (specific interactions, such as avidin-biotin, protein G-Antibody, DNA hybridization with its complementary DNA and also, aptamers), physiosorption (physical adsorption of bio-reports by low energy intermolecular interactions, such as hydrogen, Van Der Waals, hydrophilic, and electrostatic bonds), or a combination of all these three factors [45]. Da Silva et al. [46] immobilized biotinylated bacterial cells to slides coated with polymerized biotinylated functionalized pyrrole using avidin as the interacting molecules. However, immobilization of (especially) alive cells to a surface using avidin-biotin interaction can be prone to enzymatic/non-enzymatic disruption [47], and therefore, leads to release of the target cells from the surface (Figure 2). Aptamers, which are a group of synthetic/natural proteins/oligonucleotides with the ability to bind specifically to a target molecule, have been used for the design of biosensors since the 1980s [48] due to (a) high affinity to the target; (b) in vitro production; and, (c) their small size, that overall allows for the production of low cost and highly dense immobilized kits in comparison with other specific techniques [45]. Since the 1990s, many studies have been performed on the use of a single, or a cocktail, of these aptamers as effective means for design of microbial detection sensors capable of making strong bonds to different components of microbial cells, such as cell wall, cell membrane proteins, flagella, extracellular enzymes, or genomic contents (RNA/DNA) [49–54]. Application of a cocktail of these molecules improves the target attachment to the biosensors [49]. However, in case of either a single aptamer or a cocktail of these aptamer, these factor should not affect the conformational structure and functionality of the bio-reporters [45]. Application of heterobifunctional cross-linkers, which work as bridges to create large spacing between the cells and the surface, facilitates cell-Ab interactions on the electrode [31].

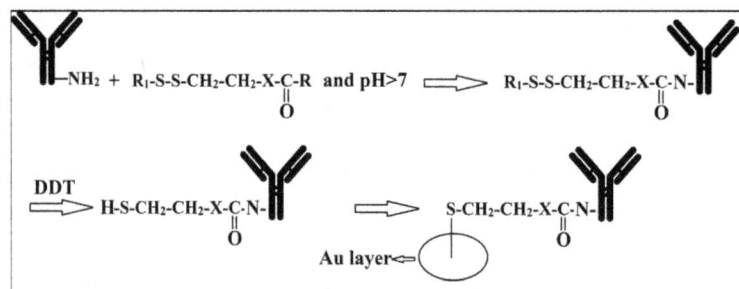

Figure 2. Antibody absorption on the surface of an electrode using thiolation cross-linkers. $R1$ and $R2$ are leaving groups; X is the residue remaining after antibody immobilization; DTT: dithiothreitol.

3. Detection of Intact Cells

Phenotypic analysis refers to the preliminary step of the identification of microorganism's observable traits. Conventional methods distinguish cellular subjects based on morphological features, such as: size, shape, motility and number of flagella, sporulation, shape and position of spores in the cytoplasm, encapsulation, inclusion bodies, staining features, and surface and ultrastructural characteristics. Cellular colonies are also distinguished based on: form, size, elevation, margin, opacity, consistency, and pigmentation [55]. Microscopic studies for detection of intact microbial cells, which primarily need staining before observation, are time-consuming and less specific [56,57], and it is not possible to use this technique in LoCs unless in combination with a reporter system, such as fluorescence.

The biosensors used for the detection of intact bacterial cells commonly employ different optical, electrical and chemical detector techniques. The optical technologies mostly rely on the traditional sandwich technique used for immunoassay analysis, in which the bacterial cells are specifically captured by the antibodies that are immobilized on the kit, and the captured cells are recognized by a second specific antibody conjugated to a fluorescent/luminescent dye [9,58,59] (Figure 3).

Myong song and his colleagues for instance, used laser induced fluorescence (LIF) for recognition of the antibody-bacterial cell (*Bacillus globigii*) reactions in a CMOS system [60].

It is possible to apply Fibre Optic Biosensors (FOBs) in the sandwich structure, in which the antibodies are immobilized on an optical fiber and cells consequently attached to them. The binding of a secondary labeling factor to this overall complex will change the optical signal passing through the fiber [61]. Altintas and his colleagues, for instance, were recently able to use a combination of a microfluidic device and a horseradish Peroxidase (HRP)-conjugated biosensor to design an automated labour free device for the detection of *E. coli*. The electron required for the reduction of H_2O_2 into H_2O by peroxidase is supplied by TMB (3,3′,5,5′-Tetramethylbenzidine), and all the electrical changes can be detected by a sensitive gold electrode (Figure 4) [62]. Yao et al. used a bacteriophage as the specific biosensor for the detection of *E. coli* in a CMOS based integrated sensor system [63]. Nikkhoo and her colleagues were able to improve the sensitivity of the bacteriophage receptor with the employment of integrated ion-sensitive field-effect transistors (ISFETs) implemented in conventional 0.18 μm CMOS with additional post-processes, most notably a PVC-based potassium-sensitive membrane atop the chip [64]. Mejri et al. indicated that the application of a bacteriophage for the detection of a bacterial species is more specific and accurate than the use of antibodies since a phage is able to generate successive dual signals of opposite trends over time [65]. Specifically, the initial increases in the impedance due to the attachment of the bacterial cells to the phages is followed by a sudden decrease in this value due to bacterial cell lysis (as a result of lytic activity of phages) [65].

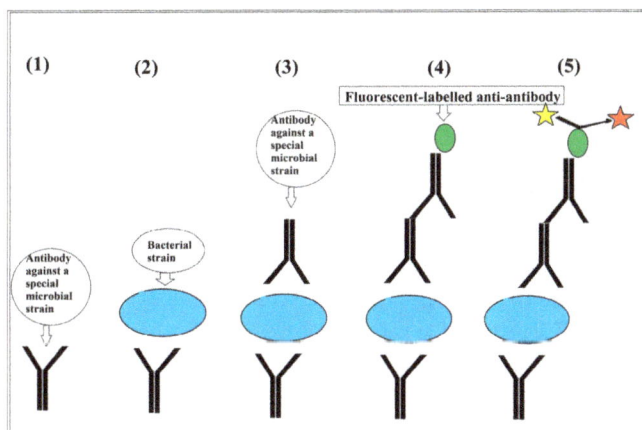

Figure 3. Sandwich technique: in this method, a specific antibody may be used to fix the particle on the surface of a chip. Attachment of another specific antibody on the free site of the particle can be recognised by the addition of a fluorescent-labelled anti-antibody (such as anti-IGM and Anti-IgG).

The development of different high affinity molecules, such as antibodies, phage, and other natural/synthetic factors for the detection of specific moieties, such as antigenic epitopes, receptors and specific hydrocarbon, respectively, on the surface of bacteria have been improving optical based devices for the detection of microorganisms [66]. Furthermore, scientists have been focusing on a way to resolve the limitation of the sample preparation step using the application of so-called label free optical methods, such as Surface Plasmon Resonance (SPR) (Figure 5) [67]. In SPR systems, plane-polarized light passes through a glass prism to reach a transducer surface, which, in response, produces electrical pulses. Attachment of specific receptors (such as an antibody, lectin, and bacteriophage) to this surface makes this transducer specific to a special type of analyte; any transducer-analyte binding will be associated with changes in the electrical pulse [67]. Bouguelia et al., for instance, were able to use this system for the label-free and real-time monitoring of single cell multiplying bacteria on a biochip [68]. Yoon et al. employed radiofrequency (RF)/microwave microstrip bandpass filter circuits in which the bacteria present in water samples were trapped atop the planar filter [69]. The bacterial cells so trapped change the relative permittivity of the insulating material, thus altering the filter's frequency response. However, this system is only able to detect the presence of microorganisms and is not able to differentiate between different types of microorganisms. Therefore, such systems are not yet of immediate benefit in routine microbiology laboratories for detection of various microorganisms. While immobilization of an array of specific antibodies on such chips and analysis of the outgoing data from this system can be applied for the specific detection of single or multiple known microorganisms even in a combined contamination or complex environmental or clinical sample, this chip is applicable for detection of only the bacteria that are specific for those antibodies.

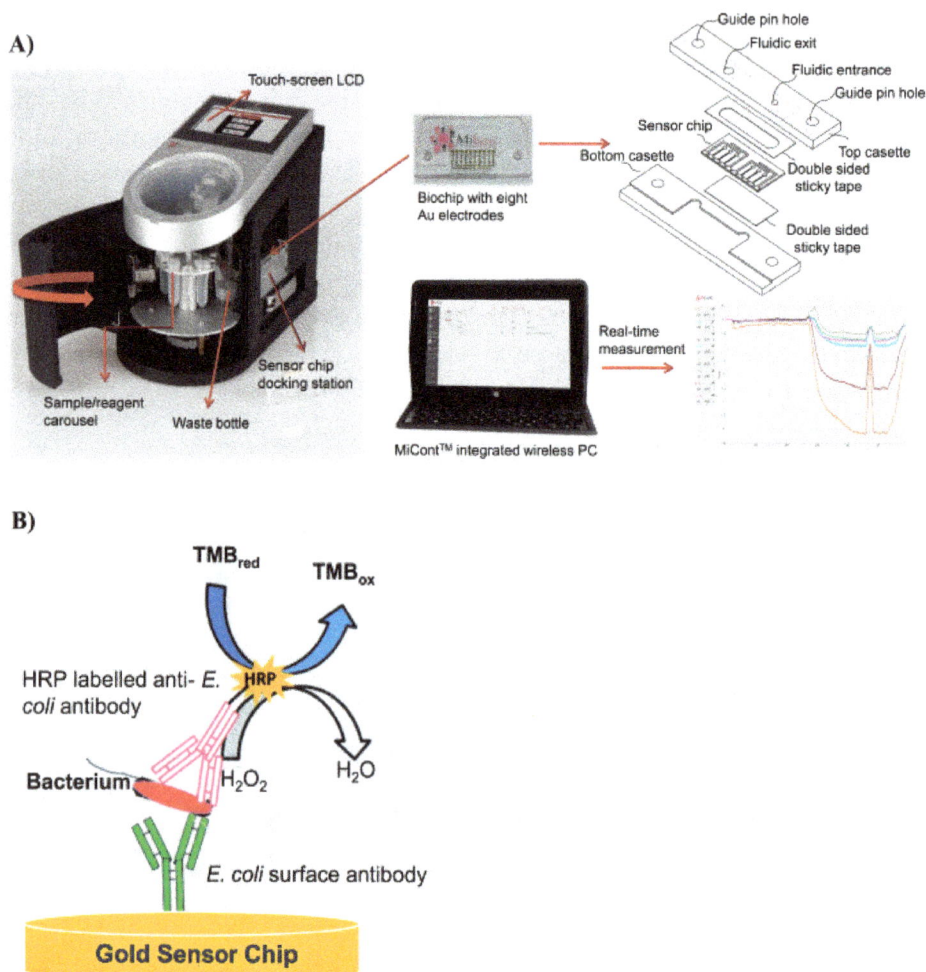

Figure 4. The design of a microfluidic device based on a horseradish Peroxidase (HRP)-conjugated biosensor for detection of *E. coli*. Here, the target bacterial cells bind to their specific antibodies on the gold sensor chip. The bacterial cells bonded to the biosensors are recognized by the addition of another strain-specific horseradish Peroxidase (HRP)-conjugated antibody. A washing step between each sample or antibody addition will remove all unbounded cells or (HRP)-conjugated antibody, and therefore, any enzymatic reaction in the media occurs as a result of the presence of the antibodies bonded to the cell [62]. The enzymatic reaction of H_2O_2 to H_2O reduction is associated with oxidation of TMB (3,3′,5,5′-Tetramethylbenzidine), which can be detected by very sensitive gold electrode (This Figure is used from Elsevier with permission).

Application of mechanical biosensors, such as cantilever technology [9,70] and quartz crystal microbalance (QCM), are new strategies to improve the sensitivity and final costs of LoC devices [71]. In combination with a specific receptor and a cantilever sensor with oscillation at a certain resonance frequency, micro-cantilever-based devices measure changes in the resonance frequency caused by the binding of specific cells to the receptors [72]. With a similar strategy, QCM detects the micro-changes in the total mass of the receptors after bonding of a cell to its specific receptor, itself attached to a piezoelectric biosensor [73]. Miniaturization of such equipment will reduce the massive size of the device and, therefore, will make these devices affordable for LoC purposes [74]. However, it must be mentioned that all such devices designed to-date are able to detect only a single bacterial strain, and therefore, application of these devices needs prior information about cell properties, especially their antigenic features.

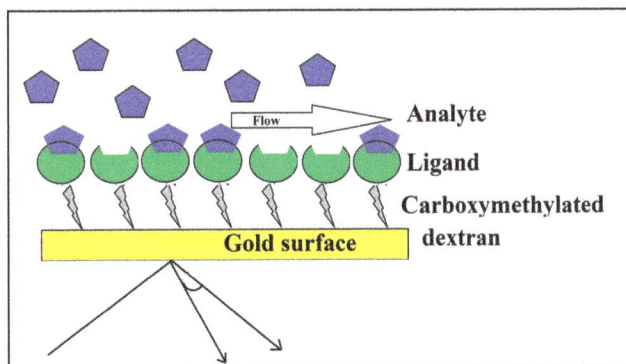

Figure 5. Surface Plasmon Resonance (SPR); the analyte binds to its specific ligand on the surface of a glass prism. This interaction is detected by passing a plane-polarized light through this film to a transducer surface which produces electrical pulses.

In addition to their physical features, microorganisms are detectable based on their metabolic and biochemical reactions as well as their physiological characteristics [56,57]. To use these features for identification, microorganisms are grown in microbial media and are subject to special tests based on the enzymatic activities, microbial growth requirements, microbial resistance to special chemicals or physical factors and microbial fatty acid composition [56,57]. Conventionally, biochemical tests are used for the differentiation of bacteria based on their genetic abilities to metabolize and convert special groups of substrates [56,57]. The cultivable bacteria are grown in a series of different media, each containing a specific substrate. Thus, their ability to use different substrate(s) (carbon, nitrogen, and sulphur), their ability to produce any sort of fermentative products, their sensitivity to metabolic inhibitors (such O_2, H_2O_2, temperature, pH and osmotic pressure) and antibiotics, and their response to the production of degrading enzymes are investigated based on observations of changes in the appearance, texture, or color of the media (Figure 6) [56,57]. The final decision for detection of a cultivable bacterium is made based on an analysis of the microbial response to a combination of different metabolic and growth abilities [57]. Indeed, microbial responses to a complex of detection criteria are exclusive to a specific strain and can be used as a signature for that strain. The culture-based analysis for bacterial detection is reported based on observations of the physical/chemical changes in the media (such as changes in the color, turbidity, the production of gas, and so on) after an overnight incubation for the growth of bacteria, which makes these tests too time consuming and occasionally unreliable (Figure 7) [75].

The employment of LoC devices incorporating electromechanical biosensors for data analysis provides the possibility of designing very sensitive, cost effective and fast approaches to sense any slight changes in the microbial growing media, which are related to microbial activities [6,7,76]. The BacT/Alert system (Figure 8), for instance, is a CO_2 production-based instrument that is used for detection of blood infections caused by different types of bacteria [75]. In this system, the production of CO_2 in aquatic media produce H_2CO_3- and H^+, and the interaction of free hydrogen with a pH sensor leads to the generation of a voltage signal. These systems however, are only able to detect the presence of microorganisms in a given sample, while the type and diversity of microorganisms have to be detected by further microbial analysis.

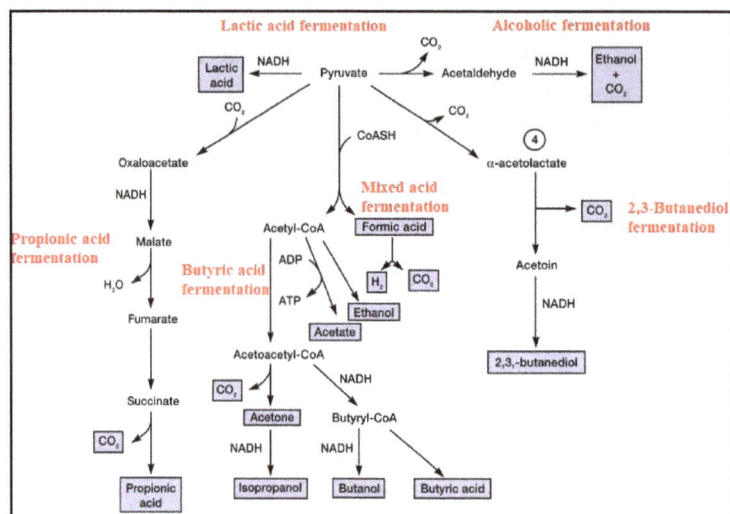

Figure 6. Schematic description of microbial fermentation.

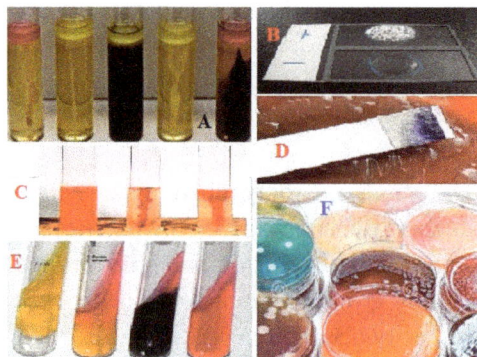

Figure 7. Metabolic activities of microorganism are the most important factor for the detection of microorganisms in traditional bacteriology; **A** = SIM (Sulfide-Indole-Motility) media; **B** = Catalase; **C** = SIM media with different bacterial inoculation; **D** = Oxidase test; **E** = TSI (Triple Sugar Iron Agar); **F** = different plate agars.

To cope with these problems, electrochemical biosensors have recently been included in the design of new versions of commercial LoCs [77,78]. The advantages of these systems are their capacity to implement label-free, cost-effective, real-time, miniaturized and highly sensitive devices [77,78]. The strategy behind the electrochemical sensors is to measure the electrical changes including voltage (potentiometery) or current (amperometry) or impedance (impedometry) [76–78]. Impedimetric biosensors rely on the electrophysiological nature of microbial cells and therefore work based on changes in the electron transfer resistance and capacitance following cell attachment to the electrodes or cell lysis [79,80]. Unlike the highly conductivity nature (up to 1 S/m) of the cytoplasm, the bacterial cell membrane is highly insulated (with an average conductivity of 10^{-7} S/m) [81]. Therefore, while cell attachment reduces the impedance of a particle, cell lysis can cause increases in this index [82,83]. This is the reason why impedimetric studies of the whole bacterial cell in deionized (DI) water lead to better responses in comparison to those in PBS (phosphate buffer solution) [79,80] which is more conductive and hence blurs impedimetric contributions from the cell material under test. Impedimetric biosensors can be made specific for certain types of microbial cells following association of the electrodes to specific receptors, such as antibodies and bacteriophages [79,80]. The impedimetric value of these interactions is categorized as either Faradaic, when the interactions are detected by a mediator such as Ferricyanide and Hexa-ammineruthenium III/II, or non-Faradaic, in the absence of a mediator.

The results of microbial detection based on impedimetric analysis depend on the size and shape of the analyte, the absorbance affinity of the chosen bio-receptors, the way of presenting data (such as absolute/imaginary/real impedance and raw/percent R_{ct} changes) as well as the materials used for electrodes and the base layer [84,85]. This technology has been used for the detection of microorganisms since the 1970s, and there are many reports of using this factor for microbial analysis in food products, clinical specimens and environmental samples [84,85]. While the application of impedance factor as the only detective factor is not specific for detection of microbial species, the integration of polyclonal antibody and electrochemical impedance spectroscopy (EIS) converts this system very specific for certain microbe(s) [86]. In the experience performed by Maalouf et al., the antibody-conjugated EIS was more sensitive than SRP for the detection of *E. coli* (10 cfu/mL versus 10^7 cfu/mL, respectively) [86].

The impedance technology for detection of microorganisms can be boosted by integration of dielectrophoresis (DEP) to the chip, which enables the system to concentrate the microbial cells from a diluted microbial sample [87,88]. In this system, a microfluidic device is designed in such a way that microbial cells are repelled by both a liquid flow force and a negative DEP to be deposited on a positive DEP situated at the downstream of the device. The microbial cells that are concentrated on the positive DEP are detected by an impedance analyzer (Figure 8) [88]. This system can be specific for the detection of a certain microorganisms in a microbial mixture if the positive DEP is covered by specific antibodies [89]. Suehero et al., for instance, applied this strategy to direct a mixture of microorganisms to the antibody-conjugated microelectrodes. A rinse step following the binding of the bacterial species specific to the antibodies removes all of the non-target bacteria, and therefore, this system was specific for detection of certain bacterial strains in a mixture of microorganism [89]. In another strategy to improve the sensitivity of impedance-based detectors, Jiang and his colleagues designed a microfluidic device in which the microbial solution is passed through a filter where the microbial cells trapped in the filter are detected by an integrated electrode located below the filter (Figure 9) [90]. The sensitivity of this device was claimed to 10 bacterial cells per millilitre.

Amperometric-based technologies measure the changes in the current of a microbial media, which are directly proportional to the concentration of reactants or microbial cells [77]. The amperometric index of microbial media changes as a result of microbial metabolism which leads to the release or absorption of ionic metabolites or elements [77]. Potentiometric devices function based on recognition of analytes by specific ion electrodes, which is trackable based on changes in the potential of the microbial growth media [91]. Potentiometric technologies have been used for whole cell detection of a few bacterial cell types, such as sulfate-reducing bacteria [SRB] [92] and *Staphylococcus aureus* [93] by potential stripping analysis (PSA) and electromotive force (EMF) technologies, respectively.

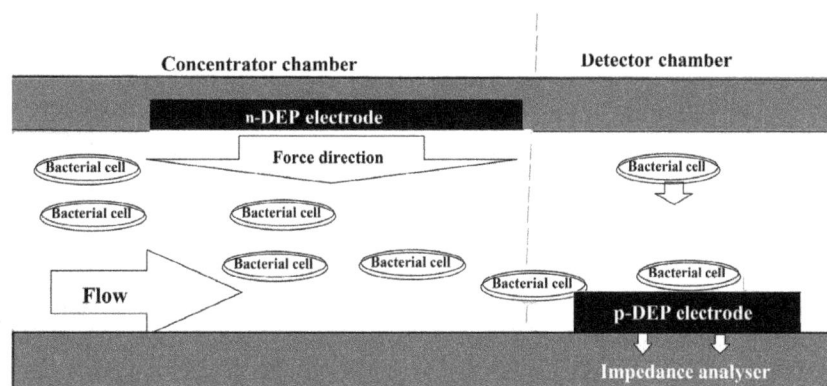

Figure 8. A microfluidic system to direct and concentrate bacterial cells on an electrode based on both flow and dielectrophoresis. The cells collected at the positive side of electrophoresis screen are detected by an impedance analyzer.

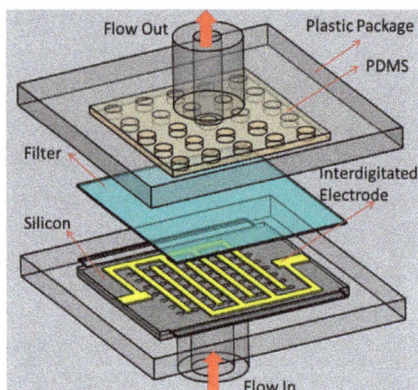

Figure 9. Cross-sectional view of the integrated the electrochemical impedance spectroscopy for detection of bacteria; in this system, bacterial suspension are flown into a nanopore filter above the electrodes where the cells trapped in the filter and the rest of the suspension is flown out [90] (This Figure is used from Elsevier with permission).

Commercially available bactometer devices use a combination of conductance, capacitance, and impedance to perform several microbial analyses per hour [94], and are able to perform microbial analysis on different types of microorganisms, including bacteria, fungi, viruses, algae and protozoa. The changes in the output electrical signals throughout microbial growth depend on the nature of media, microbial primary inoculation, microbial doubling rate, and the presence of optimal physicochemical condition, including pH, temperature [85]. The threshold time for each microorganism is the time that the electrical index curve starts to accelerate [95] (Figure 10), and starts in a concentration of 10^7 bacteria (2–11) and 10^4–10^5 yeast (2–10) per millimeter. Since the electrical changes of a microbial environment are influenced by the metabolic activity of microorganisms, the nature of microbial media is a significant factor for more sensitive detection of microorganisms [95].

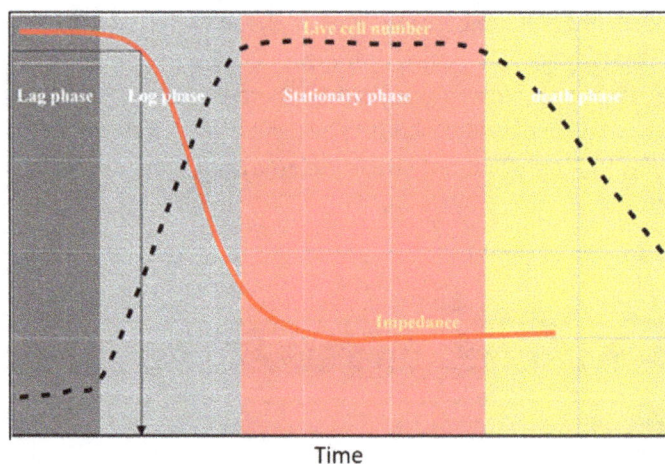

Figure 10. Bacterial growth curve. The lag phase after inoculation of a media with bacteria is followed by a logarithmic growth of the bacteria (dashed line). The shortage of nutrients and the production of toxic compounds in the media lead to a decline in the microbial growth during the stationary phase and to death at the last steps. The number of cells in this media is inversely correlated with the media impedance (solid line).

Overall, real-time analysis of microbial growth and the possibility of miniaturization of these devices are the most outstanding advantages of the application of electrochemical biosensors for point of care studies [85]. Furthermore, There is the possibility of measurement of the number of cells in a given sample based on measurement of the output electricity [85]. Besides, based on changes in the electrical indices of a given microorganism in the presence of a growth inhibitor, this technology is able to determine the susceptibility of bacteria to growth inhibitors such as antibiotics which can be used for antibiogram tests (a routine test performed by microbiologists in medical laboratories to determine the level of microbial sensitivity to an antibiotic) [96]. These technologies are especially useful to accelerate the detection of slow metabolic activities, as can be seen in slow-growing microorganism, such as *Mycobacterium* sp. [97,98] or in some anaerobic microbial activities, such as microbial desulfurization [92]. However, any innovative devices for bacterial detection based on the electrochemical activity must resolve the interference caused by other bacterial strains present in bacterial samples, especially in environmental samples or many clinical samples such as stool and sputum.

On the other hand, while application of the sandwich method based on specific reagents (such as antibodies, bacteriophages, receptors, enzymes and so on) in conjunction with an electrochemical detection factor makes the detection devices very specific for a particular type of microorganism [99,100], these approaches fail to detect the presence of multiple microorganisms in a sample. Furthermore, in the case of usage of multiple specific antibodies for multiple microbial detections, these (usually gold-made-) chips are disposable and therefore, are too expensive for the design of point of care LoCs [99,100]. However, the design and application of newly emerged technologies related to the structure of the base matrix and electrodes, such as the nanoporous membrane-based impedimetric immunosensor [101], reduced graphene oxide paper [102], platinum wires [103], nickel foam [104], can improve the sensitivity and specificity of these devices. Immobilization of antibodies on the surface of the nanoporous membrane using an adherent, such as hyaluronic acid, makes a very specific binding surface upon which that attachment of bacteria changes the ion permeability through the membrane, with its effect being traceable by electrical sensors [105].

The integration of aptamers (small peptide or oligonucleotide molecules with the ability to bind specifically to a target molecule) with a detection system is another strategy to achieve more specific detection of microorganisms even in a mixture of microorganisms [93]. Application of different detection factors in such systems can provide a detection chip for multiple pathogenic microorganisms. Wu et al., for instance, used multi-color (luminescent) up-conversion nanoparticles (UCNPs) in combination with specific aptamers for each bacteria strain (overall three strains) to design a sensitive chip for simultaneous detection of three bacteria in a complex (food) environment (Figure 11) [106]. In this case, however, an increase in the numbers of strains intensifies the luminescence interferences, and therefore, decreases the sensitivity and specificity of the kit. Furthermore, the number of strains detectable in this system is limited to the number of known aptamers specific for this microorganism.

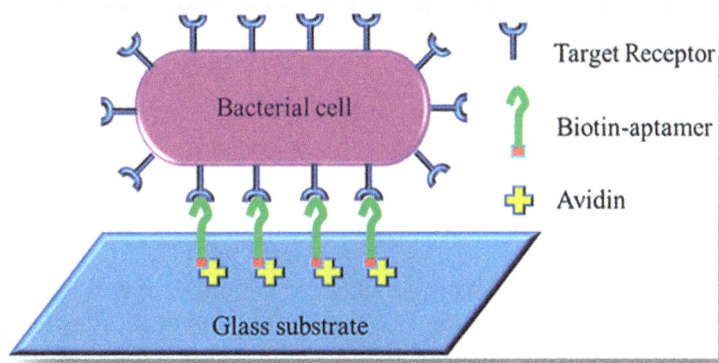

Figure 11. A simple schematic description of the employment of aptamers for the detection of microorganisms. In this condition, the complex of Biotin-aptamer works as a specific intermediate between the microbial strain, which is free in the media) and Avidin, which is stabilized on the surface. The aptamer complex is usually bounded to a recognition marker, such as a fluorescent).

4. Genomic Based Analysis

Microbial genomic studies are currently the most popular molecular identification and classification method of microorganisms in different types of samples based on the selection of one or more specific molecular target(s), such as DNA, RNA and protein, without the requirement to grow in their culture media [107]. Genetic microbial detection assays can be classified into genome GC% (Guanine + Cytosine) contents, pattern amplification (polymerase chain reaction; PCR, and loop mediated isothermal amplification; LAMP), DNA/RNA hybridization, genome polymorphism (such as RFLP: restriction fragment length polymorphism and AFLP: amplified fragment length polymorphisms) and gene sequencing [108–110]. In addition to the sample preparation and genome extraction, the majority of these methods need a nucleic acid amplification and an amplicon analysis step [110]. The sample preparation and DNA extraction steps are very challenging steps due to the removal of inhibitors from genomic samples and accurate measurement of the initial genomic contents per specific amounts of sample, which is used in quantitative methods [110]. These samples are always passed through an enrichment step, which can be performed by overnight incubation of bacteria in a specific sample (only limited to cultivable bacteria), centrifugation, and immunomagnetic separation [108–110]. These two latest samples can be applied for the removal of sample matrix from the microorganisms, which indeed is useful for removal of inhibitors from the genome [108–110]. A combination of conventional alkaline lysis protocol with DNA binding to silica or its derivatives has provided the possibility to design a cost effective automated microfluidic device for DNA isolation [111]. Bavykin et al., for instance, have developed a technique in which cell lysis is carried out in the presence of a high chaotropic agent, such as guanidine thiocyanate (GuSCN), followed by cell lysate injection through a universal syringe-operated silica micro-column to capture the RNA/DNA contents [112].

4.1. Guanine-Cytosine Contents

The genome GC% contents are the percentage of the sum of guanine and cytosine in DNA or RNA of a cell. While the GC% is markedly variable in different types of microorganisms, 17–75% [113], it cannot be used as the only criteria to distinguish microorganisms. Indeed, GC% is a useful factor to detect microorganisms along with a combination of other genetic/phenotype characterizations.

4.2. Sequence Amplification Techniques

PCR is a technique in which one or more specific fragments of the genome are amplified using specific primers, which are finally detected based on the size of the fragment in conventional PCR, polymerase chain reaction, or fluorescent reporters in real-time PCR [114]. In conventional PCR (Figure 12), the amplified fragment moves under a special electrical charge that separates the fragments based on their size and thus yields one or more identifiable band(s) onto the gel; the correct band among all of the other unspecific bands is determined in comparison with the size marker, which is run onto the gel at the same time [110].

Because of the ocular observation-based analysis, this technique cannot be fit in the LoC technologies and automation unless the primers are linked to a (fluorescent/radioactive/enzyme) label. Real-time PCR or quantitative PCR (q-PCR) (Figure 13) is a more sophisticated technology to characterize the amplifying reaction in time and to collect the data throughout the amplification process [115].

In this system, target specific fluorescent–labelled probes are bound to the target al.ong with the primers, and any single target replication through each single PCR reaction leads to the release of the fluorescent reporter, which is associated with illumination of a fluorescent light [115]. Since each fluorescence is parallel to one gene amplification, the numbers of fluorescent pulses is a reliable factor to determine the copy numbers of genes based on comparison with the reference samples, which contain a defined number of the gene of interest and are run along with the samples of interests [115]. Employment of probes linked to different types of fluorescent reporters enables researchers to use qPCR for simultaneous detection of multiple microorganisms in a sample [116]. The ability to monitor the illumination emitted from the reporters makes qPCR suitable for the design of automated detection systems. Oblath et al., introduced a microfluidic chip with the ability to perform both DNA extraction and real-time PCR (Figure 14) [117]. The microbial samples were lysed by a heat treatment before addition to micro-wells, and were filtered by an AOM nanofilter to concentrate the DNA samples on the filters. The genomic materials that remained on the surface of the filters underwent a RT-PCR reaction. However, it needs to be mentioned that the huge amount of cell debris through the DNA/RNA extraction can affect the quality of RT-PCR performance, therefore a suitable genome extraction microchip should be designed in a way to separate this cell debris. Furthermore, it must be mentioned that the PCR procedures are very expensive and time-consuming, and therefore, is not presently amenable with the spirit of real-time microfluidic devices which need to be inexpensive [118].

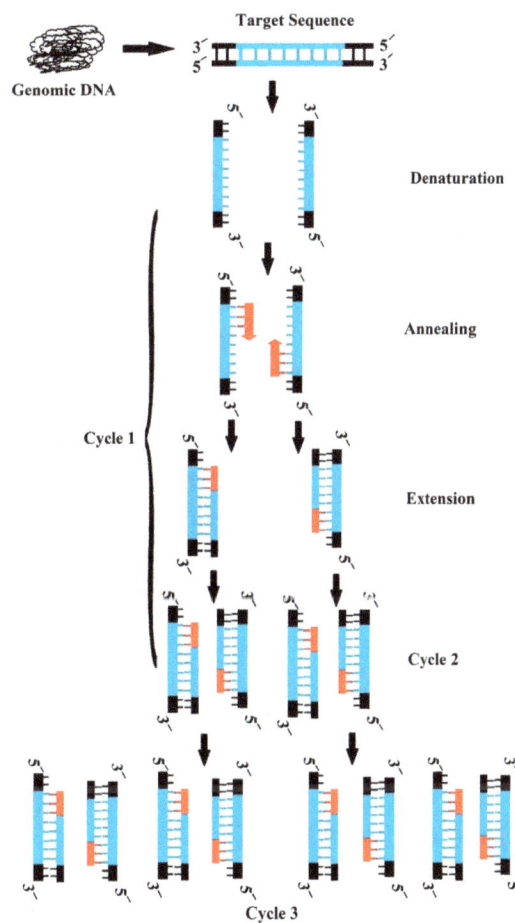

Figure 12. The principle of polymerase chain reaction (PCR); through denaturation period, increase in the temperature of the media (more than 90 °C) lead to separation of the DNA duplex in the form of single strands. Reduction of temperature to around 50 °C (depending to the primer) in the annealing step provides the condition for the single strand nucleic acids, including the primers, to re-hybrid in the form of double strand DNA. Thereafter, though extension step, an amplifier enzyme is able to use the 3′ end of the primer to extend the primer based on the complementary template DNA. These steps are usually repeated more than 30 times (depending on the concentration of the template DNA) to increase the numbers of the fragment sufficient for visibility on a gel. Since the primers are specifically designed for a special fragment of the genome, the amplification is specifically performed on that area of the genome.

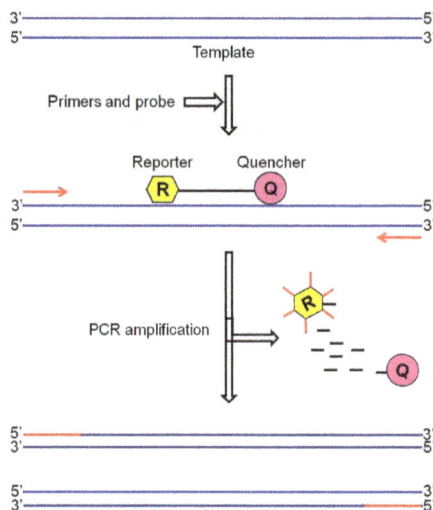

Figure 13. The real-time PCR; in this system, a specific probe labelled to two types of fluorophores, referred to as R for reporter (donor) and Q for quencher (or receptor), sits in the interval of two primers. When the distance between two fluorophores is between 1 and 10 nm, any the excited energy of the R is donated to Q in the form of Fluorescence Resonance Energy Transfer (FRET). Through amplification of the DNA, the R is cut by the amplifier enzyme, which lead to the separation of R and D fluorophores and emission of fluorescence. The level of emission is linearly proportional to the number of probe and genome.

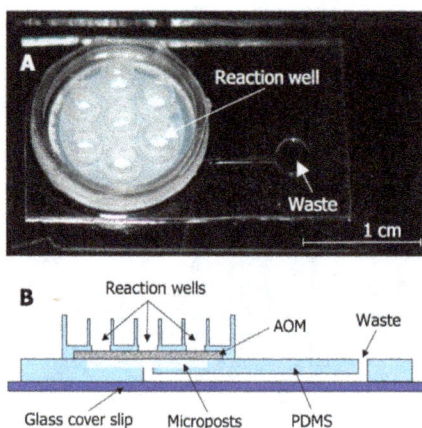

Figure 14. The microbial cell lysate is passed through an AOM filter to separate DNA on the surface, which finally undergoes a RT-PCR reaction [117]. (This Figure is used from Royal Society of Chemistry with permission).

These concerns, however, are fairly improved by the invention of the LAMP method (Figure 15), in which the genome is primarily amplified under an isothermal condition (60–65 °C) within a period of 30–60 min [109]. Amplification of six different regions of the genome as a result of the application of 4–6 primers at the first screening step of LAMP makes this detection system more sensitive and specific than PCR [119,120]. Principally, the presence of a genome complementary sequence in each inner primer leads to the formation of stem-loop concatamers at the ends of each replicated fragment, which make a free 5′ end usable as a primer for further elongation on the strands. These repetitive reactions are finally led to the production of large quantity of stem-loop containing DNAs [121]. The genome amplification in LAMP is determined based on increase in the turbidity, as a result of reaction of pyrophosphates (released through the amplification) with magnesium (present in the tube), or change in the level of fluorescence illuminated from the fluorescent dyes (such as SYTO 9) intercalating with DNA [122]. In terms of automation of these devices, in addition to using fluorescence, release of H+ through amplification, which is directly proportional to the number of nucleotide used for each amplification, can be used as an ideal parameter as a proof of amplification and measurement of the numbers of amplification. In addition to the rapidity, low cost, high sensitivity and high specificity, LAMP is effectively applicable for the detection of microorganisms in the low quantity and quality samples, which is useful for direct detection of microorganisms even in the presence of clinical sample matrices [119,120]. Guo et al. designed an integrated double layer microfluidic chip (Figure 16) in which the DNA content isolated via silica beads were amplified by LAMP followed by a Calcein staining and UV light [123].

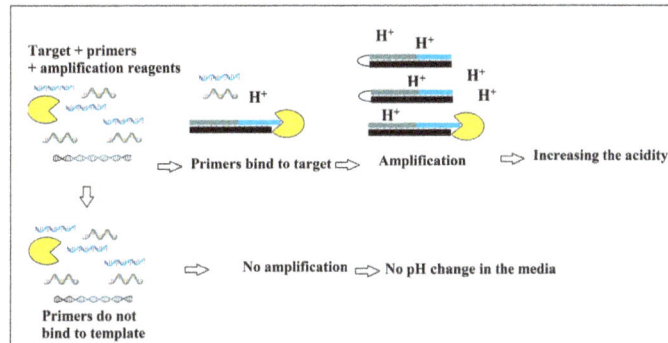

Figure 15. Loop mediated isothermal amplification (LAMP) method (description in the text).

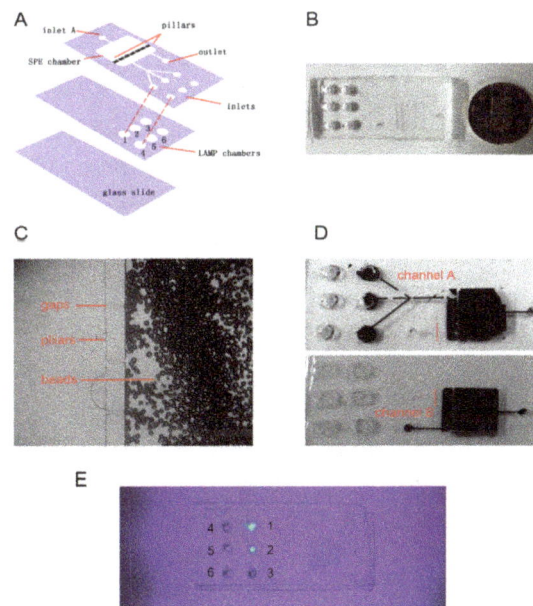

Figure 16. This microfluidic device consisted of two layers; (**A**) Schematic design of the device; (**B**) the real assembled device. The cells, DNA extraction reagents and silica beads are directed to the upper layer through the inlets to the SPE (Solid phase extraction) chamber. Following DNA extraction in the SPE chamber, the DNA samples are washed from the beads using distilled water to the second layer (LAMP chamber). The gaps on the micropillars are small enough to prevent bead transfer to the second layer; Subpanel (**C**) Shows the SPE chamber in which the beads stuck by the row of micro-pillars and the micro-gaps; (**D**): the solution flow in the chip is showed by an injection of black ink from inlet A flowed to channel **A** or **B**; the positive results can be seen by a, green fluorescence under the UV light (**E**) [123]. (This Figure is used from Elsevier with permission).

4.3. Genetic Polymorphism

Polymorphism in genetics denotes the presence of more than one alternate form (alleles) of inherited genetic contents of individuals in the population. Genetic polymorphism can be detected using different technologies, such as RFLP (Restriction Fragment Length Polymorphism) [124], T-RFLP (Terminal-Restriction Fragment Length Polymorphism) [125], AFLP (Amplified fragment length polymorphism) [126], RAPD (Random Amplified Polymorphic DNA) [127] and SSCP (Single-strand conformation polymorphism analysis) [128]. Restriction enzymes are a group of DNA endonucleases with the ability to cut the sugar-phosphate backbone of DNA at or near a very specific target sequence known as restriction sites, leading to produce several DNA fragments of individuals with a certain size [129]. Any mutation in a specific restriction site of a homologous DNA sequence resist the target against the corresponding restriction enzyme, which cause the production of fragments of different length [130]. RFLP is a combination of DNA digestion with one or more restriction enzyme(s), separation of fragments onto a gel through electrophoresis, and finally, detection of presence of specific fragments by a DNA probe [131,132].

AFLP is a more selective polymorphism technique in which, after digestion of a given DNA, the fragments are ligated to adaptors and then undergo pre-selective and selective PCR amplifications to reduce the fragment numbers [133]. The remaining fragments are run onto gel electrophoresis and the fragment profile is interpreted by software. Application of different types of adaptors enables this technique to study the polymorphism of hundreds of individuals in the same batch even without having any information about the microbial genomic sequence [133].

T-RFLP, however, works based on amplification of the gene target (mostly 16SrRNA) by fluorescent labelled primers, enzymatic digestion of the PCR products and, finally, the generation of the gel profile [134]. Due to the presence of fluorescent label at the $5^/$ end of fragments, a UV exposure will cause different peaks of various sizes and heights to appear, which is representative of the profile of a microbial community in a given sample [134]. T-RFLP, therefore, is a very useful and quick technique to investigate the microbial changes in an environment after treatment with a specific factor. However, it must be mentioned that this technique is not able to detect microorganisms in a sample, but only gives an idea of the levels of differences in whole microbial diversity of an environment [135].

RAPD is a PCR-based polymorphism in which the given DNA is amplified with a universal primer (approximately 10 bp length), and the gel profile of individuals will be different as a result of mutations in the primer annealing sites (Figure 17) [136,137]. Because of using a universal primer, no specific knowledge of the microbial genome is required for the creation of microbial RAPD profile [136,137]. SSCP is an effective screening method to distinguish a specific genomic variation (like a specific strain) among a wide range of microorganisms. In this technique, following amplification of the genome, the PCR products are denaturised in the form of single strand DNAs.

The presence of any mutation in the genome sequence leads to changes in the conformation of these single strand DNAs, which can be detected by separation onto polyacrylamide gel (PAGE) or Capillary Gel Electrophoresis (CGE)(Figure 17) [138].

While each polymorphism technique shows slight principal differences, the application of radioactive/fluorescent labels for detection of genome fragments enables automation of these technologies. However, these techniques are mostly used for detection of the presence of genomic changes in a strain in comparison to its normal sample, and to investigate the changes in the overall microbial profile of different samples. Therefore, these techniques are not recommended for the design of bio-chip devices intended for precise microbial detection in a sample.

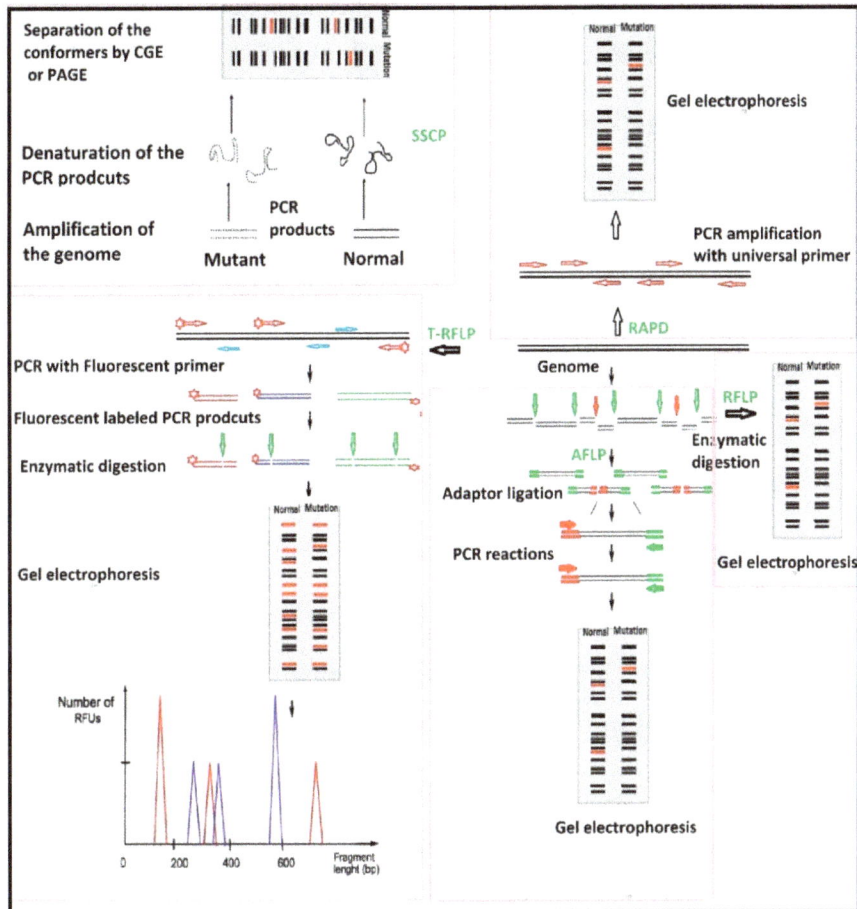

Figure 17. A schematic comparison of RFLP, AFLP, RAPD, T-RFLP and SSCP.

4.4. Hybridization-Based Technologies

Hybridization-based microbial detection technologies work based on the genome melting (de-hybridization) in high temperature and annealing (re-hybridization) of the complementary strands at lower temperatures [139]. Based on these techniques, a probe (a labelled short length single strand RNA/DNA complementary to a specific target sequence) binds specifically to its target on the genome through an annealing process, and the hybrid is detected by the (enzyme, radioactive or fluorescent) label on the probe [139].

FISH (Fluorescent In Situ Hybridization) (Figure 18) is a very sophisticated technique in which a fluorescent-labelled segment of RNA/DNA can be used to detect the presence of a specific complement of the genome in a sample and to locate the precise position of the complementary segment in the cell [140]. Application of several probes labeled to distinct fluorophore enables researchers to co-detect and co-locate multiple targets in the cells [140].

A group of microbiological diagnostic biosensors is designed based on immobilization of a probe (a single strand DNA) complementary to the target DNA on the surface of a highly sensitive transducer [76]. The DNA hybridization with the target DNA can be detected by a fluorescence [141] or based on changes in a physical index such as electrical factors or acoustic waves [142]. Xi et al., designed a microfluidic device in which peptide nucleic acid (PNA) molecular beacons (MBs) were used as bacterial detectors based on DNA hybridization. A DNA MB is a single strand probe flanked with two complementary sequences to make a hair pin loop structure that, on one end, is labelled with a fluorophore and, on the second end, is attached to a quencher (Figure 19) [143]. As in real-time PCR, this structure causes the fluorescence produced by the fluorophore to be absorbed by the quencher.

Hybridization with the target, however, increases the distances between the fluorophore and the quencher, and leads to illumination. Furthermore, based on alteration of the probe backbone with an electrically neutral compound (peptide nucleic acid; PNA) instead of a sugar phosphate backbone, they were able to improve its hybridization kinetics [143].

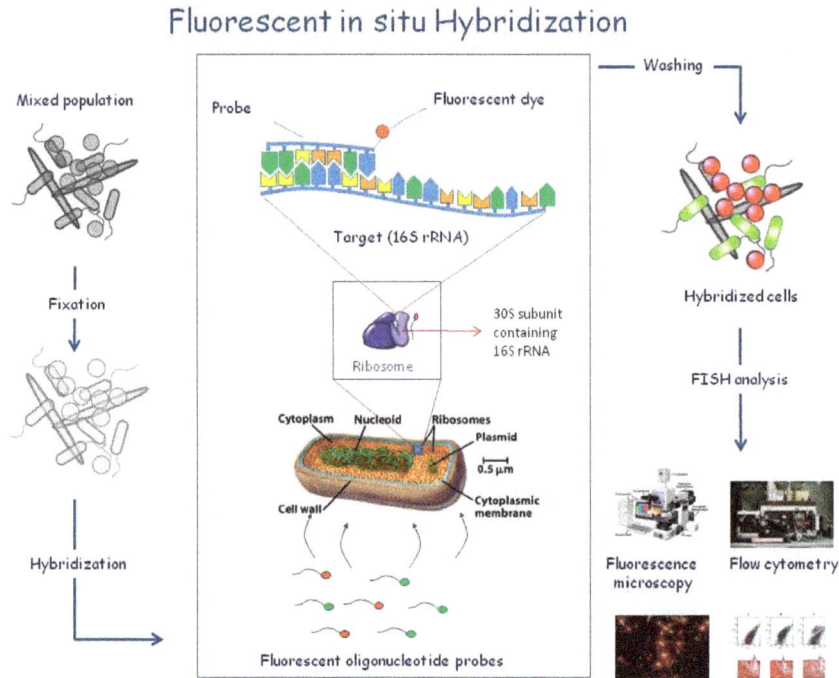

Figure 18. Fish (Florescent In Situ Hybridization).

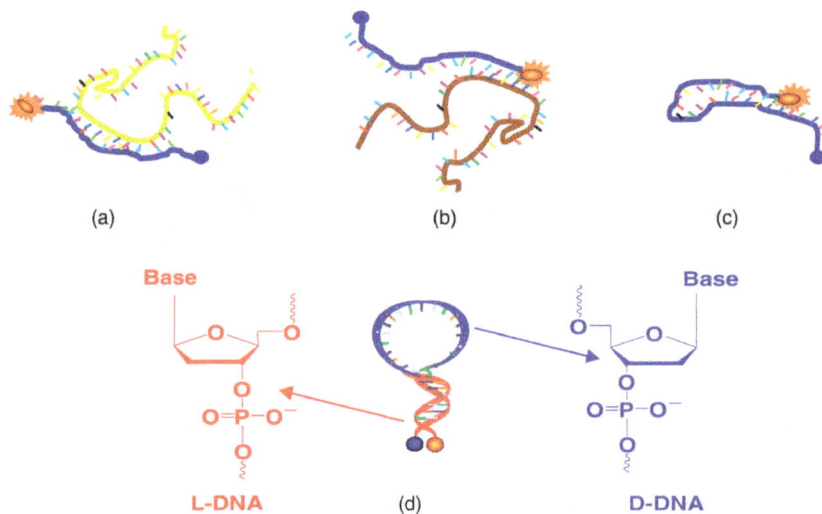

Figure 19. The structure and function of molecular beacons (MBs); MBs are target-specific single strand DNAs flanked with two complementary sequences labeled with a fluorophore (at one end) and a quencher (at the opposite end). (**a**) and (**b**) The MB-DNA target hybridization creates a distance between quencher and fluorophore, but (**c**) The hydrogen bonds between two complementary sequences at two ends make a hairpin structure in which any fluorescence released by the fluorophore is captured by quencher. (**d**) The production of D-DNA in the hairpin and L-DNA in the Double helix DNA.

The DNA microarray is arguably the best fabricated and commercialized system for diagnoses based on DNA hybridization in which a group of each, out of millions, types of identical DNA probes are bound to specific spots of a solid surface, and their hybridization with a fluorescence-labeled DNA target is detected by a fluorescent scanner [144,145]. The integrity of hybridized DNA is detected by software base on the location of light emission.

While the majority of LoCs based on the DNA hybridization technique has been designed based on a light emission technology, the background light emission can affect their sensitivity and specificity. It is believed that the employment of electrical biosensors to detect the signals from DNA hybridization can improve the reliability of LoCs. While several types of electrical DNA microarrays have been designed, the majority of them have been limited to the research level due to the high noise levels exhibited by such devices [14]. Attachment of DNA probes onto the electrical circuits in a way to evoke a sensible electrical signal is a very challenging step in the design of DNA electric biosensors [14,76]. The chip bases can be covered with Streptavidin protein, which has the ability to bind the biotinylated DNA probe [15]. Using immobilization of DNA probes on a GMR (Giant MagnetoResistives), for instance, it is possible to design a CMOS biochip in which biotinylated DNA target/probe hybrid can bind to streptavidin-containing magnetic nanoparticles, and this complex can change the magnetic field disturbances [15,146]. Alternatively, mercapto-propyl-trimethoxsilane onto the surface of biosensors exhibit the ability to bind covalently to $5^{/}$ thiol-modified DNA probes [147]. Integration of CMOS with a successful electrical biosensor for detection of DNA hybridization is the gold key to design a successful LoC device with the ability of microbial diagnosis in a very complex sample.

4.5. Sequencing Dependent Techniques

Gene sequencing, on the other hand, is a more accurate approach in which the sequences of usually a fragment of microbial genome (RNA or DNA) is determined, and the sequence is compared with the already present gene sequences in online universal databases, such as NCBI (National Centre for Biotechnology Information) [148]. Since the genetic elements of microorganisms are too variable, only a group of consensus genes, such as ribosomal DNA [149], gene *recA*, α and β submits of RNA polymerase (rpoA and rpoB) [150], Gyrase B subunit (*gyrB*) [151], which have been conserved or slightly changed throughout the evolution of microorganisms, are used to identify any phylogenetic relationships between microorganisms. The levels of similarities between two samples show the degree of relativeness between these samples. In addition to the accuracy, this technology enables researchers to identify and discover new uncultivable microorganisms in conventional cultures [151].

The existing commercial genome sequencers are classified into Sanger (an exemplar of so-called first generation sequencers), Illumina, Pyrosequencing, and Ion Torrernt (exemplars of second generation sequencers) plus Nanopore and Pacific Biosciences technologies (exemplars of third generation sequencers). The principal ability, advantages, and disadvantages of each sequencer have been described by Abbasian and colleagues [151].

Data collection in the second and third generation sequencing systems are significantly faster than the first generation techniques since these systems are designed based on DNA fragmentation, continuous cycles of enzymatic reactions, which are performed simultaneously on millions of fragments, and monitoring based on change in chemistry (such as pH and inorganic phosphor) or the emission of a special fluorescence [152]. A pre-amplification of the target genome by a specific primer increases the numbers of gene targets, and therefore, is the best way to reduce magnificently the time required for gene genome sequencing and data analysis, which are the main key reasons for limitation of the rate of microbial detection in these systems [153]. While all devices designed for next generation sequencing can be potentially used in LOC technologies, the small size of nanopore sequencers makes this technology perfect for the design of a fully automated miniaturized device for detection of microorganisms [154]. However, it is worth to remind that while the genome based methods show high sensitivity and specificity for microbial detection, their application are very questionable in terms of good laboratory practice (GLP) and biosafety assessments [155]. The limitations of these approaches for distinguishing the microorganisms with higher phylogenetic relationships and unreliability to the quality of gene sequences present in genome databases are mentioned as some of the shortages in the genome based studies. Furthermore, these methods are normally unable to distinguish viable microorganisms from the destroyed cells. However, treatment of the samples with a nuclease enzyme or an impermeable nucleic acid binding chemical to digest extracellular genomes or to prevent the DNA amplification may increase the possibility of detection of the genome obtained from viable cells.

5. Other Methods

In addition to genomic analysis, molecular based technologies for identification of microorganisms can be designed mainly based on the detection of genomic materials (RNA or DNA) and protein or fatty acid (FA) profiles of a cell and comparison of the results with information databases to analyze the similarity percentage [110]. The main advantages of these techniques are high sensitivity and very small sample requirements for detection of the microorganism of interest in a pure or mixed sample [110]. In addition to using only one or a few targets for detection of microorganisms, recently it has become possible to detect a single microorganism using whole gene materials (genomics), RNA profile (transcriptomics), protein profile (proteomics), FA profile or metabolic activity (metabolome) [156,157]. Furthermore, metagenomics is a technology to identify whole (known and unknown) microbial strains in a given environment based on the whole genetic materials of the sample [148]. The emergence of new technologies, such as advanced pattern amplification systems, next generation sequencing tools, genome/protein arrays, in-silico microbial metabolome, mass spectral protein and different automated chromatography systems have improved the sensitivity of microbial identification in a given sample [148,158,159].

5.1. Protein Based Methods

The protein based methods work either by immunological assays or gel electrophoresis. The immunological based techniques, such as ELISA (Enzyme Linked Immuno-Sorbant Assay), antigenic participation and agglutination assays and western blotting, which commonly are employed to identify microorganisms based on their surface epitopes, are very specific and all are limited to the presence of a specific antibody for the given protein as well as the sensitivity and specificity of the antigen-antibody reactions [160,161].

Since ELISA technology can be integrated with CMOS technology, we mainly focus on this subject. While routine ELISA kits used in diagnostic laboratories are able to detect only one specific microorganism or antigen in a given sample, CMOS technology offers the possibility of diagnosis of several microorganisms/antigens in only one kit [16]. In the other words, the application of CMOS-based biochips for immunological detection of pathogens improves the ability of identification of one or more microorganisms or their products within a complex matrix containing different microorganism [16]. Furthermore, in comparison to genome based assays that require genome extraction and gene replication, the immunological assays are short enough for direct detection of pathogens in urgent cases [162]. Therefore, the integration of antibodies with silicon dioxide passivated interdigitated electrodes is a very promising strategy to detect the presence of a specific An-Ab complex in a very quick (1–2 h) and sensitive device using changes in electrical indices [96]. In such devices, attachment of antigen on the antibody, which is coated on the surface of an interdigitated electrodes (IDE), led to changes the voltage and impedance values of the microelectrodes. While many attentions have been paid for the innovation of an integrative electrical device with the ability to detect Ab/Ag complex, its commercialization requires several technical upgrades and set up. Safavieh et al. [163], for instance, was able to design a HIV diagnostic paper microchip with a sensitivity of 107 virion per mL in which graphene-modified silver electrodes (GSEs) were conjugated with anti-HIV antibodies to sense the production of any Antigen (HIV)-Antibody complex in the clinical samples. Song et al., could also design a portable biochip system based on integration of a miniature CMOS sensor microarray and laser induced fluorescence (LIF) technology for single-bacteria detection in which the enzymatic product in ELISA test is excited by a laser beam, and the emitted fluorescent light is focused on a photodiode element of CMOS chip by passing through objective lens and optical mirror [60].

In addition, association of Gold Nanoparticle Probe (GNP) with antibodies and biotin enables these particles to bind at the same time to the specific microorganisms and Streptavidium-HRP (horseradish peroxidase), respectively, which generates visual light in the reaction with Tetramethyl Benzidine (TMB) [164,165]. Therefore, due to the generation of visual signals and the ability to read the result of microbial detection by naked eyes, this system does not require a specialist operator to interpret the results. To improve the sensitivity of this system, Ren et al. designed a magnetic LFIA magnetic system (mLFIA) in which the microbial cells are also captured with a magnetic nanoparticle, made of a Fe_3O_4/Au core equipped with antibodies against the specific bacteria. The creation of cell/Ironic nanoparticle facilitates the movement of these complex toward the GNP particles using a magnetic force [166]. While LFIA technology is a rapid, simple, labor free and cost effective approach for microbial screening in pathogenic and environmental samples, these devices are unable to detect a complex of antigens/microorganisms in a given sample that can be used as a multipurpose diagnostic device [166,167].

5.2. Bacteriocins

In addition to antibodies, it is possible to employ other molecules with specific affinity to microbial surface compounds. Bacteriocins are a group of compounds produced by specific bacterial strains to kill specifically other bacterial strains [168]. Therefore, the philosophy behind the application of these compound is to study the cell lysis of specific bacteria following treatment with specific bacteriocins. Nikkhou et al., for instance, employed two bacteriocins, Lysostaphin and colicin, for specific detection of *Staphylococcus aureus* and *E. coli*, respectively, onto a CMOS system. In this strategy, cell lysis caused by the bacteriocins was detected using potassium selective electrodes installed on CMOS systems to detect the increases in the potassium levels as a result of potassium efflux after cell lysis [169]. This system is applicable, however, for a narrow range of bacteria due to very specificity action of each bacteriocin on a range of bacteria [168]. Therefore, specific bacteriocins are required for detection of each bacterial species. Furthermore, since some bacteriocins can destroy a range of bacterial cells from different species (not only one strain), it is not possible to detect specifically the microbial strain. For instance, while Nikkhoo and et al. used colicin for detection of *E. coli*, a few bacteriocins exhibit inhibitory action on a range of bacteria; such as colicin against *Listeria monocytogenes*, *Salmonella* sp., [170] and subtilisin against some species of *Staphylococcus*, *Streptococcus*, *Enterococcus*, *Enterobacter aerogenes*, *Listeria monocytogenes*, *Pseudomonas*, *Porphyromonas*, *Kocuria*, *Escherichia*, *Shigella* [171]. Furthermore, the ability of bacteria to resist against bacteriocins [172–174] limits the application of bacteriocins for microbial detection.

6. Future Works

The design of hybrid CMOS/LOC devices for the detection of microorganisms in a given sample is a challenging subject for many researchers. These devices have the ability to detect microorganisms at the species, and even strain, levels. Operation speeds allowing 10–20 times shorter experimental periods and miniaturized profiles allowing for 100–1000 times less sample and reagents that are required for microbial detection are primary advantages of these technologies. All steps of an experiment, including sample collection, preparation, detection and data analysis are potentially targeted for design of a fully automated diagnosis. While many efforts have been devoted to develop this technology into a fast real-time means of microorganism mixture detection in (clinical, environmental, and industrial) samples, no significant developments have yet been achieved. The type, size, and nature of a microorganisms are all critical factors for the selection or design of a method for detection of that microorganism. Therefore, while the detection of microorganisms based on their morphological and biochemical features is too complicated and time-consuming, the engineers prefer to design genetic/immunogenic based biosensors. Since antibodies are specific to a special microbial strain, the antigenic-based microbial detection approaches are mainly limited to the presence of specific antibodies in the media. However, the application of universal 16SrRNA primers makes the genomic-based approaches very flexible, even for the detection of unknown microorganism. Therefore, since there is no idea about the type of microorganisms in complex samples, such as environmental samples, and in case of bioterrorism attacks, the genomic based approaches will be more useful for fast and effective microbial detection.

It must be noted that the commercialization of a device (see Table 1) depends mainly on the quality control and safety of the product, which, in turn, is determined by the test validation assessments to determine its precision, sensitivity, and robustness. While there is no limitation for the sensing options used in a detection method, the Fertilizer Safety Office suggests the application of an integrated approach in which both conventional and molecular techniques are used for to improve its sensitivity, specificity, and accuracy. Furthermore, since LoC devices usually need to be accompanied by a variety of ancillary machineries, such as pumps for microfluidic devices and amplifiers for electrodes, the bulkiness and final cost of these devices are the biggest challenges that bioengineers need to address to realize the technologies' full potential. Integration of LoC devices with a universal pre-existing portable device, such as a smart phone, can be an applicable strategy to resolve these kinds of problems.

With the development of smartphones in our everyday activities and the feasibility of application of the smartphone apps by the users, many researchers and programmers have tried to bring these technologies into biology. However, the first commercial applications in this filed limited to the software that introduce the microorganisms, their effects in clinical/industrial/environmental microbiology and the best ways for treatment of the infectious disease [1]. Latest in this decade, engineers tried to integrate modern technologies to sense the presence of the whole cells or their components and metabolites by smartphones [2–4]. Park et al., for instance, reported a paper microfluidic device on which they loaded anti-*E. coli* antibody-conjugated beads. Any specific antigen (*E. coli*)-Antibody interaction on these papers led to immunoagglutination, which could be detected and quantified by analysis of Mie scattering from the digital pictures that were taken with a smartphone [2,5]. Later on, other immunochromatographic tests, such as fluorescence and luminescence, were joined with smartphones for the detection of enzymatic or antibody-antigen reactions [3]. Besides, Arts et al., were able to design a Luminescent Antibody Sensor (LUMABS) platform with the ability to detect antibodies directly in solution by a smartphone. LUMABS is made of a blue luciferase protein attached to a green fluorescent acceptor protein by a semi-flexible linker [6]. The presence of antibody in a solution disrupts the linker interaction between these two components, leading to the drops of bioluminescence resonance energy transfer (BRET) efficiency. They showed that this technique is able to detect antibodies at picomolar levels. As well, they showed that the manipulating of the structure of these LUMBAS enable specific detection of microorganisms in a solution. While all these designs depend on the employment of a specific/nonspecific factor for detection of microorganisms, Pei-Shih Liang et al., developed a smartphone-utilized biosensor with the ability to detect microorganisms on the surface of meat using Mie scatter, but without any requirement to antibodies or any other reagent [5]. They irradiated an 880 nm near infrared LED vertically to the surface of contaminated meat, and the signals scattered from the surface was collected by the digital camera of a smartphone and the gyro sensor at various angles. These types of technologies, however, defeat to distinguish microorganisms, and therefore, are only able to detect the presence of microbial cells on the subjects. While the integration of smartphone to the diagnostic systems is still in the first steps, it looks like this dream will come true very soon. For instance, Oxford Nanopore sequencer, which is able to sequence the genomic contents of microorganism, is a dedicated commercial approach with the ability to convert genomic codes into electronic data, which can be processed by smartphones [7].

Table 1. Comparison between the commercialized Sensing Technologies.

Name of the Device/Kit	Designed by	Working Principle	Advantages	Limitations	Reference
Microbe sensor (BM-300C)	Sharp	Works based on a chemical reaction referred to Millard reaction in which reducing sugars and proteins are heated to produce Melanoidin with strong fluorescence in the presence of UV. This device heats the sample to proceed the Millard reaction inside the cells, and the presence of any Melanoidin is detected by UV.	- Fast detection period (10 min.) - Continuous measurement for ongoing monitoring of microbe counts - Compatible with Smartphone	Not usable for detection of types of microorganisms	(http://www.sharp-world.com/corporate/news/130926.html)
SYTO® BC bacteria stain	Molecular Probes, Inc. (Eugene, OR, USA)	The quantity of bacteria can be detected by a SYTO® BC stain, which is a high-affinity green fluorescent nucleic acid stain, and a Cytometer	Measure the numbers of both gram positive and gram negative bacteria.	- Not able to distinguish gram positive and gram negative bacteria - Not able to detect the type of microorganisms	https://www.thermofisher.com/order/catalog/product/B7277
Milliflex® Rapid Microbiology Detection and Enumeration system	Millipore Sigma	Cells are trapped on a micro-filter and the number of cells are measured by Bioluminescence-ATP (adenosine triphosphate)	- Automated device for quantification of whole microbial cells	- Requires a long incubation time to grow microorganisms - Not able to differentiate microorganisms	http://www.emdmillipore.com/CA/en/product/Milliflex-Rapid-Microbiology-Detection-and-Enumeration-system,MM_NF-C10711
Celsis® systems	Charles River	The proprietary adenosine triphosphate (ATP) bioluminescence technology	- Automated device for quantification of whole microbial cells	- Requires a long incubation time to grow microorganisms - Not able to differentiate microorganisms	https://www.criver.com/products-services/qc-microbial-microbial-detection/microbial-detection-instruments?region=3601
Bactometer Microbial Analyzer	Biomerieux	These systems use Colorimetry od grown bacteria in defined media	- High discrimination between species - Antibiotic susceptibility testing	- Not real time - Not able to detect unknown microorganisms	http://www.biomerieux-usa.com/clinical/vitek-2-healthcare
PCR (Polymerase Chain Reaction)	Several companies (such as Eppendorf, Bio-Rad and Ampicon)	Detection of microorganisms based on amplification of target gene	Highly accurate	- Not real time - Not able to distinguish dead and alive cells - Not for qualification	[151]

Table 1. *Cont.*

Name of the Device/Kit	Designed by	Working Principle	Advantages	Limitations	Reference
Real-time PCR	Several companies (such as Eppendorf, Bio-Rad and Ampicon)	Detection of microorganisms based on amplification of target gene	- Highly accurate - Quantitative	- Not real time - Not able to distinguish dead and alive cells - Not able to separate different microorganism unless with combination with other techniques	[151]
Next generation Gene Sequencing systems	Several companies-454 Life Science, Ion Torrent-Illumina-Oxford NanoporeSOLiD,	Detection of microorganisms based on amplification of target gene	- Highly accurate - Quantitative - Able to detect the presence of different microorganisms in one sample	- Expensive - Not real time - Required bioinformatic analysis	[151]

Conflicts of Interest: The authors declare no conflict of interest.

References

1. Smith, K.F.; Goldberg, M.; Rosenthal, S.; Carlson, L.; Chen, J.; Chen, C.; Ramachandran, S. Global rise in human infectious disease outbreaks. *J. R. Soc. Interface* **2014**, *11*, 20140950. [CrossRef] [PubMed]

2. Greig, J. 2016 Foodborne Outbreak Updates. In Proceedings of the IAFP 2016 Annual Meeting, St. Louis, MO, USA, 31 July–3 August 2016.

3. Guo, L.; Feng, J.; Fang, Z.; Xu, J.; Lu, X. Application of microfluidic "lab-on-a-chip" for the detection of mycotoxins in foods. *Trends Food Sci. Technol.* **2015**, *46*, 252–263. [CrossRef]

4. Gardeniers, J.; van den Berg, A. Lab-on-a-chip systems for biomedical and environmental monitoring. *Anal. Bioanal. Chem.* **2004**, *378*, 1700–1703. [CrossRef] [PubMed]

5. Liu, Y.; Zhou, H.; Hu, Z.; Yu, G.; Yang, D.; Zhao, J. Label and label-free based surface-enhanced Raman scattering for pathogen bacteria detection: A review. *Biosens. Bioelectron.* **2017**, *94*, 131–140. [CrossRef] [PubMed]

6. Andersson, H.; van den Berg, A. Microfluidic devices for cellomics: A review. *Sens. Actuators B Chem.* **2003**, *92*, 315–325. [CrossRef]

7. Reyes, D.R.; Iossifidis, D.; Auroux, P.-A.; Manz, A. Micro total analysis systems. 1. Introduction, theory, and technology. *Anal. Chem.* **2002**, *74*, 2623–2636. [CrossRef] [PubMed]

8. Charbon, E. Towards large scale CMOS single-photon detector arrays for lab-on-chip applications. *J. Phys. D Appl. Phys.* **2008**, *41*, 094010. [CrossRef]

9. Lim, J.W.; Ha, D.; Lee, J.; Lee, S.K.; Kim, T. Review of micro/nanotechnologies for microbial biosensors. *Front. Bioeng. Biotechnol.* **2015**, *3*. [CrossRef] [PubMed]

10. Zhang, X.; Ju, H.; Wang, J. *Electrochemical Sensors, Biosensors and Their Biomedical Applications*; Academic Press: Cambridge, MA, USA, 2011.

11. Mannoor, M.S.; Zhang, S.; Link, A.J.; McAlpine, M.C. Electrical detection of pathogenic bacteria via immobilized antimicrobial peptides. *Proc. Natl. Acad. Sci. USA* **2010**, *107*, 19207–19212. [CrossRef] [PubMed]

12. Zhang, D.; Liu, Q. Biosensors and bioelectronics on smartphone for portable biochemical detection. *Biosens. Bioelectron.* **2016**, *75*, 273–284. [CrossRef] [PubMed]

13. Martins, R.; Nathan, A.; Barros, R.; Pereira, L.; Barquinha, P.; Correia, N.; Costa, R.; Ahnood, A.; Ferreira, I.; Fortunato, E. Complementary metal oxide semiconductor technology with and on paper. *Adv. Mater.* **2011**, *23*, 4491–4496. [CrossRef] [PubMed]

14. Caillat, P.; David, D.; Belleville, M.; Clerc, F.; Massit, C.; Revol-Cavalier, F.; Peltie, P.; Livache, T.; Bidan, G.; Roget, A. Biochips on CMOS: An active matrix address array for DNA analysis. *Sens. Actuators B Chem.* **1999**, *61*, 154–162. [CrossRef]

15. Han, S.-J.; Xu, L.; Yu, H.; Wilson, R.J.; White, R.L.; Pourmand, N.; Wang, S.X. CMOS integrated DNA microarray based on GMR sensors. In Proceedings of the IEDM'06. International Electron Devices Meeting, San Francisco, CA, USA, 11–13 December 2006.

16. Lee, J.; Kwak, Y.H.; Paek, S.-H.; Han, S.; Seo, S. CMOS image sensor-based ELISA detector using lens-free shadow imaging platform. *Sens. Actuators B Chem.* **2014**, *196*, 511–517. [CrossRef]

17. Sze, S.M. *Semiconductor Devices: Physics and Technology*; John Wiley & Sons: New York, NY, USA, 2008.

18. Otten, M.; Ott, W.; Jobst, M.A.; Milles, L.F.; Verdorfer, T.; Pippig, D.A.; Nash, M.A.; Gaub, H.E. From genes to protein mechanics on a chip. *Nat. Methods* **2014**, *11*, 1127–1130. [CrossRef] [PubMed]

19. Kim, T.-H.; Park, J.; Kim, C.-J.; Cho, Y.-K. Fully integrated lab-on-a-disc for nucleic acid analysis of food-borne pathogens. *Anal. Chem.* **2014**, *86*, 3841–3848. [CrossRef] [PubMed]

20. Liu, Q.; Wu, C.; Cai, H.; Hu, N.; Zhou, J.; Wang, P. Cell-based biosensors and their application in biomedicine. *Chem. Rev.* **2014**, *114*, 6423–6461. [CrossRef] [PubMed]

21. Lei, K.F. Review on impedance detection of cellular responses in micro/nano environment. *Micromachines* **2014**, *5*, 1–12. [CrossRef]

22. Kim, D.; Herr, A.E. Protein immobilization techniques for microfluidic assays. *Biomicrofluidics* **2013**, *7*, 041501. [CrossRef] [PubMed]

23. Pack, S.P.; Kamisetty, N.K.; Nonogawa, M.; Devarayapalli, K.C.; Ohtani, K.; Yamada, K.; Yoshida, Y.; Kodaki, T.; Makino, K. Direct immobilization of DNA oligomers onto the amine-functionalized glass surface for DNA microarray fabrication through the activation-free reaction of oxanine. *Nucleic Acids Res.* **2007**, *35*, e110. [CrossRef] [PubMed]

24. Fabre, B.; Hauquier, F. Boronic Acid-Functionalized Oxide-Free Silicon Surfaces for the Electrochemical Sensing of Dopamine. *Langmuir* **2017**, *33*, 8693–8699. [CrossRef] [PubMed]

25. Sarvi, F.; Yue, Z.; Hourigan, K.; Thompson, M.C.; Chan, P.P. Surface-functionalization of PDMS for potential micro-bioreactor and embryonic stem cell culture applications. *J. Mater. Chem. B* **2013**, *1*, 987–996. [CrossRef]

26. Laib, S.; MacCraith, B.D. Immobilization of biomolecules on cycloolefin polymer supports. *Anal. Chem.* **2007**, *79*, 6264–6270. [CrossRef] [PubMed]

27. Xin, L.; Cao, Z.; Lau, C.; Kai, M.; Lu, J. G-rich sequence-functionalized polystyrene microsphere-based instantaneous derivatization for the chemiluminescent amplified detection of DNA. *Luminescence* **2010**, *25*, 336–342. [CrossRef] [PubMed]

28. Fixe, F.; Dufva, M.; Telleman, P.; Christensen, C.B.V. Functionalization of poly (methyl methacrylate) (PMMA) as a substrate for DNA microarrays. *Nucleic Acids Res.* **2004**, *32*, e9. [CrossRef] [PubMed]

29. Rashid, J.I.A.; Yusof, N.A. The strategies of DNA immobilization and hybridization detection mechanism in the construction of electrochemical DNA sensor: A review. *Sens. Bio-Sens. Res.* **2017**, *16*, 19–31. [CrossRef]

30. Xie, Y.; Hill, C.A.; Xiao, Z.; Militz, H.; Mai, C. Silane coupling agents used for natural fiber/polymer composites: A review. *Compos. Part A Appl. Sci. Manuf.* **2010**, *41*, 806–819. [CrossRef]

31. Park, I.-S.; Kim, N. Thiolated Salmonella antibody immobilization onto the gold surface of piezoelectric quartz crystal. *Biosens. Bioelectron.* **1998**, *13*, 1091–1097. [CrossRef]

32. Khrenov, V.; Klapper, M.; Koch, M.; Müllen, K. Surface functionalized ZnO particles designed for the use in transparent nanocomposites. *Macromol. Chem. Phys.* **2005**, *206*, 95–101. [CrossRef]

33. Kopetz, S.; Cai, D.; Rabe, E.; Neyer, A. PDMS-based optical waveguide layer for integration in electrical–optical circuit boards. *AEU Int. J. Electron. Commun.* **2007**, *61*, 163–167. [CrossRef]

34. Irawan, R.; Tjin, S.C.; Fang, X.; Fu, C.Y. Integration of optical fiber light guide, fluorescence detection system, and multichannel disposable microfluidic chip. *Biomed. Microdevices* **2007**, *9*, 413–419. [CrossRef] [PubMed]

35. Niles, W.D.; Coassin, P.J. Cyclic olefin polymers: Innovative materials for high-density multiwell plates. *Assay Drug Dev. Technol.* **2008**, *6*, 577–590. [CrossRef] [PubMed]

36. Viswanathan, P.; Johnson, D.W.; Hurley, C.; Cameron, N.R.; Battaglia, G. 3D surface functionalization of emulsion-templated polymeric foams. *Macromolecules* **2014**, *47*, 7091–7098. [CrossRef]

37. Marques, M.E.; Mansur, A.A.; Mansur, H.S. Chemical functionalization of surfaces for building three-dimensional engineered biosensors. *Appl. Surface Sci.* **2013**, *275*, 347–360. [CrossRef]

38. Hu, F.; Neoh, K.; Cen, L.; Kang, E. Antibacterial and antifungal efficacy of surface functionalized polymeric beads in repeated applications. *Biotechnol. Bioeng.* **2005**, *89*, 474–484. [CrossRef] [PubMed]

39. Drury, J.L.; Mooney, D.J. Hydrogels for tissue engineering: Scaffold design variables and applications. *Biomaterials* **2003**, *24*, 4337–4351. [CrossRef]

40. Piletsky, S.A.; Matuschewski, H.; Schedler, U.; Wilpert, A.; Piletska, E.V.; Thiele, T.A.; Ulbricht, M. Surface functionalization of porous polypropylene membranes with molecularly imprinted polymers by photograft copolymerization in water. *Macromolecules* **2000**, *33*, 3092–3098. [CrossRef]

41. Iwasa, F.; Tsukimura, N.; Sugita, Y.; Kanuru, R.K.; Kubo, K.; Hasnain, H.; Att, W.; Ogawa, T. TiO$_2$ micro-nano-hybrid surface to alleviate biological aging of UV-photofunctionalized titanium. *Int. J. Nanomed.* **2011**, *6*, 1327–1341.

42. Ribeiro, A.J.; Zaleta-Rivera, K.; Ashley, E.A.; Pruitt, B.L. Stable, covalent attachment of laminin to microposts improves the contractility of mouse neonatal cardiomyocytes. *ACS Appl. Mater. Interfaces* **2014**, *6*, 15516–15526. [CrossRef] [PubMed]

43. Wong, I.; Ho, C.-M. Surface molecular property modifications for poly (dimethylsiloxane) (PDMS) based microfluidic devices. *Microfluid. Nanofluid.* **2009**, *7*, 291. [CrossRef] [PubMed]

44. Zhang, H.; Chiao, M. Anti-fouling coatings of poly (dimethylsiloxane) devices for biological and biomedical applications. *J. Med. Boil. Eng.* **2015**, *35*, 143–155. [CrossRef] [PubMed]

45. Rusmini, F.; Zhong, Z.; Feijen, J. Protein immobilization strategies for protein biochips. *Biomacromolecules* **2007**, *8*, 1775–1789. [CrossRef] [PubMed]

46. Da Silva, S.; Grosjean, L.; Ternan, N.; Mailley, P.; Livache, T.; Cosnier, S. Biotinylated polypyrrole films: An easy electrochemical approach for the reagentless immobilization of bacteria on electrode surfaces. *Bioelectrochemistry* **2004**, *63*, 297–301. [CrossRef] [PubMed]

47. Bogusiewicz, A.; Mock, N.I.; Mock, D.M. Release of biotin from biotinylated proteins occurs enzymatically and nonenzymatically in human plasma. *Anal. Biochem.* **2004**, *331*, 260–266. [CrossRef] [PubMed]

48. Aizawa, M.; Kato, S.; Suzuki, S. Electrochemical typing of blood using affinity membranes. *J. Membr. Sci.* **1980**, *7*, 1–10. [CrossRef]

49. Kim, Y.S.; Chung, J.; Song, M.Y.; Jurng, J.; Kim, B.C. Aptamer cocktails: Enhancement of sensing signals compared to single use of aptamers for detection of bacteria. *Biosens. Bioelectron.* **2014**, *54*, 195–198. [CrossRef] [PubMed]

50. Queirós, R.B.; de-Los-Santos-Álvarez, N.; Noronha, J.; Sales, M.G.F. A label-free DNA aptamer-based impedance biosensor for the detection of *E. coli* outer membrane proteins. *Sens. Actuators B Chem.* **2013**, *181*, 766–772. [CrossRef]

51. Yi-Xian, W.; Zun-Zhong, Y.; Cheng-Yan, S.; Yi-Bin, Y. Application of aptamer based biosensors for detection of pathogenic microorganisms. *Chin. J. Anal. Chem.* **2012**, *40*, 634–642.

52. So, H.M.; Park, D.W.; Jeon, E.K.; Kim, Y.H.; Kim, B.S.; Lee, C.K.; Choi, S.Y.; Kim, S.C.; Chang, H.; Lee, J.O. Detection and Titer Estimation of Escherichia coli Using Aptamer-Functionalized Single-Walled Carbon-Nanotube Field-Effect Transistors. *Small* **2008**, *4*, 197–201. [CrossRef] [PubMed]

53. Paniel, N.; Baudart, J.; Hayat, A.; Barthelmebs, L. Aptasensor and genosensor methods for detection of microbes in real world samples. *Methods* **2013**, *64*, 229–240. [CrossRef] [PubMed]

54. Lian, Y.; He, F.; Wang, H.; Tong, F. A new aptamer/graphene interdigitated gold electrode piezoelectric sensor for rapid and specific detection of *Staphylococcus aureus*. *Biosens. Bioelectron.* **2015**, *65*, 314–319. [CrossRef] [PubMed]

55. Donelli, G. *Advances in Microbiology, Infectious Diseases and Public Health*; Springer: New York, NY, USA, 2016; Volume 1.

56. Mahon, C.R.; Lehman, D.C.; Manuselis, G., Jr. *Textbook of Diagnostic Microbiology*; Elsevier Health Sciences: Amsterdam, The Netherlands, 2014.

57. Baker, F.J.; Silverton, R.E. *Introduction to Medical Laboratory Technology*; Butterworth-Heinemann: Oxford, UK, 2014.

58. Myers, F.B.; Lee, L.P. Innovations in optical microfluidic technologies for point-of-care diagnostics. *Lab Chip* **2008**, *8*, 2015–2031. [CrossRef] [PubMed]

59. Noble, R.T.; Weisberg, S.B. A review of technologies for rapid detection of bacteria in recreational waters. *J. Water Health* **2005**, *3*, 381–392. [CrossRef] [PubMed]

60. Song, J.M.; Culha, M.; Kasili, P.M.; Griffin, G.D.; Vo-Dinh, T. A compact CMOS biochip immunosensor towards the detection of a single bacteria. *Biosens. Bioelectron.* **2005**, *20*, 2203–2209. [CrossRef] [PubMed]

61. Pospíšilová, M.; Kuncová, G.; Trögl, J. Fiber-optic chemical sensors and fiber-optic bio-sensors. *Sensors* **2015**, *15*, 25208–25259. [CrossRef] [PubMed]

62. Altintas, Z.; Akgun, M.; Kokturk, G.; Uludag, Y. A fully automated microfluidic-based electrochemical sensor for real-time bacteria detection. *Biosens. Bioelectron.* **2018**, *100*, 541–548. [CrossRef] [PubMed]

63. Yao, L.; Hajj-Hassan, M.; Ghafar-Zadeh, E.; Shabani, A.; Chodavarapu, V.; Zourob, M. CMOS capacitive sensor system for bacteria detection using phage organisms. In Proceedings of the 2008 IEEE Canadian Conference on Electrical and Computer Engineering, Niagara Falls, ON, Canada, 4–7 May 2008.

64. Nikkhoo, N.; Gulak, P.G.; Maxwell, K. Rapid detection of *E. coli* bacteria using potassium-sensitive FETs in CMOS. *IEEE Trans. Biomed. Circuits Syst.* **2013**, *7*, 621–630. [CrossRef] [PubMed]

65. Mejri, M.; Baccar, H.; Baldrich, E.; Del Campo, F.; Helali, S.; Ktari, T.; Simonian, A.; Aouni, M.; Abdelghani, A. Impedance biosensing using phages for bacteria detection: Generation of dual signals as the clue for in-chip assay confirmation. *Biosens. Bioelectron.* **2010**, *26*, 1261–1267. [CrossRef] [PubMed]

66. Bogas, D.; Nyberg, L.; Pacheco, R.; Azevedo, N.F.; Beech, J.P.; Gomila, M.; Lalucat, J.; Manaia, C.M.; Nunes, O.C.; Tegenfeldt, J.O. 1. Optical mapping in microbiology: The beginning. *BioTechniques* **2017**, *62*, 255–267. [CrossRef] [PubMed]

67. Eletxigerra, U.; Martinez-Perdiguero, J.; Barderas, R.; Pingarrón, J.M.; Campuzano, S.; Merino, S. Surface plasmon resonance immunosensor for ErbB2 breast cancer biomarker determination in human serum and raw cancer cell lysates. *Anal. Chim. Acta* **2016**, *905*, 156–162. [CrossRef] [PubMed]

68. Bouguelia, S.; Roupioz, Y.; Slimani, S.; Mondani, L.; Casabona, M.G.; Durmort, C.; Vernet, T.; Calemczuk, R.; Livache, T. On-chip microbial culture for the specific detection of very low levels of bacteria. *Lab Chip* **2013**, *13*, 4024–4032. [CrossRef] [PubMed]

69. Yoon, J. A Microstrip-Based Radio-Frequency Biosensor for the Detection of Bacteria in Water. Master of Science Thesis, University of Nevada, Reno, NV, USA, 2003.

70. Hsieh, S.; Hsieh, S.-L.; Hsieh, C.-W.; Lin, P.-C.; Wu, C.-H. Label-free glucose detection using cantilever sensor technology based on gravimetric detection principles. *J. Anal. Methods Chem.* **2013**, *2013*. [CrossRef] [PubMed]

71. Guo, X.; Lin, C.-S.; Chen, S.-H.; Ye, R.; Wu, V.C. A piezoelectric immunosensor for specific capture and enrichment of viable pathogens by quartz crystal microbalance sensor, followed by detection with antibody-functionalized gold nanoparticles. *Biosens. Bioelectron.* **2012**, *38*, 177–183. [CrossRef] [PubMed]

72. Garipcan, B.; Caglayan, M.; Demirel, G. New generation biosensors based on ellipsometry. In *New Perspectives in Biosensors Technology and Applications*; InTech: London, UK, 2011.

73. Hao, R.-Z.; Song, H.-B.; Zuo, G.-M.; Yang, R.-F.; Wei, H.-P.; Wang, D.-B.; Cui, Z.-Q.; Zhang, Z.; Cheng, Z.-X.; Zhang, X.-E. DNA probe functionalized QCM biosensor based on gold nanoparticle amplification for *Bacillus anthracis* detection. *Biosens. Bioelectron.* **2011**, *26*, 3398–3404. [CrossRef] [PubMed]

74. Malic, L.; Brassard, D.; Veres, T.; Tabrizian, M. Integration and detection of biochemical assays in digital microfluidic LOC devices. *Lab Chip* **2010**, *10*, 418–431. [CrossRef] [PubMed]

75. Carroll, K.C.; Butel, J.; Morse, S. *Jawetz Melnick & Adelbergs Medical Microbiology 27 E*; McGraw Hill Professional: New York, NY, USA, 2015.

76. Su, L.; Jia, W.; Hou, C.; Lei, Y. Microbial biosensors: A review. *Biosens. Bioelectron.* **2011**, *26*, 1788–1799. [CrossRef] [PubMed]

77. Grieshaber, D.; MacKenzie, R.; Voeroes, J.; Reimhult, E. Electrochemical biosensors-sensor principles and architectures. *Sensors* **2008**, *8*, 1400–1458. [CrossRef] [PubMed]

78. Ronkainen, N.J.; Halsall, H.B.; Heineman, W.R. Electrochemical biosensors. *Chem. Soc. Rev.* **2010**, *39*, 1747–1763. [CrossRef] [PubMed]

79. Pohanka, M.; Skládal, P. Electrochemical biosensors—Principles and applications. *J. Appl. Biomed.* **2008**, *6*, 57–64.

80. Kurkina, T.; Vlandas, A.; Ahmad, A.; Kern, K.; Balasubramanian, K. Label-free detection of few copies of DNA with carbon nanotube impedance biosensors. *Angew. Chem. Int. Ed.* **2011**, *50*, 3710–3714. [CrossRef] [PubMed]

81. Pethig, R.; Markx, G.H. Applications of dielectrophoresis in biotechnology. *Trends Biotechnol.* **1997**, *15*, 426–432. [CrossRef]

82. Yang, L. Electrical impedance spectroscopy for detection of bacterial cells in suspensions using interdigitated microelectrodes. *Talanta* **2008**, *74*, 1621–1629. [CrossRef] [PubMed]

83. Suehiro, J.; Hatano, T.; Shutou, M.; Hara, M. Improvement of electric pulse shape for electropermeabilization-assisted dielectrophoretic impedance measurement for high sensitive bacteria detection. *Sens. Actuators B Chem.* **2005**, *109*, 209–215. [CrossRef]

84. Lan, W.-J.; Maxwell, E.J.; Parolo, C.; Bwambok, D.K.; Subramaniam, A.B.; Whitesides, G.M. Based electroanalytical devices with an integrated, stable reference electrode. *Lab Chip* **2013**, *13*, 4103–4108. [CrossRef] [PubMed]

85. Ahmed, A.; Rushworth, J.V.; Hirst, N.A.; Millner, P.A. Biosensors for whole-cell bacterial detection. *Clin. Microbiol. Rev.* **2014**, *27*, 631–646. [CrossRef] [PubMed]

86. Maalouf, R.; Fournier-Wirth, C.; Coste, J.; Chebib, H.; Saïkali, Y.; Vittori, O.; Errachid, A.; Cloarec, J.-P.; Martelet, C.; Jaffrezic-Renault, N. Label-free detection of bacteria by electrochemical impedance spectroscopy: Comparison to surface plasmon resonance. *Anal. Chem.* **2007**, *79*, 4879–4886. [CrossRef] [PubMed]

87. Yang, L.; Bashir, R. Electrical/electrochemical impedance for rapid detection of foodborne pathogenic bacteria. *Biotechnol. Adv.* **2008**, *26*, 135–150. [CrossRef] [PubMed]

88. Hamada, R.; Takayama, H.; Shonishi, Y.; Mao, L.; Nakano, M.; Suehiro, J. A rapid bacteria detection technique utilizing impedance measurement combined with positive and negative dielectrophoresis. *Sens. Actuators B Chem.* **2013**, *181*, 439–445. [CrossRef]

89. Suehiro, J.; Ohtsubo, A.; Hatano, T.; Hara, M. Selective detection of bacteria by a dielectrophoretic impedance measurement method using an antibody-immobilized electrode chip. *Sens. Actuators B Chem.* **2006**, *119*, 319–326. [CrossRef]

90. Jiang, J.; Wang, X.; Chao, R.; Ren, Y.; Hu, C.; Xu, Z.; Liu, G.L. Smartphone based portable bacteria pre-concentrating microfluidic sensor and impedance sensing system. *Sens. Actuators B Chem.* **2014**, *193*, 653–659. [CrossRef]

91. Simonian, A.L.; Rainina, E.I.; Wild, J.R. Microbial biosensors based on potentiometric detection. In *Enzyme and Microbial Biosensors: Techniques and Protocols*; Humana Press: New York, NY, USA, 1998; pp. 237–248.

92. Wan, Y.; Wang, Y.; Wu, J.; Zhang, D. Graphene oxide sheet-mediated silver enhancement for application to electrochemical biosensors. *Anal. Chem.* **2010**, *83*, 648–653. [CrossRef] [PubMed]

93. Hernández, R.; Vallés, C.; Benito, A.M.; Maser, W.K.; Rius, F.X.; Riu, J. Graphene-based potentiometric biosensor for the immediate detection of living bacteria. *Biosens. Bioelectron.* **2014**, *54*, 553–557. [CrossRef] [PubMed]

94. Thorpe, T.C.; Wilson, M.; Turner, J.; DiGuiseppi, J.; Willert, M.; Mirrett, S.; Reller, L. BacT/Alert: An automated colorimetric microbial detection system. *J. Clin. Microbiol.* **1990**, *28*, 1608–1612. [PubMed]

95. Brosel-Oliu, S.; Uria, N.; Abramova, N.; Bratov, A. Impedimetric sensors for bacteria detection. In *Biosensors-Micro and Nanoscale Applications*; InTech: London, UK, 2015.

96. Abeyrathne, C.D.; Huynh, D.H.; Mcintire, T.W.; Nguyen, T.C.; Nasr, B.; Zantomio, D.; Chana, G.; Abbott, I.; Choong, P.; Catton, M. Lab on a Chip sensor for rapid detection and antibiotic resistance determination of *Staphylococcus aureus*. *Analyst* **2016**, *141*, 1922–1929. [CrossRef] [PubMed]

97. Cabibbe, A.M.; Miotto, P.; Moure, R.; Alcaide, F.; Feuerriegel, S.; Pozzi, G.; Nikolayevskyy, V.; Drobniewski, F.; Niemann, S.; Reither, K. Lab-on-chip-based platform for fast molecular diagnosis of multidrug-resistant tuberculosis. *J. Clin. Microbiol.* **2015**, *53*, 3876–3880. [CrossRef] [PubMed]

98. Foudeh, A.M.; Didar, T.F.; Veres, T.; Tabrizian, M. Microfluidic designs and techniques using lab-on-a-chip devices for pathogen detection for point-of-care diagnostics. *Lab Chip* **2012**, *12*, 3249–3266. [CrossRef] [PubMed]

99. Wang, S.; Ge, L.; Song, X.; Yu, J.; Ge, S.; Huang, J.; Zeng, F. Based chemiluminescence ELISA: Lab-on-paper based on chitosan modified paper device and wax-screen-printing. *Biosens. Bioelectron.* **2012**, *31*, 212–218. [CrossRef] [PubMed]

100. Mark, D.; Haeberle, S.; Roth, G.; Von Stetten, F.; Zengerle, R. Microfluidic Lab-on-a-chip platforms: Requirements, characteristics and applications. In *Microfluidics Based Microsystems*; Springer: New York, NY, USA, 2010; pp. 305–376.

101. Joung, C.-K.; Kim, H.-N.; Lim, M.-C.; Jeon, T.-J.; Kim, H.-Y.; Kim, Y.-R. A nanoporous membrane-based impedimetric immunosensor for label-free detection of pathogenic bacteria in whole milk. *Biosens. Bioelectron.* **2013**, *44*, 210–215. [CrossRef] [PubMed]

102. Haque, A.-M.J.; Park, H.; Sung, D.; Jon, S.; Choi, S.-Y.; Kim, K. An electrochemically reduced graphene oxide-based electrochemical immunosensing platform for ultrasensitive antigen detection. *Anal. Chem.* **2012**, *84*, 1871–1878. [CrossRef] [PubMed]

103. Prabhulkar, S.; Alwarappan, S.; Liu, G.; Li, C.-Z. Amperometric micro-immunosensor for the detection of tumor biomarker. *Biosens. Bioelectron.* **2009**, *24*, 3524–3530. [CrossRef] [PubMed]

104. Wan, Y.; Zhang, D.; Wang, Y.; Hou, B. A 3D-impedimetric immunosensor based on foam Ni for detection of sulfate-reducing bacteria. *Electrochem. Commun.* **2010**, *12*, 288–291. [CrossRef]

105. Jha, A.K.; Xu, X.; Duncan, R.L.; Jia, X. Controlling the adhesion and differentiation of mesenchymal stem cells using hyaluronic acid-based, doubly crosslinked networks. *Biomaterials* **2011**, *32*, 2466–2478. [CrossRef] [PubMed]

106. Wu, S.; Duan, N.; Shi, Z.; Fang, C.; Wang, Z. Simultaneous aptasensor for multiplex pathogenic bacteria detection based on multicolor upconversion nanoparticles labels. *Anal. Chem.* **2014**, *86*, 3100–3107. [CrossRef] [PubMed]

107. Logares, R.; Sunagawa, S.; Salazar, G.; Cornejo-Castillo, F.M.; Ferrera, I.; Sarmento, H.; Hingamp, P.; Ogata, H.; Vargas, C.; G. Lima-Mendez, G. Metagenomic 16S rDNA Illumina tags are a powerful alternative to amplicon sequencing to explore diversity and structure of microbial communities. *Environ. Microbiol.* **2014**, *16*, 2659–2671. [CrossRef] [PubMed]

108. Elizaquível, P.; Aznar, R.; Sánchez, G. Recent developments in the use of viability dyes and quantitative PCR in the food microbiology field. *J. Appl. Microbiol.* **2014**, *116*, 1–13. [CrossRef] [PubMed]

109. Kanitkar, Y.H.; Stedtfeld, R.D.; Steffan, R.J.; Hashsham, S.A.; Cupples, A.M. Loop-Mediated Isothermal Amplification (LAMP) for Rapid Detection and Quantification of Dehalococcoides Biomarker Genes in Commercial Reductive Dechlorinating Cultures KB-1 and SDC-9. *Appl. Environ. Microbiol.* **2016**, *82*, 1799–1806. [CrossRef] [PubMed]

110. Murray, P.R.; Rosenthal, K.S.; Pfaller, M.A. *Medical Microbiology*; Elsevier Health Sciences: Amsterdam, The Netherlands, 2015.

111. Lakshmi, R.; Baskar, V.; Ranga, U. Extraction of superior-quality plasmid DNA by a combination of modified alkaline lysis and silica matrix. *Anal. Biochem.* **1999**, *272*, 109–112. [CrossRef] [PubMed]

112. Bavykin, S.G.; Akowski, J.P.; Zakhariev, V.M.; Barsky, V.E.; Perov, A.N.; Mirzabekov, A.D. Portable system for microbial sample preparation and oligonucleotide microarray analysis. *Appl. Environ. Microbiol.* **2001**, *67*, 922–928. [CrossRef] [PubMed]

113. Brocchieri, L. The GC content of bacterial genomes. *J. Phylogen. Evolut. Biol.* **2013**, 1–3. [CrossRef]

114. Abbasian, F.; Saberbaghi, T. Metagenomic study of human gastrointestinal tracts in health and diseases. *Gastroenterol. Hepatol. Res.* **2013**, *2*, 885–896.

115. Ginzinger, D.G. Gene quantification using real-time quantitative PCR: An emerging technology hits the mainstream. *Exp. Hematol.* **2002**, *30*, 503–512. [CrossRef]

116. Ren-Kuibai, C.-L.P.; Hsu, C.-H. Quantitative PCR Analysis of Mitochondrial. *Mitochondr. Pathog. Genes Apoptosis Aging Dis.* **2014**, *1011*, 304–309.

117. Oblath, E.A.; Henley, W.H.; Alarie, J.P.; Ramsey, J.M. A microfluidic chip integrating DNA extraction and real-time PCR for the detection of bacteria in saliva. *Lab Chip* **2013**, *13*, 1325–1332. [CrossRef] [PubMed]

118. Medina-Plaza, C.; de Saja, J.A.; Fernandez-Escudero, J.A.; Barajas, E.; Medrano, G.; Rodriguez-Mendez, M.L. Array of biosensors for discrimination of grapes according to grape variety, vintage and ripeness. *Anal. Chim. Acta* **2016**, *947*, 16–22. [CrossRef] [PubMed]

119. Sattabongkot, J.; Tsuboi, T.; Han, E.-T.; Bantuchai, S.; Buates, S. Loop-mediated isothermal amplification assay for rapid diagnosis of malaria infections in an area of endemicity in Thailand. *J. Clin. Microbiol.* **2014**, *52*, 1471–1477. [CrossRef] [PubMed]

120. Nixon, G.; Garson, J.A.; Grant, P.; Nastouli, E.; Foy, C.A.; Huggett, J.F. Comparative study of sensitivity, linearity, and resistance to inhibition of digital and nondigital polymerase chain reaction and loop mediated isothermal amplification assays for quantification of human cytomegalovirus. *Anal. Chem.* **2014**, *86*, 4387–4394. [CrossRef] [PubMed]

121. Notomi, T.; Mori, Y.; Tomita, N.; Kanda, H. Loop-mediated isothermal amplification (LAMP): Principle, features, and future prospects. *J. Microbiol.* **2015**, *53*, 1–5. [CrossRef] [PubMed]

122. Fernández-Soto, P.; Mvoulouga, P.O.; Akue, J.P.; Abán, J.L.; Santiago, B.V.; Sánchez, M.C.; Muro, A. Development of a highly sensitive loop-mediated isothermal amplification (LAMP) method for the detection of *Loa loa*. *PLoS ONE* **2014**, *9*, e94664. [CrossRef] [PubMed]

123. Guo, Z.; Yu, T.; He, J.; Liu, F.; Hao, H.; Zhao, Y.; Wen, J.; Wang, Q. An integrated microfluidic chip for the detection of bacteria—A proof of concept. *Mol. Cell. Probes* **2015**, *29*, 223–227. [CrossRef] [PubMed]

124. Andoh, A.; Imaeda, H.; Aomatsu, T.; Inatomi, O.; Bamba, S.; Sasaki, M.; Saito, Y.; Tsujikawa, T.; Fujiyama, Y. Comparison of the fecal microbiota profiles between ulcerative colitis and Crohn's disease using terminal restriction fragment length polymorphism analysis. *J. Gastroenterol.* **2011**, *46*, 479–486. [CrossRef] [PubMed]

125. Sjöberg, F.; Nowrouzian, F.; Rangel, I.; Hannoun, C.; Moore, E.; Adlerberth, I.; Wold, A.E. Comparison between terminal-restriction fragment length polymorphism (T-RFLP) and quantitative culture for analysis of infants' gut microbiota. *J. Microbiol. Methods* **2013**, *94*, 37–46. [CrossRef] [PubMed]

126. Hoffman, J.; Clark, M.; Amos, W.; Peck, L. Widespread amplification of amplified fragment length polymorphisms (AFLPs) in marine Antarctic animals. *Polar Biol.* **2012**, *35*, 919–929. [CrossRef]

127. Katara, J.; Deshmukh, R.; Singh, N.K.; Kaur, S. Diversity Analysis of Bacillus thuringiensis Isolates Recovered from Diverse Habitats in India using Random Amplified Polymorphic DNA (RAPD) Markers. *J. Biol. Sci.* **2013**, *13*, 514. [CrossRef]

128. Schmalenberger, A.; Tebbe, C.C. Profiling the Diversity of Microbial Communities with Single-Strand Conformation Polymorphism (SSCP). In *Environmental Microbiology*; Springer: New York, NY, USA, 2014; pp. 71–83.

129. Loenen, W.A.; Dryden, D.T.; Raleigh, E.A.; Wilson, G.G. Type I restriction enzymes and their relatives. *Nucleic Acids Res.* **2013**, *42*, 20–44. [CrossRef] [PubMed]

130. Botstein, D.; White, R.L.; Skolnick, M.; Davis, R.W. Construction of a genetic linkage map in man using restriction fragment length polymorphisms. *Am. J. Human Gen.* **1980**, *32*, 314–331.

131. Fontecha, G.A.; García, K.; Rueda, M.M.; Sosa-Ochoa, W.; Sánchez, A.L.; Leiva, B. A PCR-RFLP method for the simultaneous differentiation of three *Entamoeba* species. *Exp. Parasitol.* **2015**, *151*, 80–83. [CrossRef] [PubMed]

132. Salvatore, D.; Di Francesco, A.; Poglayen, G.; Rugna, G.; Santi, A.; Morandi, B.; Baldelli, R. Molecular characterization of *Leishmania infantum* strains by kinetoplast DNA RFLP-PCR. *Vet. Ital.* **2016**, *52*, 71–75. [PubMed]

133. Esteve-Zarzoso, B.; Hierro, N.; Mas, A.; Guillamón, J.M. A new simplified AFLP method for wine yeast strain typing. *LWT Food Sci. Technol.* **2010**, *43*, 1480–1484. [CrossRef]

134. Zhao, D.; Huang, R.; Zeng, J.; Yan, W.; Wang, J.; Ma, T.; Wang, M.; Wu, Q.L. Diversity analysis of bacterial community compositions in sediments of urban lakes by terminal restriction fragment length polymorphism (T-RFLP). *World J. Microbiol. Biotechnol.* **2012**, *28*, 3159–3170. [CrossRef] [PubMed]

135. Sawamura, H.; Yamada, M.; Endo, K.; Soda, S.; Ishigaki, T.; Ike, M. Characterization of microorganisms at different landfill depths using carbon-utilization patterns and 16S rRNA gene based T-RFLP. *J. Biosci. Bioeng.* **2010**, *109*, 130–137. [CrossRef] [PubMed]

136. Jadhav, V.V.; Jamle, M.M.; Pawar, P.D.; Devare, M.N.; Bhadekar, R.K. Fatty acid profiles of PUFA producing Antarctic bacteria: Correlation with RAPD analysis. *Ann. Microbiol.* **2010**, *60*, 693–699. [CrossRef]

137. Rossetti, L.; Giraffa, G. Rapid identification of dairy lactic acid bacteria by M13-generated, RAPD-PCR fingerprint databases. *J. Microbiol. Methods* **2005**, *63*, 135–144. [CrossRef] [PubMed]

138. Swapna, M.; Sivaraju, K.; Sharma, R.; Singh, N.; Mohapatra, T. Single-strand conformational polymorphism of EST-SSRs: A potential tool for diversity analysis and varietal identification in sugarcane. *Plant Mol. Biol. Rep.* **2011**, *29*, 505–513. [CrossRef]

139. Payseur, B.A.; Rieseberg, L.H. A genomic perspective on hybridization and speciation. *Mol. Ecol.* **2016**, *25*, 2337–2360. [CrossRef] [PubMed]

140. Frickmann, H.; Zautner, A.E.; Moter, A.; Kikhney, J.; Hagen, R.M.; Stender, H.; Poppert, S. Fluorescence in situ hybridization (FISH) in the microbiological diagnostic routine laboratory: A review. *Crit. Rev. Microbiol.* **2017**, *43*, 263–293. [CrossRef] [PubMed]

141. Miller, M.B.; Tang, Y.-W. Basic concepts of microarrays and potential applications in clinical microbiology. *Clin. Microbiol. Rev.* **2009**, *22*, 611–633. [CrossRef] [PubMed]

142. Sassolas, A.; Leca-Bouvier, B.D.; Blum, L.J. DNA biosensors and microarrays. *Chem. Rev.* **2008**, *108*, 109–139. [CrossRef] [PubMed]

143. Xi, C.; Boppart, S.; Raskin, L. Use of molecular beacons for the detection of bacteria in microfluidic devices. In Proceedings of the SPIE's Photonics West, San Jose, CA, USA, 25–31 January 2003.

144. Heller, M.J. DNA microarray technology: Devices, systems, and applications. *Annu. Rev. Biomed. Eng.* **2002**, *4*, 129–153. [CrossRef] [PubMed]

145. Nguyen, D.V.; Bulak Arpat, A.; Wang, N.; Carroll, R.J. DNA microarray experiments: Biological and technological aspects. *Biometrics* **2002**, *58*, 701–717. [CrossRef] [PubMed]

146. Cardoso, F.; Costa, T.; Germano, J.; Cardoso, S.; Borme, J.; Gaspar, J.; Fernandes, J.; Piedade, M.; Freitas, P. Integration of magnetoresistive biochips on a CMOS circuit. *IEEE Trans. Magn.* **2012**, *48*, 3784–3787. [CrossRef]

147. Alessandrini, A.; De Renzi, V.; Berti, L.; Barak, I.; Facci, P. Chemically homogeneous, silylated surface for effective DNA binding and hybridization. *Surf. Sci.* **2005**, *582*, 202–208. [CrossRef]

148. Abbasian, F.; Lockington, R.; Mallavarapu, M.; Naidu, R. The Integration of Sequencing and Bioinformatics in Metagenomics. *Rev. Environ. Sci. Biotechnol.* **2015**, *14*, 357–383. [CrossRef]

149. Langille, M.G.; Zaneveld, J.; Caporaso, J.G.; McDonald, D.; Knights, D.; Reyes, J.A.; Clemente, J.C.; Burkepile, D.E.; Thurber, R.L.V.; Knight, R. Predictive functional profiling of microbial communities using 16S rRNA marker gene sequences. *Nat. Biotechnol.* **2013**, *31*, 814–821. [CrossRef] [PubMed]

150. Shu, Q.; Jiao, N. Developing a novel approach of rpoB gene as a powerful biomarker for the environmental microbial diversity. *Geomicrobiol. J.* **2013**, *30*, 108–119. [CrossRef]

151. Morata, V.; Gusils, C.; Gonzalez, S. Classification of the bacteria-traditional. *Enycl. Food Microbiol.* **1999**, *1*, 173–183.

152. Van Dijk, E.L.; Auger, H.; Jaszczyszyn, Y.; Thermes, C. Ten years of next-generation sequencing technology. *Trends Gen.* **2014**, *30*, 418–426. [CrossRef] [PubMed]

153. Buermans, H.; den Dunnen, J. Next generation sequencing technology: Advances and applications. *Biochim. Biophys. Acta Mol. Basis Dis.* **2014**, *1842*, 1932–1941. [CrossRef] [PubMed]

154. Loman, N.J.; Watson, M. Successful test launch for nanopore sequencing. *Nat. Methods* **2015**, *12*, 303–304. [CrossRef] [PubMed]

155. Marzano, M.; Manzari, C.; Filannino, D.; Pizzi, R.; D'Erchia, A.M.; Lionetti, C.; Picardi, E.; Sgaramella, G.; Pesole, G.; Lanati, A. *Good Laboratory Practices and LIMS System: The Challenge for a Next Generation Sequencing and Bioinformatic Research Laboratory*; PeerJ Preprints: San Diego, CA, USA, 2017.

156. Barrett, T.; Wilhite, S.E.; Ledoux, P.; Evangelista, C.; Kim, I.F.; Tomashevsky, M.; Marshall, K.A.; Phillippy, K.H.; Sherman, P.M.; Holko, M. NCBI GEO: Archive for functional genomics data sets—Update. *Nucleic Acids Res.* **2013**, *41*, D991–D995. [CrossRef] [PubMed]

157. Choi, D.S.; Kim, D.K.; Kim, Y.K.; Gho, Y.S. Proteomics, transcriptomics and lipidomics of exosomes and ectosomes. *Proteomics* **2013**, *13*, 1554–1571. [CrossRef] [PubMed]

158. Scalbert, A.; Brennan, L.; Manach, C.; Andres-Lacueva, C.; Dragsted, L.O.; Draper, J.; Rappaport, S.M.; van der Hooft, J.J.; Wishart, D.S. The food metabolome: A window over dietary exposure. *Am. J. Clin. Nutr.* **2014**, *99*, 1286–1308. [CrossRef] [PubMed]

159. Chen, D.; Yu, J.; Tao, Y.; Pan, Y.; Xie, S.; Huang, L.; Peng, D.; Wang, X.; Wang, Y.; Liu, Z. Qualitative screening of veterinary anti-microbial agents in tissues, milk, and eggs of food-producing animals using liquid chromatography coupled with tandem mass spectrometry. *J. Chromatogr. B* **2016**, *1017*, 82–88. [CrossRef] [PubMed]

160. Santana, A.E.; Taborda, C.P.; Severo, J.S.; Rittner, G.M.G.; Muñoz, J.E.; Larsson, C.E., Jr.; Larsson, C.E. Development of enzyme immunoassays (ELISA and Western blot) for the serological diagnosis of dermatophytosis in symptomatic and asymptomatic cats. *Med. Mycol.* **2017**, *56*, 95–102. [CrossRef] [PubMed]

161. Huang, H.; Phipps-Todd, B.; McMahon, T.; Elmgren, C.L.; Lutze-Wallace, C.; Todd, Z.A.; Garcia, M.M. Development of a monoclonal antibody-based colony blot immunoassay for detection of thermotolerant Campylobacter species. *J. Microbiol. Methods* **2016**, *130*, 76–82. [CrossRef] [PubMed]

162. Araci, I.E.; Brisk, P. Recent developments in microfluidic large scale integration. *Curr. Opin. Biotechnol.* **2014**, *25*, 60–68. [CrossRef] [PubMed]

163. Safavieh, M.; Kaul, V.; Khetani, S.; Singh, A.; Dhingra, K.; Kanakasabapathy, M.K.; Draz, M.S.; Memic, A.; Kuritzkes, D.R.; Shafiee, H. Paper microchip with a graphene-modified silver nano-composite electrode for electrical sensing of microbial pathogens. *Nanoscale* **2017**, *9*, 1852–1861. [CrossRef] [PubMed]

164. Daraee, H.; Eatemadi, A.; Abbasi, E.; Fekri Aval, S.; Kouhi, M.; Akbarzadeh, A. Application of gold nanoparticles in biomedical and drug delivery. *Artif. Cells Nanomed. Biotechnol.* **2016**, *44*, 410–422. [CrossRef] [PubMed]

165. Kashid, S.B.; Tak, R.D.; Raut, R.W. Antibody tagged gold nanoparticles as scattering probes for the pico molar detection of the proteins in blood serum using nanoparticle tracking analyzer. *Colloids Surf. B Biointerfaces* **2015**, *133*, 208–213. [CrossRef] [PubMed]

166. Ren, W.; Cho, I.-H.; Zhou, Z.; Irudayaraj, J. Ultrasensitive detection of microbial cells using magnetic focus enhanced lateral flow sensors. *Chem. Commun.* **2016**, *52*, 4930–4933. [CrossRef] [PubMed]

167. Cao, Y.; Yang, X.; Wang, X. Lateral Flow Immunoassay Method of Simultaneously Detecting Hemoglobin s, Hemoglobin c, and Hemoglobin a in Newborns, Infants, Children, and Adults. U.S. Patent US20160116489A1, 29 May 2015.

168. Hegarty, J.W.; Guinane, C.M.; Ross, R.P.; Hill, C.; Cotter, P.D. Bacteriocin production: A relatively unharnessed probiotic trait? *F1000Research* **2016**, *5*, 2587. [CrossRef] [PubMed]

169. Nikkhoo, N.; Cumby, N.; Gulak, P.G.; Maxwell, K.L. Rapid Bacterial Detection via an All-Electronic CMOS Biosensor. *PLoS ONE* **2016**, *11*, e0162438. [CrossRef] [PubMed]

170. Budič, M.; Rijavec, M.; Petkovšek, Ž.; Žgur-Bertok, D. Escherichia coli bacteriocins: Antimicrobial efficacy and prevalence among isolates from patients with bacteraemia. *PLoS ONE* **2011**, *6*, e28769. [CrossRef] [PubMed]

171. Shelburne, C.E.; An, F.Y.; Dholpe, V.; Ramamoorthy, A.; Lopatin, D.E.; Lantz, M.S. The spectrum of antimicrobial activity of the bacteriocin subtilosin A. *J. Antimicrob. Chemother.* **2007**, *59*, 297–300. [CrossRef] [PubMed]

172. Guterman, S.K. Colicin B: Mode of action and inhibition by enterochelin. *J. Bacteriol.* **1973**, *114*, 1217–1224. [PubMed]

173. Davies, J.K.; Reeves, P. Genetics of resistance to colicins in Escherichia coli K-12: Cross-resistance among colicins of group A. *J. Bacteriol.* **1975**, *123*, 102–117. [PubMed]

174. Wayne, R.; Frick, K.; Neilands, J. Siderophore protection against colicins M, B, V, and Ia in *Escherichia coli*. *J. Bacteriol.* **1976**, *126*, 7–12. [PubMed]

5

Anti-RSV Peptide-Loaded Liposomes for the Inhibition of Respiratory Syncytial Virus [†]

Sameer Joshi [1], Atul A. Chaudhari [1], Vida Dennis [1], Daniel J. Kirby [2] [iD], Yvonne Perrie [3] and Shree Ram Singh [1,*] [iD]

[1] Center for NanoBiotechnology Research, Alabama State University, Montgomery, AL 36016, USA; sjoshi@alasu.edu (S.J.); atulvet@gmail.com (A.A.C.); vdennis@alasu.edu (V.D.)

[2] Aston Pharmacy School, Life and Health Sciences, Aston University, Birmingham B4 7ET, UK; d.j.kirby1@aston.ac.uk

[3] Strathclyde Institute of Pharmacy and Biomedical Sciences, University of Strathclyde, 161 Cathedral Street, Glasgow G4 0RE, UK; yvonne.perrie@strath.ac.uk

* Correspondence: ssingh@alasu.edu

[†] This paper is an extended version of our paper published in conference proceedings of TechConnect, 2017.

Abstract: Although respiratory syncytial virus (RSV) is one of the leading causes of acute respiratory tract infection in infants and adults, effective treatment options remain limited. To circumvent this issue, there is a novel approach, namely, the development of multifunctional liposomes for the delivery of anti RSV-peptides. While most of the peptides that are used for loading with the particulate delivery systems are the penetrating peptides, an alternative approach is the development of liposome-peptide systems, which are loaded with an RSV fusion peptide (RF-482), which has been designed to inhibit the RSV fusion and block infection. The results of this work have revealed that the liposomes themselves can serve as potential RSV inhibitors, whilst the anti-RSV-peptide with liposomes can significantly increase the RSV inhibition when compared with the anti-RSV peptide alone.

Keywords: liposomes; respiratory syncytial virus; peptide; hydrophilic

1. Introduction

Respiratory syncytial virus (RSV), as well as Rhinovirus (HRV), are the primary causes of acute lower respiratory tract (LRTI) infections [1]. An RSV infection is particularly noticeable during winter for populations, including the fetus, infants [2], young children, and adults [3–5]. A focused cause of bronchitis and pneumonia is RSV. It is widely recognized that there is a need for a vaccine against RSV, as the natural infection is not capable of providing life-long immunity and patients are prone to suffer from repeated RSV infection [6].

RSV, which belongs to the *Pneumoviridae* family, is a distinct serotype that has two major antigenic circulating subgroups (A & B), of which the A is dominant [5,7,8]. RSV has an RNA genome consisting of 15,191 base pairs. The virus can be identified with 11 proteins, including 2-non-structural proteins (NS-1 and NS-2), 3-surface proteins (glycoprotein-G, fusion protein-F, and small hydrophobic protein-HP), two overlapping frames of M2 mRNA producing 2 distinct transcription factors (M-1 and M-2), and four other structural proteins (matrix protein-M, nucleocapsid-N, phosphoprotein-P, and large protein-L) [5]. The viruses of the *Pneumoviridae* family fuse their membrane with the plasma membrane of the host, which results in a cell fusion if it is added to the cell in large quantity [9]. The entry of the RSV virus into the host cells occurs with the aid of the fusion protein-F, which has two hepated-repeated regions that form a hairpin-like structure, which facilitates the entrance of the virus into the cells [10].

The root cause analysis of any disease or disorder is at the foundation of the treatment design. An infection of RSV can start with a mild upper respiratory tract infection (URTI) and may lead to a potentially precarious lower respiratory tract infection [5]. RSV transmission occurs from person to person contact, directly or indirectly; an RSV infected person, upon sneezing or coughing, can leave viral droplets suspended in the air, which have the potential for transmission of the infection by entering the healthy individual through the mouth, nose, or eyes [11].

The first line treatment of RSV infection is the use of bronchodilators, such as α and/or β adrenergic agonist [5]. For pediatrics, since corticosteroids are not approved for treating RSV-infected individuals that are less than 1 year old, because of safety concerns [12], the use of vaporub and non-aspirin formulations, such as paracetamol, are the treatments of choice prior to clinical attention. Of the very few options available for the treatment of RSV, Ribavirin, a broad spectrum antiviral drug, is used. Although this too comes with limitations and drawbacks [13]. Despite several concept studies that have claimed the effectiveness of ribavirin in significantly reducing the RSV load and minimizing the disease severity, the disadvantages of the mutagenicity, teratogenicity, and carcinogenicity have subsequently resulted in FDA denial [14]. An active rather than passive prophylaxis would be a better choice but, unfortunately, there is no current vaccine that has been developed for the RSV infection. The formalin-inactivated vaccine was launched in the 1960s but was later withdrawn because of inadequate immunogenic responses, as well as an atypical T_H2-type response, which increases the chances of reinfection with similar or deadly infections [12].

The use of nanoparticulate systems, such as liposomes, can provide adjuvant action by an enhanced antigen delivery or by inducing an innate immune response [15]. Commercially, liposomes are the most successful carrier systems across the globe. Liposomes are vesicles with an aqueous core, which are surrounded by a phospholipid bilayer. From the numerous pre-clinical and clinical studies, it is clear that those liposomes are not only carriers of the chemical and biological materials, but they are also non-toxic and good at retaining efficacy. Indeed, the potential of the liposomes to act as a latent carrier of active pharmacological agents is well established [16–18], whilst liposomes have also shown their ability to carry antigens and to serve as immunomodulators [19]. Recently, a novel approach to inhibit the RSV has been the use of gold nanoparticles (GNPs) [20]. These GNPs can be functionalized with nucleic acid, antibodies, drugs, and peptides, and these functionalized GNPs then can be applied in the diagnosis or treatment of the diseases [21]. The anti-RSV fusion peptide RF-482 (sequence: VFPSDEFDASISQVNEKINQSLAFIRKSDLLHNVNAGKK) is used to functionalize the GNPs, and a significant inhibition of the RSV was observed after the loading of the GNPs [20]. The liposomes can be loaded with peptides, however insufficient research has been performed using liposomes as carrier systems for the inhibition of RSV [22–24]. Therefore, considering the global need and structural attributes of liposomes, this work describes an approach that employs the liposomes for the loading of RF-482, for the enhanced inhibition of RSV.

2. Materials and Methods

2.1. Materials

The lipid 1,2-disteroylphosphatidylcholine (DSPC) was purchased from Avanti Polar Lipids, Inc. (Alabaster, AL, USA). The cholesterol was obtained from Sigma-Aldrich Co. (St Louis, MO, USA). The minimal essential medium (MEM), Dulbecco's Modified Eagle's media (DMEM), Hank's balanced salt solution (HBSS), fetal bovine serum (FBS), 7-amino actinomycin D (7-AAD), L-glutamine (100 mM), antibiotics, TrypLE™, Lipofectamine 2000, real-time probe, primers, SuperScript® II Reverse Transcriptase, and TaqMan® Master Mix 2× were purchased from Life Technologies (Thermo Fisher Scientific, Waltham, MA, USA). Both the DNA and ribonucleic acid (RNA) isolation kits were purchased from Qiagen (Germantown, MD, USA). The human epithelial type 2 (HEp-2) cells were obtained from

American Type Culture Collection (ATCC®, Manassas, VA, USA). The cell toxicity was analyzed using the 3-(4,5-dimethylthiazol-2-yl)-2,5-diphenyltetrazolium bromide (MTT) dye-based cell proliferation assay kit from Promega Corp (Madison, WI, USA). The osmium tetroxide (4% solution) that was used for the fixative staining of the liposomes was purchased from Electron Microscopy Sciences (Hatfield, PA, USA). The ethanol was purchased from Fisher Scientific, Fair Lawn, NJ, USA. All of the reagents that were used for the experiments were of analytical grade.

2.2. Preparation of Small Unilamellar Liposomes

The liposomes were prepared using the thin-film hydration method [25]. Briefly, the lipid DSPC and cholesterol (5:2 w/w) were dissolved in an organic solvent mixture of chloroform and methanol, with the addition of the lipophilic drug, followed by a solvent evaporation to obtain a thin, dry film. The film was then hydrated with a phosphate buffered saline (PBS) (1 mM, pH 7.4, 25 °C), which resulted in the formation of the multi-lamellar vesicles (MLV). This MLV suspension was then sonicated using a probe sonication so as to obtain small unilamellar vesicles (SUV). The sonicated liposomal suspension was then centrifuged to remove titanium debris.

2.3. Loading of Peptide

The liposomal suspension was obtained after the removal of the titanium debris was used for the peptide loading. The RF-482 (500 µg) was added to the liposomal suspension (2.0 mL) and it was subjected to the mechanical shaking for 30 min. The non-associated peptide was then removed, using centrifugal filter units (Ultracel-50K, Millipore Ireland Ltd., Cork, UK). The RF-482 loaded liposomes were then used for further studies.

2.4. Particle Characteristics

Dynamic light scattering (DLS) (Zetasizer Nano-S, Malvern Instruments, Westborough, MA, USA) was used for the size determination of the empty and RF-482 loaded liposomes. The zeta potential was determined using the laser Doppler velocimetry (Zetasizer Nano-S, Malvern Instruments, Westborough, MA, USA). The samples were prepared using the PBS that was diluted 1–300 times (pH 7.4, 25 °C).

2.5. Transmission Electron Microscopy (TEM)

Osmium tetroxide has commonly been used for the fixative stain for cells in the TEM analysis. Since the liposome structure resembles cell structure, a similar fixative staining was performed in this study. Briefly, a drop of each liposome sample was placed on to the carbon film mesh copper grid, and the excess suspension was removed using a filter paper. The staining was performed using a 4% osmium tetroxide solution and the images were captured using a high-resolution TEM (EM10A/B, ZEISS, Oberkochen, Germany).

2.6. Quantification of Peptide Association

After the separation of the associated and non-associated peptide, the eluent and the liposome suspension were tested for peptide presence using the micro bicinchoninic acid (BCA) assay kit (Thermo Scientific, Rockford, IL, USA), according to the manufacturer's protocol. Briefly, the standard curves were prepared using the RF-482 peptide, for linearity and calibration. All of the samples were diluted using PBS, and since ethanol was used to separate the liposome associated protein, this was also further diluted with PBS. The standards and samples (150 µL each) were each added to wells of a 96-well plate, which was followed by the addition of 150 µL of working reagent and incubation for 2 h at 37 °C. The plate was then cooled to room temperature, and the absorbance measurements were taken at 562 nm, using an ELISA plate reader (TECAN™, Morrisville, NC, USA). Apart from the standards and samples, the ethanol, buffer, and empty liposomes were tested for BCA interference.

2.7. Cell Viability Assay

The CellTiter 96® Non-Radioactive cell proliferation assay kit (Promega) and the MTT (3-(4,5-dimethyl-thiazol-2-yl)-2,5-diphenyl-tetrazolium bromide) dye was used to assess the cell toxicity of the empty and peptide-loaded liposomes, as well as the peptide RF-482 to human HEp-2 cells. The human HEp-2 cells were propagated using MEM, which was supplemented with 10% FBS, 2 mM L-glutamine, 75 U/mL penicillin, 100 mg/mL kanamycin, and 75 mg/mL streptomycin (MEM-10). Each well of a 96-well plate was seeded with 25,000 cells and incubated overnight at 37 °C in a 5% CO_2 atmosphere for adherence. Two different concentrations of the peptide RF-482 (0.01 mg and 0.02 mg), empty liposomes and peptide-loaded liposomes were added to the wells, and the cell toxicity was assessed 72 h post incubation. The absorbance was measured at 570 nm, using an ELISA plate reader (TECAN™). The percent viability was then calculated from a comparison of the samples against the negative control.

2.8. Fluorescence Microscopy

The fluorescein isothiocyanate (FITC) labeled peptide RF-482 was used to demonstrate the association of the peptide with the liposomes. However, in order to study the RSV inhibition, 30,000 cells per well were seeded into an 8-chambered slide. The cells were incubated with peptide RF-482, empty liposomes, and peptide-loaded liposomes for 48 h, followed by fixing it in a paraformaldehyde-glutaraldehyde and buffer (PBS) wash. The nuclei were stained using DAPI and the cell membranes were stained using Cell Mask™ (Life Technologies). All of the chamber slides were imaged using the DAPI and FITC channels of the Nikon Ti Eclipse fluorescence microscope (Nikon Inc. Melville, NY, USA).

2.9. Plaque Assay

The plaque assay was one of the most common and reliable methods for the determination of the viral/antiviral activity in the cell cultures. The plaque assay was performed using the HEp-2 cells (1.5×10^5/well) that were cultured in MEM-10 for 48 h, in order to achieve a maximum confluency. Dilutions of a mixture of RSV and peptide, empty liposome, and peptide-loaded liposomes were prepared in DMEM before the infection. Post-infection, the cells were covered by immobilizing the overlaying medium (1.6% Methylcellulose) and were subsequently incubated for five days at 37 °C in 5%, CO_2 environment. On day 5, the overlaying medium was removed, and the monolayer was fixed with cold methanol at −20 °C, followed by being stained with a 0.1% crystal violet solution. The plaques were counted in order to determine the viral or antiviral activity.

2.10. TaqMan qPCR Analysis

The HEp-2 cells (1.5×10^5/well) were seeded in 12-well plates, followed by treatment with varied RSV dilutions (10^2 to 10^8), with and without the peptide, empty liposomes, or peptide-loaded liposomes. These treated cells were then incubated for 48 h at 37 °C in a 5% CO_2 environment, which was followed by being harvested for RNA extractions. A total of 1 µg RNA/sample was converted to cDNA, using reverse transcriptase according to the manufacturer's protocols (Applied Biosystems, Waltham, MA, USA). TaqMan qPCR was performed for the amplification of the RSV-F gene, using the RSV-F gene-specific primers and probe, and previously published procedures [26,27]. The RSV-F gene amplicon dilutions (10^2 to 10^8) were used as the standards. The qPCR for each sample was run in a duplicate on the Applied Biosystems® ViiA™ 7 real-time PCR (Life Technologies). The fold changes were calculated by comparing the untreated cells.

2.11. Statistical Analysis

Unless stated otherwise, the results were calculated as the mean ± standard deviation (SD). The data were analyzed by the student's *t*-test, alone, or by ANOVA, which was followed by Dunnett's post-hoc analysis for comparison. The significance was acknowledged for *p* values < 0.05.

3. Results

3.1. Confirmation of Peptide Loading

Liposomes are spherical vesicles made up of phospholipids [25]. Upon the hydration of a thin lipid film, MLVs were formed by vortexing and heating the suspension above the transition temperature (Tc) of the lipid. These MLVs could then be transformed into SUV by a variety of methods, including a probe sonication, which was used in this study. The liposomes were one of the most flexible structures that conjugate moieties, like lactose and peptide [22,28]. There was a slight change in the size of the liposome that was observed after the loading with RF-482 (Figure 1).

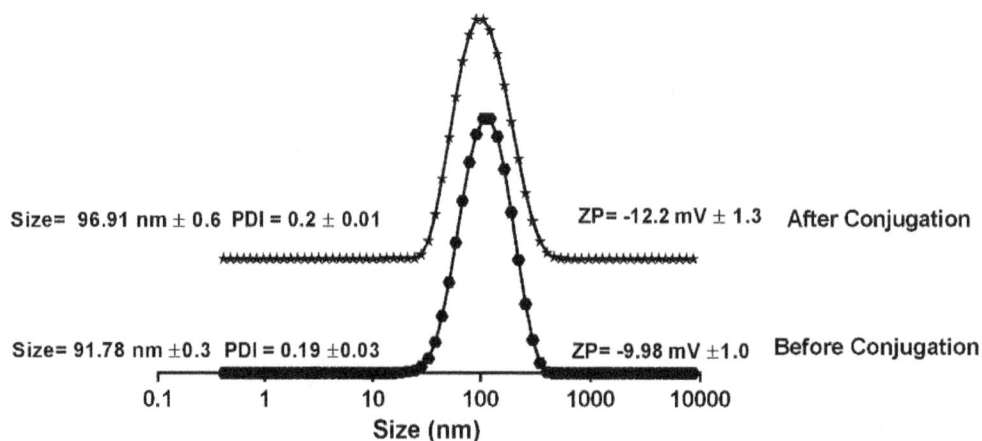

Size= 96.91 nm ± 0.6 PDI = 0.2 ± 0.01   ZP= -12.2 mV ± 1.3   After Conjugation

Size= 91.78 nm ±0.3 PDI = 0.19 ±0.03   ZP= -9.98 mV ±1.0   Before Conjugation

Figure 1. Dynamic light scattering (DLS) measurement of liposomes before and after RF-482 loading. (*n* = 3 ± SD).

The average hydrodynamic size, before and after the liposome loading, was 91.78 nm ± 0.3 (PDI 0.2 ± 0.01) and 96.91 nm ± 0.6 (PDI 0.19 ± 0.03), respectively. However, no change was observed on the zeta potential of the liposomes before and after the peptide loading. The first confirmation of the association of the peptide with the liposome was observed when the FITC-labeled peptide RF-482 was used for the loading. After the separation of the unloaded peptide, the liposomal suspension was dried, covered with phosphotungstic acid for better resolution, and observed under the fluorescence microscope (Figure 2a). Subsequently, the empty liposomes, as well as the peptide-loaded liposomes, were imaged under TEM. The presence of protein was confirmed by a cloudy environment that was observed around the peptide-loaded liposomes (Figure 2b). This was not observed in the case of the empty liposomes (Figure 2c).

It was possible that a certain amount of protein might not have crossed the membrane during the separation of the unloaded protein. This non-separated protein, therefore, might get counted as a loaded protein. To eradicate this doubt, three different concentrations of peptide RF-482 were run through the same column for the same amount of time, and the eluents were tested for recovery. It was observed that more than 97% of the protein passed through the column (results not included). This confirmed that the liposomal sample that was obtained after the separation of the unloaded protein would have had negligible amounts of the unloaded protein present along with the liposomes. Therefore, this could be the third confirmation of RF-482 loading with the liposomes.

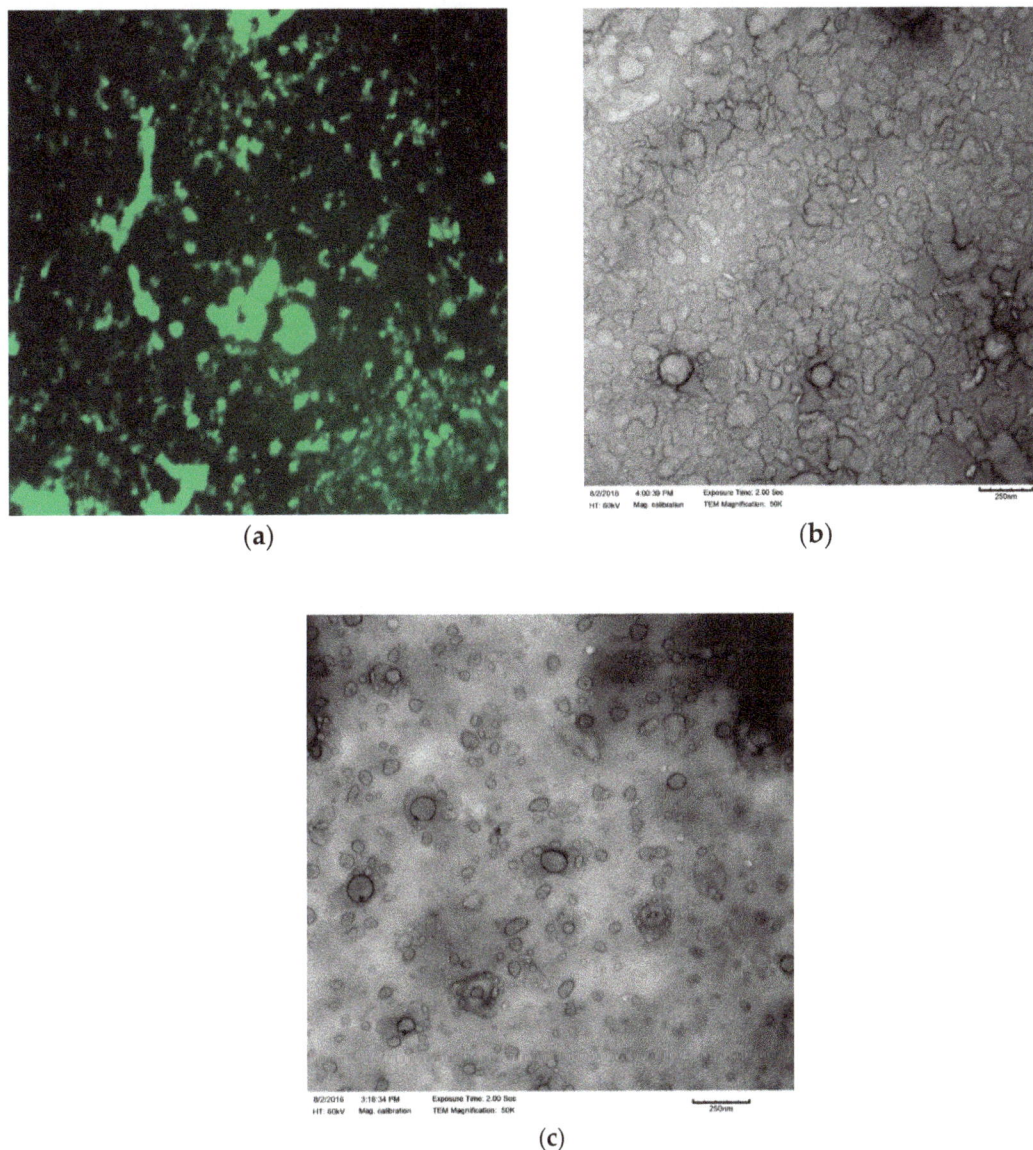

Figure 2. Fluorescence microscopy analysis of the presence of the fluorescein isothiocyanate (FITC)-labeled peptide RF-482 (green) confirms association with liposomes ($40\times$ magnification) (**a**). Transmission electron microscopy analysis. Comparison between the empty liposomes (**c**) and the RF-482 peptide-loaded liposomes (**b**).

3.2. Quantification of Loading

The exact mechanism of the protein–liposome association is still unknown. However, it was confirmed by the fluorescence microscopy as well as the TEM that the peptide RF-482 was associated with the liposomes. Since the association of the peptide–liposome was confirmed, it was necessary to quantify the amount of protein that was loaded with the liposomes. However, it was equally important to achieve the mass balance so as to study the actual amount of loading of the peptide. Hence, the BCA assay was performed for the analysis of both the eluent and liposome samples. It was confirmed that $81.7\% \pm 0.1\%$ ($n = 3$) was not loaded and that $19.1\% \pm 0.4\%$ ($n = 3$) was loaded (Figure 3). This also confirmed the 100% recovery of the initial amount of the peptide.

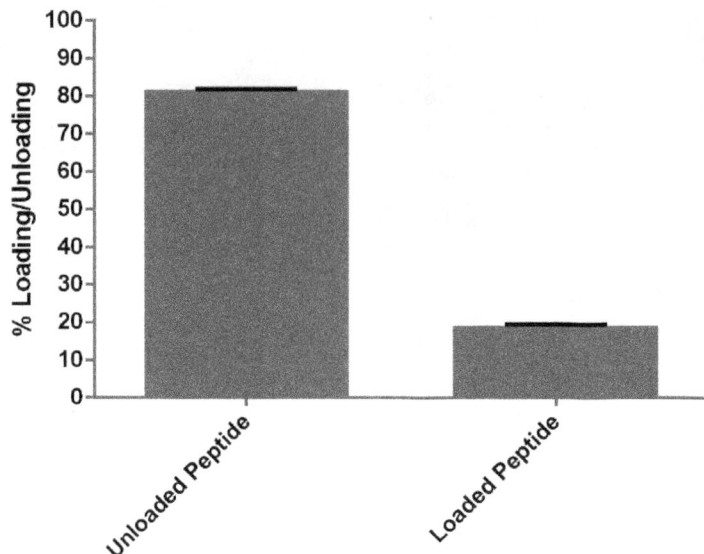

Figure 3. Liposomal loading with peptide RF-482 determined by the bicinchoninic acid (BCA) assay. Results presented as peptide loading determined after separation of the unloaded peptide ($n = 3 \pm$ standard deviation (SD)).

3.3. Cell Toxicity Analysis

The peptide, liposomes, and peptide-loaded liposomes, all at two different concentrations, were tested for their cell toxicity. There was no toxicity observed with both of the concentrations of all of the samples. After 72 h of incubation, the observed cell viability was more than 80% ($n = 3 \pm$ SD) for the chosen concentrations of the peptide RF-482. Whereas, in the case of the liposome alone, both of the concentrations of lipids were found to be completely non-toxic. However, when the peptides were loaded on to the liposomes, the toxicity was reduced slightly, as the cell viability for the chosen concentrations were observed to be ~90% ($n = 3 \pm$ SD). (Figure 4).

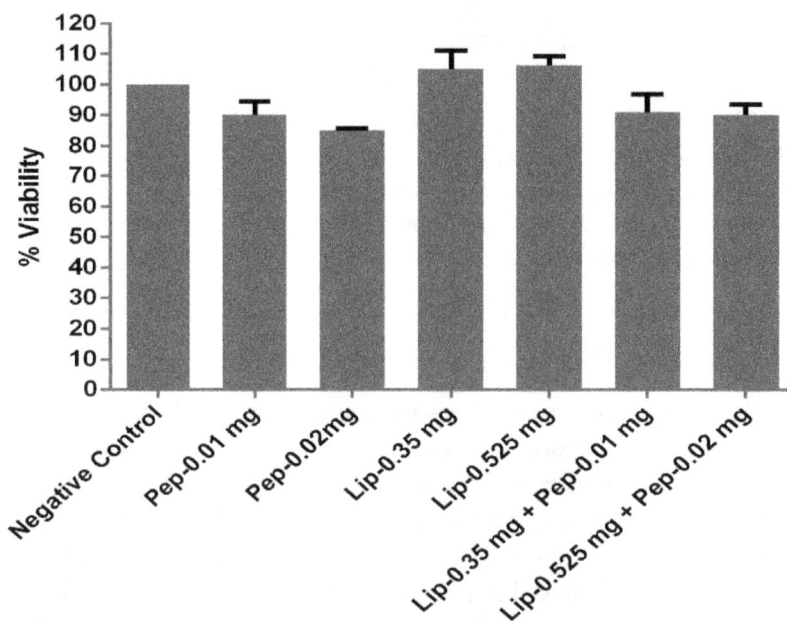

Figure 4. Toxicity profiling of peptide RF-482, liposomes, and RF-482-loaded liposomes presented through cell viability count, performed using the 3-(4,5-dimethylthiazol-2-yl)-2,5-diphenyltetrazolium bromide (MTT) assay (72 h, $n = 3 \pm$ SD).

3.4. Evaluation of Viral Inhibition

The plaque reduction assay has been known as the optimum standard for the antiviral activity analysis [29]. During the plaque assay, the monolayer of the HEp-2 cells was infected with the lytic RSV. The infected cells experienced lytic cycles and eventually appeared as plaques or, in other words, zones of cell death [30]. It was reported recently that the surfactant phospholipids that were bound to the RSV had markedly suppressed the infection though fusion inhibition [31]. We tested the anti-RSV peptide, liposomes, and liposome loaded peptide against RSV infection to HEp-2 cells. When these plaques were counted, it was observed that, although the peptide RF-482 and liposomes alone were capable of inhibiting the RSV, when they were combined, the peptide-loaded liposomes had a significant increase in RSV inhibition ($p < 0.05$, ANOVA, post hoc Dunnett's multiple comparison tests) (Figure 5). Also, it was an interesting finding that the liposomes alone could also inhibit the RSV, and significantly more so ($p < 0.05$, ANOVA, post hoc Dunnett's multiple comparison tests) than the peptide RF-482 alone.

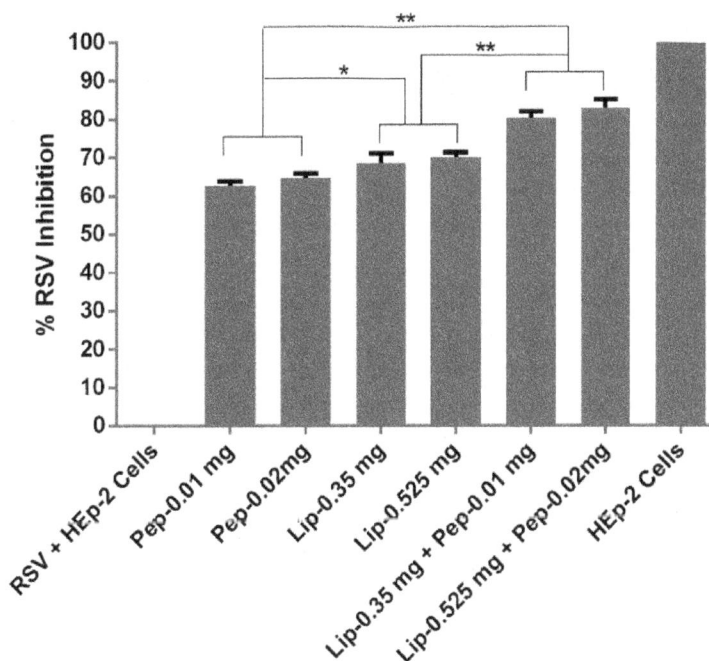

Figure 5. Screening of peptide, liposomes, and peptide encapsulated against respiratory syncytial virus (RSV). Plaques were counted, and the mean plaque was a count of each sample, which was expressed as a percentage of the mean count of the control. ($n = 3 \pm$ SD). Significant differences between the samples represented as * $p < 0.05$ and ** $p < 0.01$.

To validate the plaque assay results, qualitative and quantitative testing was performed. Immunofluorescence imaging was used as a qualitative tool, whilst PCR was employed as a quantitative tool. The cells were incubated for 48 h with the peptide RF-482, empty liposomes, and peptide-loaded liposomes, followed by being fixed in a paraformaldehyde-glutaraldehyde and being washed with a buffer (PBS). An appropriate chamber from the eight-chambered slide was observed under the fluorescence microscope for the RSV activity. The observation that was made from this analysis (Figure 6) was concurrent with the findings that were obtained from the plaque assay (Figure 5) and confirmed that the peptide RF-482, liposomes, and peptide-loaded liposomes, were capable of inhibiting the RSV.

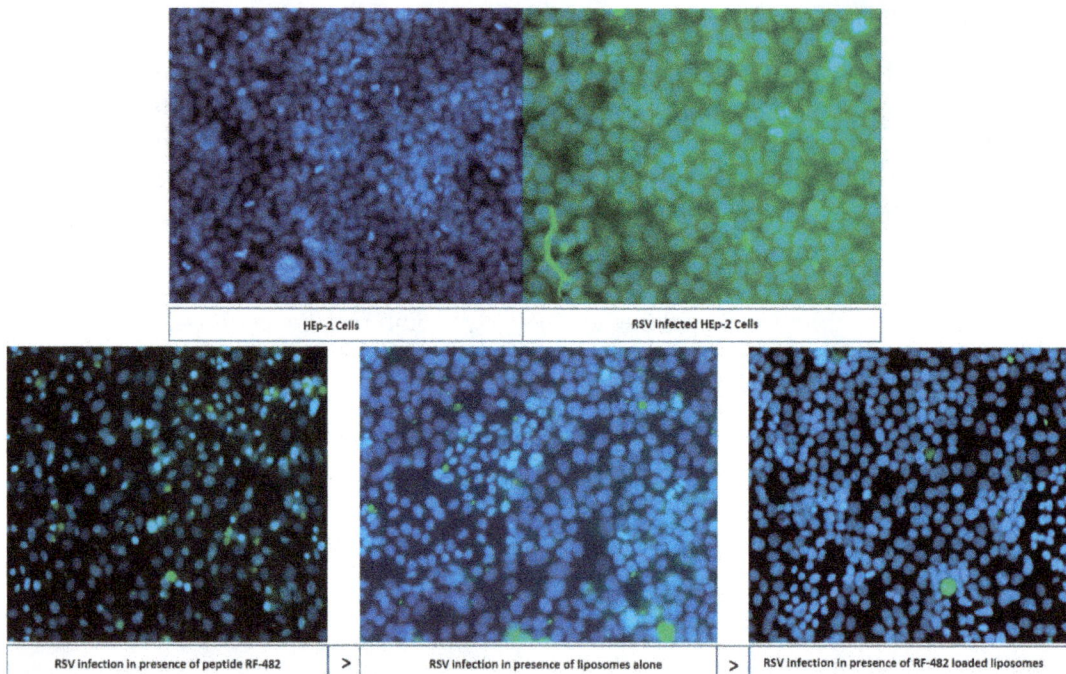

Figure 6. Fluorescence microscopy analysis. FITC (green)—RSV; and DAPI (blue)—HEP-2 cell nucleus. In liposomes and RF-482 liposomes, the blue color represents the survived cells, and the green color represents the presence of RSV.

Having the confirmation from the microscopic observation of the viral inhibition, it was necessary to quantitatively validate the plaque assay results. A significant difference (p <0.05, ANOVA, post hoc Dunnett's multiple comparison tests) was observed between a number of gene copies of the virus sample and samples having the peptide, liposomes, or peptide-loaded liposomes (Figure 7). This confirmed that the peptide RF 482, liposomes, and peptide-loaded liposomes were equally capable of inhibiting the RSV. However, although there was no significant difference (p >0.05, t-test) observed between the peptide, liposome alone, and peptide-loaded liposomes, the loaded liposome samples displayed a trend for slightly lower gene copies, which indicated slightly more inhibition than the individual components when used solely (Figure 7).

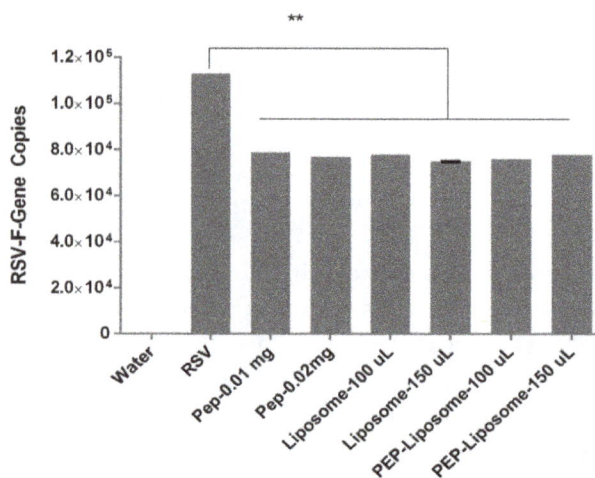

Figure 7. Screening of the RSV-F gene amplicon dilution with water as negative control. Comparison of viral gene amplicon and peptide, liposomes, and peptide-loaded liposomes. (n = 3 ± SD). Significant difference between the samples represented as ** p < 0.01.

4. Discussion

To date, the F, G, M, and small hydrophobic proteins have been considered as targets so as to avoid the RSV infection [23,32–34]. Nearly a dozen peptide-based formulations have been under clinical trials, and most of these are targeting the F-protein [35–37]. However, the role of the F-protein is vital in the spread of the virus, because targeting the G-protein could neutralize the virus, however the actual spread of the virus is only possible after inhibiting the F-protein [38–40]. Over a decade ago, studies were suggested that all of the three F-, G-, and RSV-SH protein inhibitors could be used for the complete RSV inhibition [41]. However, within the last decade, several reports indicated that only the F-protein was capable of inhibiting the RSV infection [42,43]. Therefore, some recent studies have specifically targeted the F-protein [20,44,45]. For example, Singh et al. described that the anti-RSV peptide RF-482 was an F-protein inhibitor and was used to functionalize the GNPs, and a significant inhibition was reported after the loading of the GNPs [20].

With the quest of finding an alternative carrier for delivering the anti-RSV peptide RF-482, the liposome formulations were prepared here and were tested for the inhibition of the RSV infection. RF-482 was a small fusion peptide with 39 amino acids, with a total of 611 atoms [20]. Since it was a fusion of the peptide and hydrophilic, it was expected that the RF-482 might have gotten entrapped in the hydrophilic core and that it might have been adsorbed on the surface of the liposomes. Although the exact mechanism of the peptide association with the liposomes was unrevealed, the functionalization of the liposomes was confirmed through the dynamic light scattering (Figure 1), change in the surface charge (Figure 1), fluorescence imaging (Figure 2c), and transmission electron microscopy (Figure 2a,b).

Knowing that the toxicity was an emerging problem in the RSV treatment [7], it was one of the primary objectives when designing the treatment for the RSV infection. The liposomal research to date described them as a system that could be used, not only as a delivery system, but also as adjuvants [15,46,47]. Designing the protein-based liposomal adjuvant vaccine could be an approach to attaining a maximum efficacy and low toxicity [48]. Similarly, in this scenario, the liposomes had shown their non-toxic nature for the chosen HEp-2 cells for 72 h, which reflected their potential application in designing the RSV treatment. The cytotoxicity of the various concentrations of these liposomes was tested by the MTT assay, and it was found that the liposome formulations of two different concentrations of lipids, with and without protein loading, did not render any cytotoxic effect or molecular effect on the host HEp-2 cells (Figure 4).

Amongst the several ways of viral inhibition analysis, the plaque assay, immunofluorescence microscopy, and qPCR were considered as standard tools [13,21,49]. The anti-RSV effect was therefore confirmed in the plaque assay (Figure 5), immunofluorescence imaging (Figure 6), and qRT-PCR. The anti-RSV activity of the peptide RF-282 had already been reported, but an interesting finding confirmed the inhibition of the RSV in the presence of the DSPC cholesterol liposomes, which resembled the results from Hendricks et al., where the decoy liposomes were found to be capable of inhibiting RSV from cellular binding [23]. The functionalization of the gold nanoparticles using the anti-RSV peptide was shown to have a significant effect on the RSV inhibition, compared with the peptide alone [20,21]. Similarly, our results also confirmed that the RSV percent inhibition was significantly increased ($p < 0.05$) for the RF-482 loaded liposomes compared with the RF-482 and liposomes alone (Figures 5 and 7).

Overall, our research demonstrated that the RF-482 itself was capable of inhibiting RSV, but the inhibition of RSV was significantly increased with the loading on the liposome. Moreover, recent reports showed that phosphatidylinositol (PI) inhibited the respiratory syncytial virus (RSV), as the PI bound the RSV with a high affinity, which inhibited its fusion to the epithelial cells [31,50]. There were five derivatives of the phospholipids, including the PI and the phosphocholine (PC). Although the exact mechanism of RSV inhibition by PC was not confirmed, it was possible that, similarly to the PI, the PC could have an affinity towards the RSV, inhibiting its fusion to the epithelial cells. However, the liposomal loading of the anti-RSV agents, like RF-482, were shown to have decreased the viral activity of RSV. Liposomes are multifaceted delivery systems and are capable of co-encapsulating compounds depending on their characteristics [51]. This structural attribute of the liposomes could become a carrier of multiple proteins and/or other anti-RSV compounds. Liposomes could be designed to look like a virus, by attaching multiple proteins to it [52]. Therefore, liposomes are multifaceted systems and hold the potential of entering the mainstream for designing the prophylaxis against RSV infection.

5. Conclusions

RF-482 has been reported as an inhibitor of RSV fusion and, to date, gold nanoparticles are the only reported carrier of RF-482 [21]. However, liposomes have been considered as an alternative carrier for delivering the anti-RSV peptide, RF-482. Moreover, the liposomes, as a new candidate for RSV inhibition, have been tested and the inhibitory effect of the liposomes has been shown to be better compared with the peptide alone, while the peptide-loaded liposomes have proved to be a better candidate for the RSV inhibition compared with both the peptide and liposomes alone. However, this has generated another quest to unveil the exact mechanism of the RSV infection inhibition, using the liposomes or peptide-loaded liposomes, which could lead to the commercialization of the formulation.

Author Contributions: S.J. conceived, designed, and performed the experiments. A.A.C. designed and performed the TEM analysis. S.J. wrote the paper. S.R.S., D.J.K., Y.P., and V.D. contributed to the reagents, materials, and analysis tools, and supervised the investigation, analysis, and interpretation of the data and participated in the manuscript preparation.

Funding: The research was funded by the Alabama State University (Montgomery, AL, USA) using grants from NSF-CREST (HRD-1241701) and in collaboration with the Aston University (Birmingham, UK).

Acknowledgments: We are thankful to Michael Miller (Auburn University, Auburn, AL, USA) for the kind assistance in the microscopic analysis.

Conflicts of Interest: The authors declare no conflict of interest.

References

1. Luchsinger, V.; Ampuero, S.; Palomino, M.A.; Chnaiderman, J.; Levican, J.; Gaggero, A.; Larrañaga, C.E. Comparison of virological profiles of respiratory syncytial virus and rhinovirus in acute lower tract respiratory infections in very young Chilean infants, according to their clinical outcome. *J. Clin. Virol.* **2014**, *61*, 138–144. [CrossRef] [PubMed]

2. Tregoning, J.S.; Schwarze, J. Respiratory viral infections in infants: Causes, clinical symptoms, virology, and immunology. *Clin. Microbiol. Rev.* **2010**, *23*, 74–98. [CrossRef] [PubMed]

3. Falsey, A.R.; Hennessey, P.A.; Formica, M.A.; Cox, C.; Walsh, E.E. Respiratory syncytial virus infection in elderly and high-risk adults. *N. Engl. J. Med.* **2005**, *352*, 1749–1759. [CrossRef] [PubMed]

4. Rappuoli, R.; Mandl, C.W.; Black, S.; De Gregorio, E. Vaccines for the twenty-first century society. *Nat. Rev. Immunol.* **2011**, *11*, 865–872. [CrossRef] [PubMed]

5. Borchers, A.T.; Chang, C.; Gershwin, M.E.; Gershwin, L.J. Respiratory syncytial virus—A comprehensive review. *Clin. Rev. Allergy Immunol.* **2013**, *45*, 331–379. [CrossRef] [PubMed]

6. Kamphuis, T.; Meijerhof, T.; Stegmann, T.; Lederhofer, J.; Wilschut, J.; de Haan, A. Immunogenicity and Protective Capacity of a Virosomal Respiratory Syncytial Virus Vaccine Adjuvanted with Monophosphoryl Lipid A in Mice. *PLoS ONE* **2012**, *7*, e36812. [CrossRef] [PubMed]

7. Wyde, P.R. Respiratory syncytial virus (RSV) disease and prospects for its control. *Antivir. Res.* **1998**, *39*, 63–79. [CrossRef]

8. Mufson, M.A.; Belshe, R.B.; Örvell, C.; Norrby, E. Respiratory syncytial virus epidemics: Variable dominance of subgroups A and B strains among children, 1981–1986. *J. Infect. Dis.* **1988**, *157*, 143–148. [CrossRef] [PubMed]

9. Haywood, A.M. Interaction of liposomes with viruses. *Ann. N. Y. Acad. Sci.* **1978**, *308*, 275–280. [CrossRef] [PubMed]

10. Zhao, X.; Singh, M.; Malashkevich, V.N.; Kim, P.S. Structural characterization of the human respiratory syncytial virus fusion protein core. *Proc. Natl. Acad. Sci. USA* **2000**, *97*, 14172–14177. [CrossRef] [PubMed]

11. CDC. Respiratory Syncytial Virus Infection (RSV). 2014. https://www.cdc.gov/rsv/index.html (accessed on 27 April 2016).

12. Piedimonte, G.; Perez, M.K. Respiratory Syncytial Virus Infection and Bronchiolitis. *Pediatr. Rev.* **2014**, *35*, 519–530. [CrossRef] [PubMed]

13. Bawage, S.S.; Tiwari, P.M.; Pillai, S.; Dennis, V.; Singh, S.R. Recent advances in diagnosis, prevention, and treatment of human respiratory syncytial virus. *Adv. Virol.* **2013**, *2013*, 595768. [CrossRef] [PubMed]

14. Simões, E.A.; DeVincenzo, J.P.; Boeckh, M.; Bont, L.; Crowe, J.E.; Griffiths, P.; Hayden, F.G.; Hodinka, R.L.; Smyth, R.L.; Spencer, K. Challenges and opportunities in developing respiratory syncytial virus therapeutics. *J. Infect. Dis.* **2015**, *211*, S1–S20. [CrossRef] [PubMed]

15. Schwendener, R.A. Liposomes as vaccine delivery systems: A review of the recent advances. *Ther. Adv. Vaccines* **2014**, *2*, 159–182. [CrossRef] [PubMed]

16. Gregoriadis, G.; Ryman, B.E. Lysosomal localization of β-fructofuranosidase-containing liposomes injected into rats. Some implications in the treatment of genetic disorders. *Biochem. J.* **1972**, *129*, 123–133. [CrossRef] [PubMed]

17. Gregoriadis, G.; Ryman, B.E. Fate of Protein-Containing Liposomes Injected into Rats. *Eur. J. Biochem.* **1972**, *24*, 485–491. [CrossRef] [PubMed]

18. Gregoriadis, G.; Leathwood, P.; Ryman, B.E. Enzyme entrapment in liposomes. *FEBS Lett.* **1971**, *14*, 95–99. [CrossRef]

19. Perrie, Y.; Kastner, E.; Kaur, R.; Wilkinson, A.; Ingham, A.J. A case-study investigating the physicochemical characteristics that dictate the function of a liposomal adjuvant. *Hum. Vaccines Immunother.* **2013**, *9*, 1374–1381. [CrossRef] [PubMed]

20. Singh, S.R.; Tiwari, P.M.; Dennis, V.A. Anti Respiratory Syncytial Virus Peptide Functionalized Gold Nanoparticles. U.S. Patent 8,815,295, 26 August 2014.

21. Tiwari, P.M.; Eroglu, E.; Bawage, S.S.; Vig, K.; Miller, M.E.; Pillai, S.; Dennis, V.A.; Singh, S.R. Enhanced intracellular translocation and biodistribution of gold nanoparticles functionalized with a cell-penetrating peptide (VG-21) from vesicular stomatitis virus. *Biomaterials* **2014**, *35*, 9484–9494. [CrossRef] [PubMed]

22. Vabbilisetty, P.; Sun, X.-L. Liposome surface functionalization based on different anchoring lipids via Staudinger ligation. *Org. Biomol. Chem.* **2014**, *12*, 1237–1244. [CrossRef] [PubMed]

23. Hendricks, G.L.; Velazquez, L.; Pham, S.; Qaisar, N.; Delaney, J.C.; Viswanathan, K.; Albers, L.; Comolli, J.C.; Shriver, Z.; Knipe, D.M. Heparin octasaccharide decoy liposomes inhibit replication of multiple viruses. *Antivir. Res.* **2015**, *116*, 34–44. [CrossRef] [PubMed]

24. Joshi, S.; Kirby, D.; Perrie, Y.; Singh, S.R. Novel nano-biomaterials for inhibition of respiratory syncytial virus. Proceedings of TechConnect, Washington, DC, USA, 14 May 2017; pp. 75–78.

25. Bangham, A.D.; Standish, M.M.; Watkins, J.C. Diffusion of univalent ions across the lamellae of swollen phospholipids. *J. Mol. Biol.* **1965**, *13*, 238–252. [CrossRef]

26. Mentel, R.; Wegner, U.; Bruns, R.; Gürtler, L. Real-time PCR to improve the diagnosis of respiratory syncytial virus infection. *J. Med. Microbiol.* **2003**, *52*, 893–896. [CrossRef] [PubMed]

27. Eroglu, E.; Tiwari, P.M.; Waffo, A.B.; Miller, M.E.; Vig, K.; Dennis, V.A.; Singh, S.R. A nonviral pHEMA+ chitosan nanosphere-mediated high-efficiency gene delivery system. *Int. J. Nanomed.* **2013**, *8*, 1403–1415.

28. Nahar, K.; Absar, S.; Gupta, N.; Kotamraju, V.R.; McMurtry, I.F.; Oka, M.; Komatsu, M.; Nozik-Grayck, E.; Ahsan, F. Peptide-coated liposomal fasudil enhances site specific vasodilation in pulmonary arterial hypertension. *Mol. Pharm.* **2014**, *11*, 4374–4384. [CrossRef] [PubMed]

29. Landry, M.L.; Stanat, S.; Biron, K.; Brambilla, D.; Britt, W.; Jokela, J.; Chou, S.; Drew, W.L.; Erice, A.; Gilliam, B. A standardized plaque reduction assay for determination of drug susceptibilities of cytomegalovirus clinical isolates. *Antimicrob. Agents Chemother.* **2000**, *44*, 688–692. [CrossRef] [PubMed]

30. Baer, A.; Kehn-Hall, K. Viral concentration determination through plaque assays: Using traditional and novel overlay systems. *J. Vis. Exp.* **2014**, *4*, e52065. [CrossRef] [PubMed]

31. Numata, M.; Chu, H.W.; Dakhama, A.; Voelker, D.R. Pulmonary surfactant phosphatidylglycerol inhibits respiratory syncytial virus–induced inflammation and infection. *Proc. Natl. Acad. Sci. USA* **2010**, *107*, 320–325. [CrossRef] [PubMed]

32. Connors, M.; Kulkarni, A.; Collins, P.; Firestone, C.; Holmes, K.; Morse, H.D.; Murphy, B. Resistance to respiratory syncytial virus (RSV) challenge induced by infection with a vaccinia virus recombinant expressing the RSV M2 protein (Vac-M2) is mediated by CD8+ T cells, while that induced by Vac-F or Vac-G recombinants is mediated by antibodies. *J. Virol.* **1992**, *66*, 1277–1281. [PubMed]

33. Mader, D.; Huang, Y.; Wang, C.; Fraser, R.; Issekutz, A.C.; Stadnyk, A.W.; Anderson, R. Liposome encapsulation of a soluble recombinant fragment of the respiratory syncytial virus (RSV) G protein enhances immune protection and reduces lung eosinophilia associated with virus challenge. *Vaccine* **2000**, *18*, 1110–1117. [CrossRef]

34. Li, Y.; To, J.; Verdià-Baguena, C.; Dossena, S.; Surya, W.; Huang, M.; Paulmichl, M.; Liu, D.X.; Aguilella, V.M.; Torres, J. Inhibition of the human respiratory syncytial virus small hydrophobic protein and structural variations in a bicelle environment. *J. Virol.* **2014**, *88*, 11899–11914. [CrossRef] [PubMed]

35. Drugbank. BTA9881. https://www.drugbank.ca/drugs/DB05226 (accessed on 22 September 2017).

36. ClinicalTrials.gov. Safety Study of Oral BTA9881 to Treat RSV Infection. https://clinicaltrials.gov/ct2/show/NCT00504907 (accessed on 22 September 2017).

37. Costello, H.M. *The N500 Glycan of the Respiratory Syncytial Virus F Protein is Required for Fusion, but Not for Stabilization or Triggering of the Protein*; The Ohio State University: Columbus, OH, USA, 2013.

38. Hancock, G.E.; Speelman, D.J.; Frenchick, P.J. Adjuvants for Vaccines against Respiratory Syncytial Virus. U.S. Patents 5723130A, 3 March 1998.

39. McLellan, J.S.; Ray, W.C.; Peeples, M.E. Structure and function of respiratory syncytial virus surface glycoproteins. In *Challenges and Opportunities for Respiratory Syncytial Virus Vaccines*; Springer: Berlin/Heidelberg, Germany, 2013; pp. 83–104.

40. Bukreyev, A.; Yang, L.; Collins, P.L. The secreted G protein of human respiratory syncytial virus antagonizes antibody-mediated restriction of replication involving macrophages and complement. *J. Virol.* **2012**, *86*, 10880–10884. [CrossRef] [PubMed]

41. Heminway, B.; Yu, Y.; Tanaka, Y.; Perrine, K.; Gustafson, E.; Bernstein, J.; Galinski, M. Analysis of respiratory syncytial virus F, G, and SH proteins in cell fusion. *Virology* **1994**, *200*, 801–805. [CrossRef] [PubMed]

42. Techaarpornkul, S.; Barretto, N.; Peeples, M.E. Functional analysis of recombinant respiratory syncytial virus deletion mutants lacking the small hydrophobic and/or attachment glycoprotein gene. *J. Virol.* **2001**, *75*, 6825–6834. [CrossRef] [PubMed]

43. Battles, M.B.; Langedijk, J.P.; Furmanova-Hollenstein, P.; Chaiwatpongsakorn, S.; Costello, H.M.; Kwanten, L.; Vranckx, L.; Vink, P.; Jaensch, S.; Jonckers, T.H. Molecular mechanism of respiratory syncytial virus fusion inhibitors. *Nat. Chem. Biol.* **2016**, *12*, 87–93. [CrossRef] [PubMed]

44. Perron, M.; Stray, K.; Kinkade, A.; Theodore, D.; Lee, G.; Eisenberg, E.; Sangi, M.; Gilbert, B.E.; Jordan, R.; Piedra, P.A. GS-5806 inhibits a broad range of respiratory syncytial virus clinical isolates by blocking the virus-cell fusion process. *Antimicrob. Agents Chemother.* **2016**, *60*, 1264–1273. [CrossRef] [PubMed]

45. Samuel, D.; Xing, W.; Niedziela-Majka, A.; Wong, J.S.; Hung, M.; Brendza, K.M.; Perron, M.; Jordan, R.; Sperandio, D.; Liu, X. GS-5806 inhibits pre-to postfusion conformational changes of the respiratory syncytial virus fusion protein. *Antimicrob. Agents Chemother.* **2015**, *59*, 7109–7112. [CrossRef] [PubMed]

46. Perrie, Y.; Mohammed, A.R.; Kirby, D.J.; McNeil, S.E.; Bramwell, V.W. Vaccine adjuvant systems: Enhancing the efficacy of sub-unit protein antigens. *Int. J. Pharm.* **2008**, *364*, 272–280. [CrossRef] [PubMed]

47. Allison, A.; Gregoriadis, G. Liposomes as immunological adjuvants. *Nature* **1974**, *252*, 252. [CrossRef] [PubMed]

48. Perrie, Y.; Frederik, P.M.; Gregoriadis, G. Liposome-mediated DNA vaccination: The effect of vesicle composition. *Vaccine* **2001**, *19*, 3301–3310. [CrossRef]

49. Kim, K.S.; Kim, A.-R.; Piao, Y.; Lee, J.-H.; Quan, F.-S. A rapid, simple, and accurate plaque assay for human respiratory syncytial virus (HRSV). *J. Immunol. Methods* **2017**, *446*, 15–20. [CrossRef] [PubMed]

50. Numata, M.; Kandasamy, P.; Nagashima, Y.; Fickes, R.; Murphy, R.C.; Voelker, D.R. Phosphatidylinositol inhibits respiratory syncytial virus infection. *J. Lipid Res.* **2015**, *56*, 578–587. [CrossRef] [PubMed]

51. Joshi, S.; Hussain, M.T.; Roces, C.B.; Anderluzzi, G.; Kastner, E.; Salmaso, S.; Kirby, D.J.; Perrie, Y. Microfluidics based manufacture of liposomes simultaneously entrapping hydrophilic and lipophilic drugs. *Int. J. Pharm.* **2016**, *514*, 160–168. [CrossRef] [PubMed]

52. Zurbriggen, R.; Amacker, M.; Krammer, A.R. Immunopotentiating reconstituted influenza virosomes. *Liposome Technol.* **2006**, *1*, 85–96.

6

Theoretical Insight into the Biodegradation of Solitary Oil Microdroplets Moving through a Water Column

George E. Kapellos [1,2,*] ⓘ**, Christakis A. Paraskeva** [3]**, Nicolas Kalogerakis** [2] **and Patrick S. Doyle** [1]

[1]   Department of Chemical Engineering, Massachusetts Institute of Technology, Cambridge, MA 02139, USA; pdoyle@mit.edu

[2]   School of Environmental Engineering, Technical University of Crete, 73100 Chania, Greece; nicolas.kalogerakis@enveng.tuc.gr

[3]   Department of Chemical Engineering, University of Patras, 26504 Rion Achaia, Greece; takisp@chemeng.upatras.gr

*   Correspondence: george.kapellos@gmail.com or kapellos@mit.edu

Abstract: In the aftermath of oil spills in the sea, clouds of droplets drift into the seawater column and are carried away by sea currents. The fate of the drifting droplets is determined by natural attenuation processes, mainly dissolution into the seawater and biodegradation by oil-degrading microbial communities. Specifically, microbes have developed three fundamental strategies for accessing and assimilating oily substrates. Depending on their affinity for the oily phase and ability to proliferate in multicellular structures, microbes might either attach to the oil surface and directly uptake compounds from the oily phase, or grow suspended in the aqueous phase consuming solubilized oil, or form three-dimensional biofilms over the oil–water interface. In this work, a compound particle model that accounts for all three microbial strategies is developed for the biodegradation of solitary oil microdroplets moving through a water column. Under a set of educated hypotheses, the hydrodynamics and solute transport problems are amenable to analytical solutions and a closed-form correlation is established for the overall dissolution rate as a function of the Thiele modulus, the Biot number and other key parameters. Moreover, two coupled ordinary differential equations are formulated for the evolution of the particle size and used to investigate the impact of the dissolution and biodegradation processes on the droplet shrinking rate.

Keywords: biofilm; crude oil; modeling; oil spill; droplet cloud; droplet dissolution; droplet biodegradation; Sherwood number; mass transfer; compound droplet model

1. Introduction

After a natural or accidental release of crude oil in the sea, part of the oil ends up in the form of droplets moving through the seawater column. The droplets may be created either at the sea surface during the breakup of an oil slick (i.e., floating oil layer) by sea waves [1,2], or at the seafloor during the atomization of live crude oil (i.e., gas/oil mixture) extruding at sufficiently high speed from a natural crack or a broken wellhead [3–5]. The latter case occurred, for example, after the blowout of the Deepwater Horizon rig in the Gulf of Mexico where the addition of the chemical dispersant Corexit in the leaking crude oil resulted in clouds of droplets travelling underwater along with sea currents [6,7]. At present, there are no practical means for the collection or in situ treatment of oil droplets in vast bodies of marine waters and, inevitably, their removal relies solely on natural attenuation processes, notably on dissolution and biodegradation. Specifically, it is anticipated that in the long run, most of the released oil in the sea is consumed by autochthonous oil-degrading microorganisms (bacteria, fungi, yeasts) that have developed appropriate machinery for accessing and assimilating oily substrates [8–10]. In this way, crude oil enters as a nutrient into the marine food chain. In spite of this long-term bright side, large amounts of dispersed oil droplets in the seawater column disturb the established ecosystem dynamics and pose an imminent risk of toxic effects from various crude oil components to many marine species (invertebrates, fishes, mammals, etc.) [11–14]. In particular, small oil droplets might be more toxic than crude oil itself, if consumed by fish and marine mammals [14]. It is therefore imperative to understand and quantify the physical and biological mechanisms that rule the fate of dispersed oil droplets in marine waters and, upon that knowledge, build technologies that will enable the mitigation of pertinent adverse effects.

Once the droplets entrain to the seawater column, the most critical quantity to assess is the droplet *retention time* in the underwater body until complete dissolution, degradation or relocation to the sea surface or seafloor. The retention time depends strongly on the direction of droplet motion and the rate of droplet shrinking. Dispersed droplets might be rising, settling, or drifting along sea currents. The detailed motion of the droplets depends on a number of factors, including the physical properties of the oil–water system (density, viscosity, interfacial tension), the temperature profile, the droplet size, the composition of the oil surface, the presence of marine snow and snot, and the flow direction and strength of underwater currents [15]. Under the action of buoyancy, large drops (>2 mm) and oil blobs rise towards the sea surface where they (re)coalesce with the oil slick. On the other hand, microdroplets with a size in the range of 10–100 μm have a lower rise velocity and higher probability of being carried away by underwater currents. Adsorption of chemical dispersants or naturally-occurring colloids and surfactants to the droplet surface hinders the tangential mobility of the oil–water interface, reduces the recirculating flow within the droplet, retards the overall motion of the droplet, and prevents droplet–droplet coalescence [16]. Interaction of the drifting droplets with settling marine snow (i.e., plankton and suspended microbial flocs) may lead to the formation of complex aggregates that tend to settle down on the seafloor [17,18], and stimulate chemotactic responses of other oil-degrading microbial species residing in sediments [19,20]. The probability of collision between marine snow and oil droplets depends on the concentration and size distribution of the two particulate populations. A higher concentration of larger particles creates a higher probability of aggregation and sedimentation [21]. In general, the combined effects of smaller size and interfacial contamination result in a higher probability of microdroplets forming stable droplet clouds with significant retention time in the seawater column.

The droplet shrinking rate is determined by the dissolution and biodegradation processes. The dissolution rate depends on the solubility of oil in water, the diffusivity of oil in water, and the velocity of the surrounding fluid relative to the droplet [16,22]. The solubility of most oil compounds is rather low, but may be enhanced by the action of surfactant micelles [23–27]. The biodegradation rate depends on the microbial strategy for oil uptake, the concentration of microbes, the intrinsic kinetics for oil consumption, the physical conditions (temperature, pressure, pH, salinity) and the availability of electron acceptors and mineral nutrients [8–10,28–31].

Three major microbial strategies have been identified for accessing and assimilating oily substrates; an outline is given here and a more detailed discussion is available in [31]. In a first strategy, microbes *firmly adhere* to the oil–water interface and sip oil compounds directly from the oily phase. This approach has been observed in pure cultures of super-hydrophobic, Gram-positive microbes, like *Mycobacterium* and *Rhodococcus* species. In a second strategy, microbes grow *suspended* in the bulk aqueous phase and uptake-dissolved and micellar oil compounds. This strategy has been observed, for example, in pure cultures of Gram-negative microbes, mainly of *Pseudomonas* species, that have a hydrophilic cell surface and produce biosurfactants of low molecular weight (e.g., rhamnolipids). In a third strategy, individual or clustered microbes adhere to the oil surface and actively *form biofilms* by secreting excessive amounts of biopolymers with high molecular weight. The biopolymers, mainly polysaccharides and proteins, do not dissolve into the bulk aqueous phase, but instead accumulate in the extracellular space and spontaneously assemble to form a three-dimensional matrix enmeshing the cells. The biofilm growth mode over oily substrates has been reported for several pure cultures and mixed microbial consortia. Current theoretical models for the fate of oil droplets in marine waters account only for the direct uptake strategy [32–34], neglecting any effects of biodegradation in the bulk aqueous phase or the formation of biofilm over the droplet and bioreaction therein.

In this work, a compound particle model (CPM) is developed for the biodegradation of solitary oil microdroplets moving through a water column. The compound particle is of the core-shell type and consists of an oily core that is successively surrounded by a bioreactive skin of negligible thickness and another bioreactive shell of finite thickness (Figure 1). The bioreactive skin represents a thin layer of microbes that uptake oil directly from the oily phase, whereas the bioreactive shell represents a distinct biofilm phase. In line with the abovementioned microbial strategies of oil uptake, the model accounts for all three modes of biodegradation: direct interfacial uptake, bioreaction in the bulk aqueous phase, and bioreaction in a biofilm formed around the droplet. A set of simplifying hypotheses is introduced so as to make the mathematical analysis tractable, and the governing equations solvable by analytical methods. The most important hypotheses are that the compound particle is considered to move as a non-deforming rigid sphere, the flow of the aqueous phase is dominated by viscous stresses, and the transport of dissolved oil in the biofilm phase is dominated by diffusion, whereas in the bulk aqueous phase it is dominated by advection. The analysis of the local mass balances results in a closed-form expression for the overall dissolution rate as a function of the Biot number, the Thiele modulus, the thickness of the biofilm, and the diffusivity and solubility ratios. Furthermore, from the overall mass balances, two coupled ordinary differential equations are established for the evolution of the particle size.

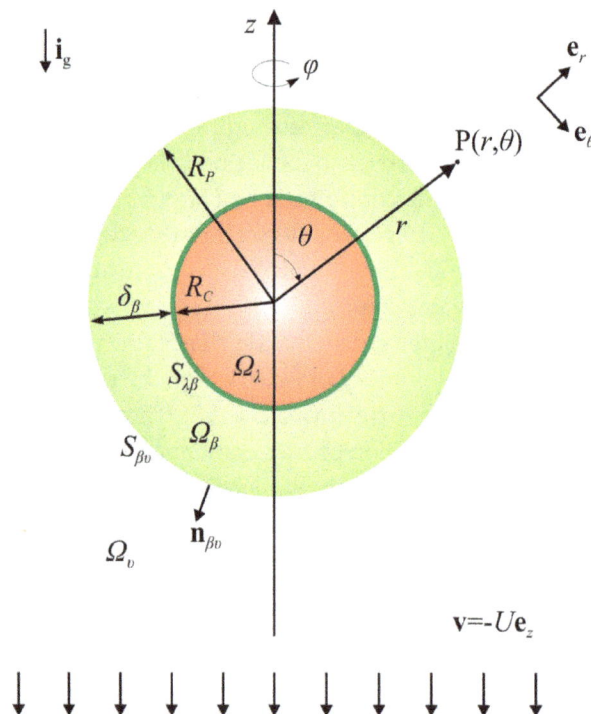

Figure 1. Geometry and coordinate system for the compound particle model (description in the text).

2. Model Formulation

With reference to Figure 1, the process under consideration is the transport and reaction of dissolved oil, denoted as the A solute, from the oil droplet (Ω_λ) to the surrounding biofilm (Ω_β) and aqueous (Ω_υ) phases. The thick line at the oil–biofilm interface ($S_{\lambda\beta}$) represents a thin layer of microbes that uptake oil compounds directly from the oily phase. The first step in the theoretical analysis is to determine the oil dissolution rate at the droplet surface, based on an appropriate formulation of the *local* mass balances (Section 2.3). The second step is to determine the droplet shrinking rate using the *overall* mass balances (Section 2.4). Before proceeding with the mathematical analysis, certain key considerations on modeling the different biodegradation modes (Section 2.1) and a set of basic hypotheses (Section 2.2) are set forth.

2.1. Considerations on Modeling the Three Major Biodegradation Modes

A few remarks are in order with regard to the theoretical modeling of each one of the three basic modes of biodegradation; that is, direct interfacial uptake, bioreaction in the bulk aqueous phase, and bioreaction in a biofilm formed around the droplet.

The first type of oil-degrading microbes is the *flatlanders*; that is superhydrophobic microbes able to firmly adhere to the oil surface and directly uptake organic compounds from the oily phase. Here, it is considered that the oil surface ($S_{\lambda\beta}$) is fully and uniformly covered by flatlanders. Partial coverage is expected to lead to more complex phenomena of fluid dynamics and solute transport and, thus, deserves to be investigated separately. The layer of flatlanders is usually found embedded in the oil side of the oil–water interface [35] and can be viewed as a *bioreactive skin* (interphase) of negligible thickness (~a few μm) on the droplet scale of observation (~100 μm). As the microbes have direct access to the oily substrate, the oil consumption rate is considered to be limited only by the intrinsic microbial kinetics. Under these conditions, this mode of biodegradation is essentially decoupled from the dissolution of oil to the surrounding phases. The physical presence of microbes on the droplet surface and the process of interfacial reaction are assumed to affect only implicitly the dissolution of oil; that is, by (possibly) changing the value of oil solubility.

The second type of oil-degrading microbes is the *drifters*; that is, hydrophilic microbes that remain suspended in the bulk aqueous phase (Ω_υ) and consume solubilized (molecular or micellar) oil. For this biodegradation mode, it is considered that the concentration of microbes is constant throughout the aqueous phase and the oil consumption rate follows first-order kinetics. In addition, solute A represents both molecular and micellar oil and, thus, the action of surfactants is taken into account only implicitly by modifying the apparent solubility of oil in the aqueous phase.

The third type of oil-degrading microbes is the *biofilm* formers; that is, microbes able to actively construct three-dimensional biofilm communities over the oil surface. The thickness of the biofilm might be appreciable and, thus, the biofilm is viewed as a distinct phase on the droplet scale of observation. Supplementary hypotheses for this mode include a uniform biofilm thickness, constant concentration of active microbes within the biofilm, and first order kinetics for the oil consumption rate. Interstitial flow is neglected and solute transport within the biofilm is dominated by diffusion.

With regard to the microbial proliferation rate, it is customary to assume a linear dependence on the concentration of active cells, that is

$$\widetilde{r}_{C,\alpha} = \widetilde{\mu}_{C,\alpha} \widetilde{B}_\alpha, \tag{1}$$

where α denotes the physical domain in which the microbes grow and takes the values $\alpha = \beta,\ \upsilon,\ \lambda\beta$. All of the primary symbols are defined in the nomenclature. The tilde (~) over a variable or parameter denotes a dimensional quantity, whereas the lack of it denotes a dimless quantity. The term dimensionless is abbreviated to *dimless* throughout the paper. The specific growth rate $\widetilde{\mu}_{C,\alpha}$ is usually considered to follow Monod kinetics.

$$\widetilde{\mu}_{C,\alpha} = \frac{\widetilde{\mu}_{m,\alpha} \widetilde{c}_{A\alpha}}{\widetilde{K}_{S,\alpha} + \widetilde{c}_{A\alpha}}. \tag{2}$$

Any possible effects of lag phase, cell maintenance, limitation by electron acceptors, nitrogen and phosphorous sources, substrate cometabolism or inhibition are neglected. Of particular interest are the limiting forms for the specific growth rate under *sufficiently high or low* concentration.

$$
\widetilde{\mu}_{C,\alpha} \cong
\begin{cases}
\widetilde{\mu}_{m,\alpha} & \text{if } \widetilde{c}_{A\alpha} \gg \widetilde{K}_{S,\alpha} \\[2mm]
\dfrac{\widetilde{\mu}_{m,\alpha}}{\widetilde{K}_{S,\alpha}}\widetilde{c}_{A\alpha} & \text{if } \widetilde{c}_{A\alpha} \ll \widetilde{K}_{S,\alpha}
\end{cases}.
\tag{3}
$$

The zeroth-order kinetics is expected to be applicable in the interfacial uptake mode because the microbial cells have access to the pure oily phase. On the other hand, the first-order kinetics is expected to be applicable in the suspended and biofilm growth modes because of the low solubility of oil compounds in aqueous phases. In all cases, the oil consumption rate is considered to be proportional to the cell proliferation rate. Therefore, the volumetric consumption rate of dissolved oil in a bulk phase, is given by

$$
\widetilde{r}_{A,\alpha} = -\frac{\widetilde{r}_{C,\alpha}}{Y_{C/A,\alpha}} = -\widetilde{k}_{1\alpha}\widetilde{c}_{A\alpha}, \quad \text{with } \widetilde{k}_{1\alpha} = \frac{\widetilde{\mu}_{m,\alpha}\widetilde{B}_{\alpha}}{\widetilde{K}_{S,\alpha}Y_{C/A,\alpha}},
\tag{4}
$$

for $\alpha = \beta, v$; and the surficial consumption rate on the droplet surface, is given by

$$
\widetilde{r}_{A,\lambda\beta} = -\frac{\widetilde{\mu}_{m,\lambda\beta}}{Y_{C/A,\lambda\beta}}\widetilde{B}_{\lambda\beta}.
\tag{5}
$$

Here, $\widetilde{B}_{\alpha}$ and $\widetilde{B}_{\lambda\beta}$ is the volumetric and surface concentration of cells, respectively. The minus sign denotes consumption rates. In this work, the cell concentration is treated as a constant for all three biodegradation modes.

2.2. Basic Hypotheses for the Hydrodynamics and Mass Transport

In addition to the previous considerations for the biodegradation process, a set of hypotheses is introduced for the flow and mass transport processes so as to simplify the mathematical description as much as possible while retaining the most important mechanisms. First, the external flow in the unbounded aqueous phase is dominated by viscous stresses and, thus, characterized by a low Reynolds number ($\mathrm{Re}_v = \widetilde{R}_P\widetilde{U}\widetilde{\rho}_v/\widetilde{\mu}_v \ll 1$). Second, the internal recirculating flow and the deformation of the particle are considered to be negligible. In all cases, the adsorption of biopolymers and microbial cells to the oil–water interface is expected to hinder the interfacial mobility and, consequently, diminish the internal flow in the oily phase. On the basis of a combination of small particle size, slow velocity and rigid-like interface, it is expected that interfacial tension dominates over viscous and gravitational forces that tend to deform the particle and the system is, thus, characterized by low capillary and Bond numbers ($\mathrm{Ca} = \widetilde{\mu}_v\widetilde{U}/\widetilde{\gamma}_{\beta v} \ll 1$, $\mathrm{Bo} = \widetilde{R}_P^2\Delta\widetilde{\rho}\widetilde{g}/\widetilde{\gamma}_{\beta v} \ll 1$; where $\Delta\widetilde{\rho} = |\widetilde{\rho}_v - \widetilde{\rho}_p|$ is the excess density and $\widetilde{\gamma}_{\beta v}$ is the interfacial tension at the particle surface) [36]. Therefore, the particle, either simple or compound, is considered to move as a rigid sphere. Third, the transport of dissolved oil in the biofilm phase is dominated by diffusion and, thus, characterized by a low Péclet number ($\mathrm{Pe}_{\beta} = \widetilde{R}_P\widetilde{U}_{\beta}/\widetilde{D}_{A\beta} \ll 1$). On the other hand, mass transport in the bulk aqueous phase is considered to be dominated by advection and characterized by a high Péclet number ($\mathrm{Pe}_v = \widetilde{R}_P\widetilde{U}/\widetilde{D}_{Av} \gg 1$). In both phases, solute diffusion is considered to obey Fick's constitutive law. Fourth, the oily phase is treated as a single compound and mass transfer therein is not taken into account (e.g., the solute A represents the total petroleum hydrocarbon in the case of crude oil). Finally, the quasi-steady state hypothesis is adopted for the analysis of the flow and mass transport problems at the local level (Section 2.3). Besides a high Péclet and a low Reynolds number, this assumption also requires a low droplet shrinking rate as compared to the characteristic velocity of the external flow. Thereafter, the evolution of the particle size is treated as a sequence of steady states in Section 2.4.

2.3. Overall Dissolution Rate: Analysis of the Local Mass Balances

Under the detailed set of considerations and hypotheses given in the previous subsections, mass transport is described in the context of the CPM by the following equations

$$\tilde{c}_{A\beta} = \tilde{c}_{A,\lambda/\beta}, \text{ at } \tilde{r} = \tilde{R}_C, \tag{6a}$$

$$0 = \tilde{D}_{A\beta}\tilde{\nabla}^2\tilde{c}_{A\beta} - \tilde{k}_{1\beta}\tilde{c}_{A\beta}, \text{ in the } \beta - \text{phase}, \tag{6b}$$

$$\tilde{\mathbf{J}}_{A\beta}\cdot\mathbf{n}_{\beta\upsilon} = \tilde{\mathbf{J}}_{A\upsilon}\cdot\mathbf{n}_{\beta\upsilon}, \text{ at } \tilde{r} = \tilde{R}_P, \tag{6c}$$

$$H_{A,\upsilon/\beta}\tilde{c}_{A\beta} = \tilde{c}_{A\upsilon}, \text{ at } \tilde{r} = \tilde{R}_P, \tag{6d}$$

$$\tilde{\mathbf{v}}_\upsilon\cdot\tilde{\nabla}\tilde{c}_{A\upsilon} = \tilde{D}_{A\upsilon}\tilde{\nabla}^2\tilde{c}_{A\upsilon} - \tilde{k}_{1\upsilon}\tilde{c}_{A\upsilon}, \text{ in the } \upsilon - \text{phase}, \tag{6e}$$

$$\tilde{c}_{A\upsilon} = 0, \text{ at } \tilde{r} \to \infty. \tag{6f}$$

It is possible to further reduce the complexity of the above set of governing equations by introducing two *educated hypotheses*. First, the tangential diffusion in the spherical shell is neglected. Strictly, this hypothesis holds for a thin shell ($\tilde{\delta}_\beta \ll \tilde{R}_P$) or fast reaction ($Da_\beta \gg 1$). Thus, Equation (6b) becomes

$$0 = \frac{\tilde{D}_{A\beta}}{\tilde{r}^2}\frac{d}{d\tilde{r}}\left(\tilde{r}^2\frac{d\tilde{c}_{A\beta}}{d\tilde{r}}\right) - \tilde{k}_{1\beta}\tilde{c}_{A\beta}, \text{ in the } \beta - \text{phase}. \tag{7}$$

Second, the continuity of the mass flux at the $\upsilon\beta$-interface is imposed in an *average sense* by demanding equality of the surface-averaged fluxes, instead of equality of the local fluxes. Therefore, Equation (6c) is expressed as follows

$$\int_{S_{\beta\upsilon}} \tilde{\mathbf{J}}_{A\beta}\cdot\mathbf{n}_{\beta\upsilon}d\tilde{S} = \int_{S_{\beta\upsilon}} \tilde{\mathbf{J}}_{A\upsilon}\cdot\mathbf{n}_{\beta\upsilon}d\tilde{S}, \text{ at } \tilde{r} = \tilde{R}_P. \tag{8a}$$

The above equation can be tidied up by considering that $S_{\beta\upsilon}$ is a spherical surface at $\tilde{r} = \tilde{R}_P$ with $d\tilde{S} = \tilde{r}^2\sin\theta d\theta d\varphi$ and $\mathbf{n}_{\beta\upsilon} = \mathbf{e}_r$. Also, the radial mass flux in the β-phase is independent of the polar and azimuthal angles, while the surface averaged flux in the right hand side of Equation (8a) defines the dissolution rate from the particle surface to the υ-phase. Thus, Equation (8a) becomes

$$-\tilde{D}_{A\beta}\left[\frac{d\tilde{c}_{A\beta}}{d\tilde{r}}\right]_{\tilde{r}=\tilde{R}_P}\tilde{S}_{\beta\upsilon} = \tilde{k}_{p/\upsilon}\tilde{S}_{\beta\upsilon}\tilde{c}_{A\upsilon}(\tilde{R}_P). \tag{8b}$$

Here, the dissolution rate has been expressed in terms of the mass transfer coefficient, $\tilde{k}_{p/\upsilon}$, the area of the compound particle surface, $\tilde{S}_{\beta\upsilon} = 4\pi\tilde{R}_P^2$, and the interfacial solute concentration at the side of the υ-phase, $\tilde{c}_{A\upsilon}(\tilde{R}_P)$, using knowledge that will be substantiated in the following paragraphs. The value of the solute concentration at the particle surface is constant, albeit not prescribed. Substitution of the boundary condition (6d) into Equation (8b), gives

$$-\tilde{D}_{A\beta}\left[\frac{d\tilde{c}_{A\beta}}{d\tilde{r}}\right]_{\tilde{r}=\tilde{R}_P} = \tilde{k}_{p/\upsilon}H_{A,\upsilon/\beta}\tilde{c}_{A\beta}(\tilde{R}_P). \tag{8c}$$

The partition coefficient of oil at the $\upsilon\beta$-interface, $H_{A,\upsilon/\beta}$, is approximated as the solubility ratio in the corresponding phases. By replacing Equations (6b) and (6c) with Equations (7) and (8c), respectively, and also by introducing dimless quantities, the mass transport problem defined in Equations (6a)–(6f) obtains the form

$$c_{A\beta}(R_C) = 1, \text{ at } r = R_C, \tag{9a}$$

$$0 = \frac{1}{r^2}\frac{d}{dr}\left(r^2\frac{dc_{A\beta}}{dr}\right) - h_T^2 c_{A\beta}, \text{ in the } \upsilon\beta - \text{phase,} \tag{9b}$$

$$-\left[\frac{dc_{A\beta}}{dr}\right]_{r=1} = \text{Bi } c_{A\beta}(1), \text{ at } r = 1, \tag{9c}$$

$$c_{A\beta}(1) = c_{A\upsilon}(1), \text{ at } r = 1, \tag{9d}$$

$$\text{Pe}_\upsilon \mathbf{v}_\upsilon \cdot \nabla c_{A\upsilon} = \nabla^2 c_{A\upsilon} - \text{Da}_\upsilon c_{A\upsilon}, \text{ in the } \upsilon - \text{phase,} \tag{9e}$$

$$c_{A\upsilon} = 0, \text{ at } r \to \infty. \tag{9f}$$

For the non-dimensionalization, the particle radius $\widetilde{R}_P$ is the reference length, the velocity $\widetilde{U}$ of the approaching fluid relative to the particle is the reference velocity, the solubility of oil in the biofilm, $\widetilde{c}_{A,\lambda/\beta}$, and in the aqueous phase, $\widetilde{c}_{A,\lambda/\upsilon}$, is the reference concentration for the respective phase. In particular, the following dimless quantities are defined

$$r = \frac{\widetilde{r}}{\widetilde{R}_P}; \ \nabla = \widetilde{R}_P \widetilde{\nabla}; \ \mathbf{v}_\upsilon = \frac{\widetilde{\mathbf{v}}_\upsilon}{\widetilde{U}}; \ c_{A\beta} = \frac{\widetilde{c}_{A\beta}}{\widetilde{c}_{A,\lambda/\beta}}; \ c_{A\upsilon} = \frac{\widetilde{c}_{A\upsilon}}{\widetilde{c}_{A,\lambda/\upsilon}}; \tag{10a}$$

$$\text{Pe}_\upsilon = \frac{\widetilde{R}_P \widetilde{U}}{\widetilde{D}_{A\upsilon}}; \ \text{Da}_\upsilon = \frac{\widetilde{k}_{1\upsilon} \widetilde{R}_P^2}{\widetilde{D}_{A\upsilon}}; \ h_T = \sqrt{\frac{\widetilde{k}_{1\beta} \widetilde{R}_P^2}{\widetilde{D}_{A\beta}}}; \ \text{Bi} = \frac{\widetilde{k}_{p/\upsilon} \widetilde{R}_P}{\widetilde{D}_{A\beta}} H_{A,\upsilon/\beta} \tag{10b}$$

The equation set defined in Equations (9a)–(9f) can be broken down into two subproblems that can be solved independently. The external mass transport problem defined by Equations (9d)–(9f) must be solved first, in order to determine the mass transfer coefficient $\widetilde{k}_{p/\upsilon}$ and the Biot number. As will be shown, the specific value of the solute concentration at the particle surface affects the concentration field in the υ-phase, but not the Biot number. Thereafter, the internal mass transport problem defined by Equations (9a)–(9c) must be solved in order to determine the overall dissolution rate at the surface of the oily core.

2.3.1. Advection-Dominated Transport in the Aqueous Phase without Bioreaction

In the absence of bioreaction ($\text{Da}_\upsilon = 0$), the external mass transport problem for the unbounded aqueous domain (Ω_υ) obtains the form

$$\text{Pe}_\upsilon \mathbf{v}_\upsilon \cdot \nabla c_{A\upsilon} = \nabla^2 c_{A\upsilon}, \tag{11a}$$

$$c_{A\upsilon}(1, \theta) = c_{A\beta}(1), \tag{11b}$$

$$c_{A\upsilon}(\infty, \theta) = 0, \tag{11c}$$

and can be solved analytically in the limits of very low ($\text{Pe}_\upsilon \ll 1$) or high Péclet number ($\text{Pe}_\upsilon \gg 1$) [37,38]. Here, the high-Péclet regime is of primary interest and, thus, the derivation of the pertinent analytical solution is outlined. In spherical coordinates, for an axisymmetric concentration field (i.e., independent of the azimuthal angle), Equation (11a) obtains the detailed form

$$v_{\upsilon,r}\frac{\partial c_{A\upsilon}}{\partial r} + \frac{v_{\upsilon,\theta}}{r}\frac{\partial c_{A\upsilon}}{\partial \theta} = \frac{1}{\text{Pe}_\upsilon}\left[\frac{\partial^2 c_{A\upsilon}}{\partial r^2} + \frac{2}{r}\frac{\partial c_{A\upsilon}}{\partial r} + \frac{1}{r^2 \sin\theta}\frac{\partial}{\partial \theta}\left(\sin\theta\frac{\partial c_{A\upsilon}}{\partial \theta}\right)\right]. \tag{12}$$

Moreover, for creeping Newtonian flow past a rigid sphere, the velocity components are [39]

$$v_{\upsilon,r}(r, \theta) = -\left(1 - \frac{3}{2r} + \frac{1}{2r^3}\right)\cos\theta, \tag{13a}$$

$$v_{\upsilon,\theta}(r, \theta) = \left(1 - \frac{3}{4r} - \frac{1}{4r^3}\right)\sin\theta. \tag{13b}$$

For advection-dominated mass transport, the change in the concentration from the value at the sphere surface ($c_{Av} = const.$) to the bulk value away from the sphere ($c_{Av} = 0$) is expected to occur within a *thin* boundary layer around the sphere. Upon this consideration, the following (dimless) independent variable is introduced

$$y \equiv r - 1, \tag{14}$$

to measure the distance from the sphere surface, within the boundary layer. On the basis that the thickness of the concentration boundary layer is small as compared to the radius of the sphere, i.e., $y \ll 1$, the velocity terms can be simplified and certain diffusion terms can be neglected in Equation (12). In particular, order of magnitude analysis shows that the terms of tangential diffusion and normal diffusion due to surface curvature are much less important than the normal diffusion term. Under the boundary layer approximation, the final form of the *reduced* advection–diffusion equation is

$$-\frac{3}{2}y^2 \cos\theta \frac{\partial c_{Av}}{\partial y} + \frac{3}{2}y \sin\theta \frac{\partial c_{Av}}{\partial \theta} = \frac{1}{\mathrm{Pe}_v} \frac{\partial^2 c_{Av}}{\partial y^2}. \tag{15}$$

A detailed derivation of the above equation and the development of an analytical solution by means of a similarity transformation is given in [22] (pp. 80–87) and [39] (pp. 414–417). The exact solution of Equation (15) can be expressed as follows

$$c_{Av}(y,\theta) = c_{A\beta}(1)\left[1 - \frac{1}{C_2}\int_0^{\chi(y,\theta)} \exp\left(-\frac{1}{3}s^3\right)ds\right], \tag{16a}$$

where C_2 is an integration constant given by

$$C_2 = \int_0^\infty \exp\left(-\frac{1}{3}s^3\right)ds \cong 1.2879, \tag{16b}$$

and $\chi(y,\theta)$ is a composite variable defined as

$$\chi(y,\theta) = \mathrm{Pe}_v^{1/3} f(\theta) y, \tag{16c}$$

with

$$f(\theta) = \frac{\sin\theta}{\left(\theta - \frac{\sin(2\theta)}{2}\right)^{1/3}}. \tag{16d}$$

The concentration field given in Equation (16) is used to determine the diffusive mass flux

$$\tilde{\mathbf{J}}_{Av} = -\frac{\widetilde{D}_{Av}\tilde{c}_{A,\lambda/v}}{\widetilde{R}_P}\nabla c_{Av}, \tag{17}$$

and, ultimately, the average mass transfer rate from the particle surface to the aqueous phase

$$\widetilde{W}_{A,p/v}^0 \equiv \int_{S_{\beta v}} \tilde{\mathbf{J}}_{Av}\cdot\mathbf{n}_{\beta v}d\tilde{S} = -2\pi\widetilde{R}_P\widetilde{D}_{Av}\tilde{c}_{A,\lambda/v}\int_0^\pi \left[\frac{\partial c_{Av}}{\partial r}\right]_{r=1}\sin\theta d\theta. \tag{18}$$

The concentration derivative is calculated using the fundamental theorem of calculus, as follows

$$\left[\frac{\partial c_{Av}}{\partial r}\right]_{r=1} = \left[\frac{\partial c_{Av}}{\partial y}\right]_{y=0} = \frac{\partial\chi(y,\theta)}{\partial y}\left[\frac{dc_{Av}}{d\chi}\right]_{\chi=0} = -\frac{\mathrm{Pe}_v^{1/3}f(\theta)}{C_2}c_{A\beta}(1), \tag{19}$$

and, after some operations, the final expression for the dissolution rate from the particle surface to the υ-phase, is given by the following expression

$$\widetilde{W}^0_{A,p/\upsilon} = \widetilde{k}^0_{p/\upsilon} \widetilde{S}_{\beta\upsilon} \widetilde{c}_{A\upsilon}\left(\widetilde{R}_P\right), \tag{20}$$

where $\widetilde{c}_{A\upsilon}(\widetilde{R}_P) = \widetilde{c}_{A,\lambda/\upsilon} c_{A\beta}(1) = H_{A,\upsilon/\beta} \widetilde{c}_{A\beta}(\widetilde{R}_P)$, and

$$\widetilde{k}^0_{p/\upsilon} = \frac{\widetilde{D}_{A\upsilon}}{2\widetilde{R}_P} \frac{I_\theta}{C_2} \mathrm{Pe}^{1/3}_\upsilon, \tag{21a}$$

$$I_\theta = \int_0^\pi f(\theta) \sin\theta d\theta \cong 1.6087. \tag{21b}$$

Here, $\widetilde{k}^0_{p/\upsilon}$ is the mass transfer coefficient and the "0" superscript denotes the absence of bioreaction in the bulk aqueous phase. At this point, it is very useful to introduce the *Sherwood number* which is defined as follows

$$\mathrm{Sh}^0_{p/\upsilon} \equiv \frac{\widetilde{k}^0_{p/\upsilon}\left(2\widetilde{R}_P\right)}{\widetilde{D}_{A\upsilon}} = 1.249\, \mathrm{Pe}^{1/3}_\upsilon, \tag{22a}$$

and represents a dimless mass transfer coefficient. The above correlation underestimates the Sherwood number for about 10% for $\mathrm{Pe}_\upsilon > 100$ [37] and, as expected, provides a wrong asymptotic value for $\mathrm{Pe}_\upsilon \to 0$. By simply adding the value of the Sherwood number that corresponds to diffusion-only (i.e., $\mathrm{Sh} = 2$ for $\mathrm{Pe} = 0$), the following improved correlation is obtained

$$\mathrm{Sh}^0_{p/\upsilon} \equiv \frac{\widetilde{k}^0_{p/\upsilon}(2\widetilde{R}_P)}{\widetilde{D}_{A\upsilon}} = 2 + 1.249\, \mathrm{Pe}^{1/3}_\upsilon. \tag{22b}$$

Levich suggested the above superposition based on the rationale that the resistances to mass transfer by diffusion and advection act in parallel [22]. The estimates of Equation (22b) agree with numerical data within approximately 7% for the entire range of Pe. At this point, two remarks are in order. First, the mass transfer coefficient that appears in the Biot number does not depend on the, yet unknown, interfacial concentration of the solute. Second, the definition given in Equation (18) and the final expression given in Equation (20) for the surface averaged mass transfer rate were introduced earlier in the derivation of the modified boundary condition given in Equation (8).

2.3.2. Advection-Dominated Transport and Homogeneous Bioreaction in the Aqueous Phase

Following the analysis presented previously for advection dominated mass transport under the boundary layer theory approximation, the reduced form of the advection–diffusion–reaction equation given in Equation (9e) is

$$-\frac{3}{2}y^2 \cos\theta \frac{\partial c_{A\upsilon}}{\partial y} + \frac{3}{2}y \sin\theta \frac{\partial c_{A\upsilon}}{\partial\theta} = \frac{1}{\mathrm{Pe}_\upsilon} \frac{\partial^2 c_{A\upsilon}}{\partial y^2} - \frac{\mathrm{Da}_\upsilon}{\mathrm{Pe}_\upsilon} c_{A\upsilon}, \tag{23}$$

with the same boundary conditions as given in Equations (11b)–(11c). To the best of our knowledge, an analytical solution is not available for the above partial differential equation. Approximate solutions have been developed using the empirical θ-expansion method of Yuge, perturbation analysis, and numerical methods [37,40–45]. For engineering applications, the following simple correlation has been proposed for the Sherwood number [42,45]

$$\mathrm{Sh}_{p/\upsilon} \equiv \frac{\widetilde{k}_{p/\upsilon}(2\widetilde{R}_P)}{\widetilde{D}_{A\upsilon}} = \frac{\mathrm{Ha}}{\tanh\mathrm{Ha}} \mathrm{Sh}^0_{p/\upsilon}, \tag{24}$$

with

$$\text{Ha} = \frac{2\sqrt{\text{Da}_v}}{\text{Sh}_{p/v}^0}, \tag{25}$$

where Ha is the Hatta modulus and $\text{Sh}_{p/v}^0$ is the Sherwood number given by Equation (22) for the case of no bioreaction in the aqueous phase. The heuristic correlation given in Equation (24) is based on the film theory approximation and has been shown to provide an acceptable fit to more accurate numerical data. It is used here to provide estimates for the dissolution rate from the particle surface to the v-phase, through the following relation

$$\widetilde{W}_{A,p/v} = \widetilde{k}_{p/v} \widetilde{S}_{\beta v} \widetilde{c}_{Av}(\widetilde{R}_P). \tag{26}$$

2.3.3. Diffusion and Reaction in the Biofilm Phase

The solution of the internal mass transfer problem

$$0 = \frac{1}{r^2} \frac{d}{dr} \left(r^2 \frac{dc_{A\beta}}{dr} \right) - h_T^2 c_{A\beta}, \text{ in the } \beta - \text{phase}, \tag{27a}$$

$$c_{A\beta}(R_C) = 1, \tag{27b}$$

$$-\left[\frac{dc_{A\beta}}{dr} \right]_{r=1} = \text{Bi } c_{A\beta}(1), \tag{27c}$$

is expressed as follows

$$c_{A\beta}(r) = \frac{C_3}{r} \cosh(h_T r) + \frac{C_4}{r} \sinh(h_T r), \tag{28}$$

where h_T is the Thiele modulus for homogeneous reaction in a spherical shell, and the integration constants are given by

$$C_3 = \frac{(1 - \delta_\beta) \cosh(h_T)[h_T + (\text{Bi} - 1)\tanh(h_T)]}{h_T \cosh(h_T \delta_\beta) + (\text{Bi} - 1)\sinh(h_T \delta_\beta)}, \tag{29a}$$

$$C_4 = -\frac{(1 - \delta_\beta) \cosh(h_T)[h_T \tanh(h_T) + \text{Bi} - 1]}{h_T \cosh(h_T \delta_\beta) + (\text{Bi} - 1)\sinh(h_T \delta_\beta)}. \tag{29b}$$

The concentration at the particle surface ($v\beta$-interface) is now given by the expression

$$c_{A\beta}(1) = \frac{h_T(1 - \delta_\beta) \sec \text{h}(h_T \delta_\beta)}{h_T + (\text{Bi} - 1)\tanh(h_T \delta_\beta)}, \tag{30}$$

and the concentration derivative at the core surface ($\lambda\beta$-interface) is

$$\left[\frac{dc_{A\beta}}{dr} \right]_{r=R_C} = -h_T \left[\frac{h_T \tanh(h_T \delta_\beta) + \text{Bi} - 1}{h_T + (\text{Bi} - 1)\tanh(h_T \delta_\beta)} \right] - \frac{1}{1 - \delta_\beta}. \tag{31}$$

The overall dissolution rate is

$$\widetilde{W}_{A,\lambda/\beta} \equiv \int_{S_{\lambda\beta}} \widetilde{\mathbf{J}}_{A\beta} \cdot \mathbf{n}_{\lambda\beta} d\widetilde{S} = \widetilde{k}_{\lambda/\beta} \widetilde{S}_{\lambda\beta} \widetilde{c}_{A,\lambda/\beta}, \tag{32}$$

where $\widetilde{S}_{\lambda\beta} = 4\pi \widetilde{R}_C^2$ is the area of the spherical core and the mass transfer coefficient is given by

$$\widetilde{k}_{\lambda/\beta} = \frac{\widetilde{D}_{A\beta}}{\widetilde{R}_P} \left[-\frac{dc_{A\beta}}{dr} \right]_{r=R_C}. \tag{33}$$

Again, it is convenient to define the Sherwood number

$$\text{Sh}_{\lambda/\beta} \equiv \frac{\widetilde{k}_{\lambda/\beta}\left(2\widetilde{R}_P\right)}{\widetilde{D}_{Av}} = 2\Lambda_{A\beta}h_T\left[\frac{h_T\tanh\left(h_T\delta_\beta\right) + \text{Bi} - 1}{h_T + (\text{Bi} - 1)\tanh\left(h_T\delta_\beta\right)}\right] + \frac{2\Lambda_{A\beta}}{1 - \delta_\beta}. \tag{34}$$

The final quantity of interest is the volume averaged concentration of solute A in the β-phase

$$\langle \widetilde{c}_{A\beta}\rangle \equiv \frac{1}{\widetilde{V}_\beta}\int_{\Omega_\beta}\widetilde{c}_{A\beta}dV, \tag{35}$$

which is later necessary in the determination of the particle size evolution. Here, $\widetilde{V}_\beta = \widetilde{V}_P - \widetilde{V}_C$ is the volume of the biofilm shell, with $\widetilde{V}_P = \pi\widetilde{D}_P^3/6$ and $\widetilde{V}_C = \pi\widetilde{D}_C^3/6$. After some algebraic operations, the final expression for the volume averaged concentration is

$$\langle \widetilde{c}_{A\beta}\rangle = \frac{4\pi\widetilde{R}_P^3}{\widetilde{V}_\beta}\frac{J_C\widetilde{c}_{A,\lambda/\beta}}{h_T^2}, \tag{36}$$

with

$$J_C \equiv h_T^2\int_{R_C}^1 r^2 c_{A\beta}(r)dr = (1 - \delta_\beta)^2\left[-\frac{dc_{A\beta}}{dr}\right]_{r=R_C} - \text{Bi}\,c_{A\beta}(1), \tag{37}$$

where the concentration and its derivative are given in Equations (30) and (31), respectively.

2.4. Evolution of the Particle Size: Analysis of the Overall Mass Balances

The knowledge of the oil dissolution rate at the oil–biofilm and biofilm–water interfaces as well as of the average oil concentration in the biofilm can be used to determine the change in the dimensions of the compound particle over time. This is achieved through the analysis of the *overall* mass balance for the λ- and β-phases.

2.4.1. Overall Mass Balance for the λ-Phase

Upon considering the λ-phase as an open system that may exchange mass with the surrounding phases, the integral form of the mass balance is

$$\frac{d}{d\widetilde{t}}\int_{\Omega_\lambda}\widetilde{\rho}_\lambda d\widetilde{V} = \int_{\Omega_\lambda}\widetilde{r}_{A,\lambda}d\widetilde{V} + \int_{S_{\lambda\beta}}\widetilde{\rho}_\lambda\left[\widetilde{\mathbf{v}}_{\lambda\beta} - \widetilde{\mathbf{v}}_\beta\right]\cdot\mathbf{n}_{\lambda\beta}d\widetilde{S}. \tag{38}$$

The term on the left hand side of the above equation represents the net accumulation of mass in the Ω_λ-region. On the right hand side, the first term represents the change in the mass because of reaction in the Ω_λ-region, and the second term represents the net influx of mass passing through the λβ-interface. Considering that the density of the λ-phase is constant, the accumulation term gives

$$\frac{d}{d\widetilde{t}}\int_{\Omega_\lambda}\widetilde{\rho}_\lambda d\widetilde{V} = \pi\widetilde{D}_{C,t}^2\frac{\widetilde{\rho}_\lambda}{2}\frac{d\widetilde{D}_{C,t}}{d\widetilde{t}}, \tag{39}$$

where $\widetilde{D}_{C,t} = 2\widetilde{R}_{C,t}$ is the diameter of the oily core at the t instant of time. The direct interfacial uptake of oil is modeled as a surface reaction occurring uniformly over the droplet surface and the reaction term obtains the form

$$\int_{\Omega_\lambda}\widetilde{r}_{A,\lambda}d\widetilde{V} = \int_{\Omega_\lambda}\widetilde{r}_{A,\lambda\beta}\delta_{\lambda\beta}d\widetilde{V} = \widetilde{r}_{A,\lambda\beta}\widetilde{S}_{\lambda\beta,t}, \tag{40}$$

where $\delta_{\lambda\beta}$ is Dirac's delta function concentrated on the λβ-interface, $\widetilde{S}_{\lambda\beta,t} = \pi\widetilde{D}_{C,t}^2$ is the surface area of the oily core, and $\widetilde{r}_{A,\lambda\beta}$ is the constant reaction rate given in Equation (5).

The last term in Equation (38) represents the diffusive mass flux of oil across the $\lambda\beta$-interface. For completely immiscible phases, this term should be nil. For the problem at hand, the dissolution of the oil droplet is considered to be *sufficiently slow* ($\widetilde{k}_{\lambda/\beta} << \widetilde{U}$) so as not to have an appreciable impact on fluid dynamics but, nonetheless, this results in a non-zero diffusive flux across the droplet surface. Therefore, for this term we have

$$\int_{S_{\lambda\beta}} \widetilde{\rho}_\lambda \left[\widetilde{\mathbf{v}}_{\lambda\beta} - \widetilde{\mathbf{v}}_\beta\right] \cdot \mathbf{n}_{\lambda\beta} d\widetilde{S} = -\int_{S_{\lambda\beta}} \widetilde{\mathbf{J}}_{A\beta} \cdot \mathbf{n}_{\lambda\beta} d\widetilde{S} = -\widetilde{W}_{A,\lambda/\beta}, \tag{41}$$

with the overall dissolution rate given by Equation (32). Substitution of Equations (39)–(41) into Equation (38), gives

$$\frac{d\widetilde{D}_{C,t}}{d\widetilde{t}} = -\frac{2\widetilde{\mu}_{m,\lambda\beta}\widetilde{B}_{\lambda\beta}}{\widetilde{\rho}_\lambda Y_{C/A,\lambda\beta}} - \frac{2\mathrm{Sh}_{\lambda/\beta}}{\widetilde{D}_{P,t}} \frac{\widetilde{D}_{Av}\widetilde{c}_{A,\lambda/\beta}}{\widetilde{\rho}_\lambda}, \tag{42}$$

where the Sherwood number varies along with the changing particle dimensions over time.

2.4.2. Overall Mass Balance for the β-Phase

The diameter of the compound particle also changes as the dissolving oily core is shrinking and its evolution is determined by the overall mass balance for the β-phase. We have that

$$\frac{d}{d\widetilde{t}} \int_{\Omega_\beta} \widetilde{\rho}_\beta d\widetilde{V} = \int_{\Omega_\beta} \widetilde{r}_\beta d\widetilde{V} + \int_{S_{\beta\lambda}} \widetilde{\rho}_\beta \left[\widetilde{\mathbf{v}}_{\beta\lambda} - \widetilde{\mathbf{v}}_\beta\right] \cdot \mathbf{n}_{\beta\lambda} d\widetilde{S} + \int_{S_{\beta\upsilon}} \widetilde{\rho}_\beta \left[\widetilde{\mathbf{v}}_{\beta\upsilon} - \widetilde{\mathbf{v}}_\beta\right] \cdot \mathbf{n}_{\beta\upsilon} d\widetilde{S}, \tag{43}$$

On the right hand side of the above equation, the second term could represent cell migration into the oily phase and the third term could represent the attachment of suspended cells to the biofilm or biofilm detachment and entrainment into the aqueous phase. However, for the problem at hand, both of these terms are considered to be nil. For the accumulation term on the left hand side, we have

$$\frac{d}{d\widetilde{t}} \int_{\Omega_\beta} \widetilde{\rho}_\beta d\widetilde{V} = \pi\widetilde{D}_{P,t}^2 \frac{\widetilde{\rho}_\beta}{2} \frac{d\widetilde{D}_{P,t}}{d\widetilde{t}} - \pi\widetilde{D}_{C,t}^2 \frac{\widetilde{\rho}_\beta}{2} \frac{d\widetilde{D}_{C,t}}{d\widetilde{t}}. \tag{44}$$

The rate of change in the biofilm mass, which is caused by the growth of cells and the synthesis of the extracellular matrix, is considered to be proportional to the microbial cell proliferation rate, i.e., $\widetilde{r}_\beta = \widetilde{r}_{C,\beta}/Y_{C/\beta} = -\widetilde{r}_{A,\beta}Y_{\beta/A}$ with $Y_{\beta/A} = Y_{C/A}/Y_{C/\beta}$. Therefore, for the first term on the right hand side of Equation (43), we have

$$\int_{\Omega_\beta} \widetilde{r}_\beta d\widetilde{V} = Y_{\beta/A} \int_{\Omega_\beta} (-\widetilde{r}_{A,\beta}) d\widetilde{V} = Y_{\beta/A}\widetilde{k}_{1\beta} \int_{\Omega_\beta} \widetilde{c}_{A\beta} d\widetilde{V} = Y_{\beta/A}\widetilde{k}_{1\beta}\widetilde{V}_{\beta,t}\langle\widetilde{c}_{A\beta}\rangle_t, \tag{45}$$

where $\langle\widetilde{c}_{A\beta}\rangle_t$ is the volume averaged concentration of oil in the β-phase at the t time instant, and is given by the expression in Equation (36). Substitution of Equations (44) and (45) into Equation (43), gives

$$\frac{d\widetilde{D}_{P,t}}{d\widetilde{t}} = \frac{\widetilde{S}_{\lambda\beta,t}}{\widetilde{S}_{\beta\upsilon,t}} \frac{d\widetilde{D}_{C,t}}{d\widetilde{t}} + \frac{2\widetilde{k}_{1\beta}Y_{\beta/A}}{\widetilde{\rho}_\beta} \frac{\widetilde{V}_{\beta,t}}{\widetilde{S}_{\beta\upsilon,t}} \langle\widetilde{c}_{A\beta}\rangle_t, \tag{46}$$

and, further, substitution of Equation (36) for the average concentration, after some operations, gives

$$\frac{d\widetilde{D}_{P,t}}{d\widetilde{t}} = \frac{\widetilde{S}_{\lambda\beta,t}}{\widetilde{S}_{\beta\upsilon,t}} \frac{d\widetilde{D}_{C,t}}{d\widetilde{t}} + \frac{4J_{C,t}}{\widetilde{D}_{P,t}} \frac{\widetilde{c}_{A,\lambda/\beta}}{\widetilde{\rho}_\lambda} \widetilde{D}_{Av}\Phi_{grt}, \tag{47}$$

with

$$\Phi_{grt} = \Lambda_{A\beta}Y_{\beta/A}\frac{\widetilde{\rho}_\lambda}{\widetilde{\rho}_\beta}. \tag{48}$$

2.4.3. Compact and Dimless Forms of the Coupled ODEs

It is very convenient to express the coupled ordinary differential equations (ODEs) given in Equations (42) and (47), into the following compact form

$$\frac{d\widetilde{D}_{C,t}}{d\widetilde{t}} = -\widetilde{k}_{srn} - \widetilde{k}_{dis}(\widetilde{t}), \tag{49a}$$

$$\frac{d\widetilde{D}_{P,t}}{d\widetilde{t}} = \frac{\widetilde{D}_{C,t}^2}{\widetilde{D}_{P,t}^2}\frac{d\widetilde{D}_{C,t}}{d\widetilde{t}} + \widetilde{k}_{grt}(\widetilde{t}), \tag{49b}$$

with

$$\widetilde{k}_{srn} = \frac{2\widetilde{\mu}_{m,\lambda\beta}\widetilde{B}_{\lambda\beta}}{\widetilde{\rho}_\lambda Y_{C/A,\lambda\beta}}, \tag{50a}$$

$$\widetilde{k}_{dis}(\widetilde{t}) = \frac{2\mathrm{Sh}_{\lambda/\beta}}{\widetilde{D}_{P,t}}\frac{\widetilde{D}_{Av}\widetilde{c}_{A,\lambda/\beta}}{\widetilde{\rho}_\lambda}, \tag{50b}$$

$$\widetilde{k}_{grt}(\widetilde{t}) = \frac{4J_{C,t}}{\widetilde{D}_{P,t}}\frac{\widetilde{c}_{A,\lambda/\beta}}{\widetilde{\rho}_\lambda}\widetilde{D}_{Av}\Phi_{grt}. \tag{50c}$$

Here, $\widetilde{k}_{srn}$ is the droplet shrinking rate caused by direct interfacial uptake, $\widetilde{k}_{dis}$ is the droplet shrinking rate caused by dissolution into the surrounding biofilm and aqueous phases, and $\widetilde{k}_{grt}$ is the biofilm expansion rate due to growth.

The Damköhler and Thiele numbers that appear in the expressions given in Equations (34) and (37) for the Sherwood number $\mathrm{Sh}_{\lambda/\beta}$ and the $J_{C,t}$ parameter, respectively, depend explicitly on the changing diameter of the compound particle. The situation might be a little bit more complex for the Péclet number in the case of a freely rising or sinking particle because the Stokes velocity also depends on the changing particle dimensions and density as follows

$$\widetilde{U}_{S,t} = \frac{\widetilde{g}}{18\widetilde{\mu}_v}\widetilde{D}_{P,t}^2\Delta\widetilde{\rho}_t, \tag{51}$$

where $\Delta\widetilde{\rho}_t = |\widetilde{\rho}_v - \widetilde{\rho}_{P,t}|$ is the excess density of the compound particle as compared to the density of the surrounding aqueous phase, $\widetilde{\rho}_{P,t} = \varphi_{\lambda,t}\widetilde{\rho}_\lambda + (1 - \varphi_{\lambda,t})\widetilde{\rho}_\beta$ is the density of the compound particle, and $\varphi_{\lambda,t} = \widetilde{D}_{C,t}^3/\widetilde{D}_{P,t}^3$ is the volume fraction of the oily core. For a given set of parameters and initial conditions, the coupled ODEs given in Equations (49a) and (49b) can be solved numerically using, for instance, the explicit Euler or the classical Runge–Kutta method. In this work, both methods have been successfully implemented with an in-house Fortran code.

One step further, it is useful to establish a dimless form for the coupled ODEs that describe the evolution of the dimensions of the compound particle. For this purpose, a scaled characteristic diffusion time is introduced as follows

$$\widetilde{\tau}_D = \frac{\widetilde{D}_{P,0}^2}{\widetilde{D}_{Av}}\frac{\widetilde{\rho}_\lambda}{\widetilde{c}_{A,\lambda/\beta}}. \tag{52}$$

Multiplication of both parts of Equations (49a) and (49b) with $\widetilde{\tau}_D/\widetilde{D}_{P,0}$ gives

$$\frac{dD_{C,t}}{d\tau} = -k_{srn} - k_{dis}(\tau), \tag{53a}$$

$$\frac{dD_{P,t}}{d\tau} = \frac{D_{C,t}^2}{D_{P,t}^2}\frac{dD_{C,t}}{d\tau} + k_{grt}(\tau), \tag{53b}$$

with

$$k_{srn} = \tilde{k}_{srn} \frac{\tilde{\tau}_D}{\tilde{D}_{P,0}}, \ k_{dis}(\tau) = \frac{2\mathrm{Sh}_{\lambda/\beta}}{D_{P,t}}, \ k_{grt}(\tau) = \frac{4J_{C,t}}{D_{P,t}} \Phi_{grt}. \tag{54}$$

For the above dimless ODEs, it is only required to define dimless ratios (diffusivity, solubility, density) and the initial values of the dimless moduli. Thereafter, at each time instant the Damköhler and the Thiele moduli are updated as follows

$$\mathrm{Da}_{v,t} = D_{P,t}^2 \mathrm{Da}_{v,0}, \ h_{T,t} = D_{P,t} h_{T,0}. \tag{55}$$

For the Péclet number, if the particle velocity is held constant we have $\mathrm{Pe}_{v,t} = D_{P,t}\mathrm{Pe}_{v,0}$, whereas if the particle is freely rising or sinking we have

$$\mathrm{Pe}_{v,t} = D_{P,t}^3 \Delta\rho_t \mathrm{Pe}_{v,0}, \tag{56}$$

with the dimless excess density given by $\Delta\rho_t = |1 - \rho_{P,t}|/|1 - \rho_{P,0}|$, and $\rho_{P,t} = \tilde{\rho}_{P,t}/\tilde{\rho}_v$. Two final remarks are in order. First, by setting $\tilde{\delta}_\beta = 0$, $\tilde{D}_{C,t} = \tilde{D}_{P,t}$, and $\tilde{c}_{A,\lambda/\beta} = \tilde{c}_{A,\lambda/v}$, the compound particle model degenerates into a single-phase shrinking particle model. Second, the radius is the preferred characteristic length in the analysis of the local mass balances because it naturally arises with the spherical coordinate system. On the other hand, the particle diameter is used in the evolution of the particle dimensions because this length is determined experimentally in particle size analyses.

3. Results and Discussion

The main outcome of the theoretical model developed in Section 2 is the expression for the overall dissolution rate for a given particle configuration and the coupled ordinary differential equations for the evolution of the dimensions of the compound particle. In this section, the effect of key system parameters on the dissolution rate and the particle size evolution are investigated.

3.1. Overall Sherwood Number

According to Equation (32), for a given particle configuration and constant oil solubility, the overall dissolution rate increases with increasing Sherwood number. As already mentioned, the Sherwood number $\mathrm{Sh}_{\lambda/\beta}$ represents the dimless mass transfer coefficient from the surface of the oily core to the surrounding biofilm shell and depends on the Biot number, the Thiele modulus, the thickness of the biofilm, and the solubility and diffusivity ratios. Among these parameters, the Biot number expresses the ratio of the external mass transfer rate (i.e., from the surface of the compound particle to the unbounded aqueous phase) over the characteristic intraparticle diffusion rate. The Biot number, in turn, depends on the Péclet and Damköhler numbers for the aqueous phase as well as on the solubility and diffusivity ratios.

Figure 2 presents the dependence of the Biot number on the Péclet number for different values of the Damköhler number, keeping the other parameters constant. As expected, the Biot number increases with increasing Péclet and Damköhler numbers because the external mass transfer rate is enhanced by the contributions of advection and bulk bioreaction, respectively. For $\mathrm{Pe}_v = 0$ and $\mathrm{Da}_v = 0$, the solute moves away from the particle surface only by diffusion and the Biot number obtains the asymptotic value of $\mathrm{Bi} = H_{A,v/\beta}/\Lambda_{A\beta}$. For most solutes, diffusion within the biofilm is hindered by the extracellular matrix [46,47] and the diffusivity ratio is expected to be $\Lambda_{A\beta} \leq 1$. Furthermore, scarce experimental evidence suggests that the solubility of hydrophobic organic compounds might be significantly higher in the biofilm than in the aqueous phase ($H_{A,v/\beta} \leq 1$). In the limit of exiguous solubility in the aqueous phase, i.e., $H_{A,v/\beta} \ll 1$, the Biot number practically becomes nil and the dissolved oil is retained within the biofilm shell.

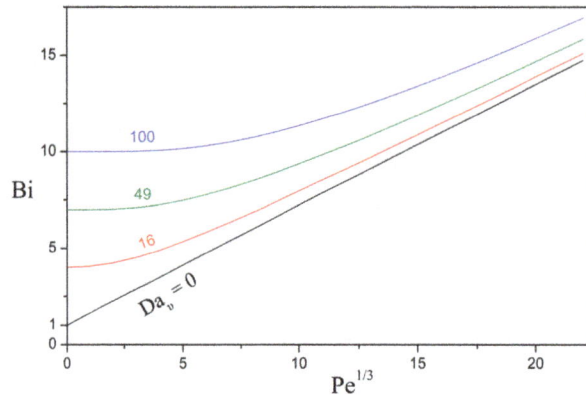

Figure 2. Impact of the Péclet and Damköhler numbers on the Biot number, for $\Lambda_{A\beta} = 1$, $H_{A,v/\beta} = 1$.

Figure 3a shows that the overall Sherwood number increases monotonically with increasing Biot and Thiele numbers, while keeping constant the other parameters. The Thiele number is a measure of the bioreaction rate over the diffusion rate within the biofilm shell. Faster bioreaction results in a steeper concentration gradient which, in turn, drives a higher rate of oil dissolution from the surface of the oily core. Of particular interest is the case of the vanishingly small Biot number which translates into the dissolved oil remaining trapped within the biofilm until complete biodegradation is achieved. Such a function would be of great practical importance and could perhaps be implemented by biofilms with hydrophobic or lipophilic biopolymers in their extracellular matrix [48]. This aspect deserves to be examined experimentally.

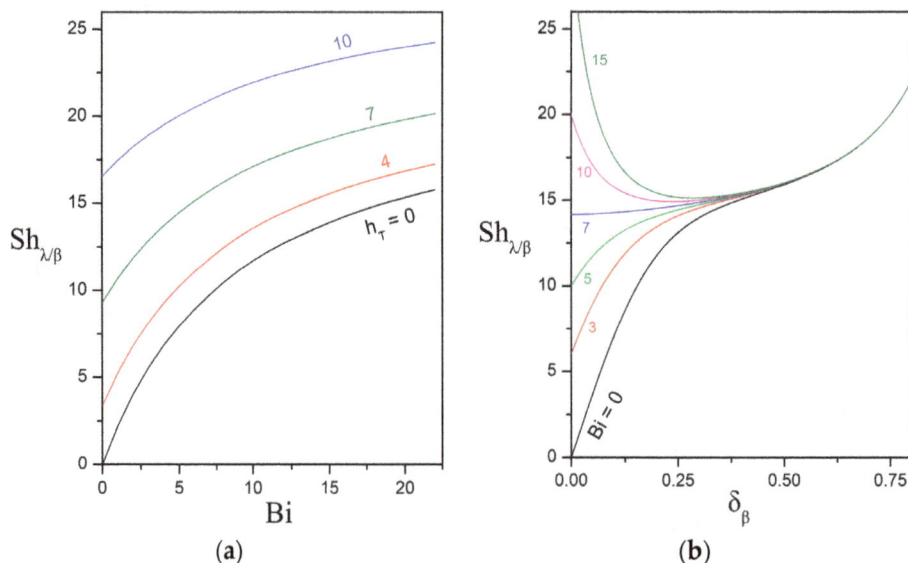

(a)   (b)

Figure 3. Dependence of the overall Sherwood number on: (**a**) the Biot and Thiele numbers for $\delta_\beta = 0.1$; and (**b**) the dimless biofilm thickness and the Biot number for $h_T = 6$. Also, $\Lambda_{A\beta} = 1$.

Figure 3b presents a very interesting finding. If the Biot number is below a critical value ($\text{Bi}_{crit} \approx 7$ in this figure), then the overall Sherwood number increases monotonically as the dimless biofilm thickness increases. On the other hand, if the Biot number is above the critical value, then the Sherwood number decreases as the biofilm thickness increases from zero up to a critical value $\delta_{\beta,crit}$ and, beyond that value, the Sherwood number re-increases with increasing biofilm thickness. In order to elucidate this behavior, a detailed examination of the expression given in Equation (34) for the overall Sherwood number is required. It has been established that

$$\text{Sh}_{\lambda/\beta} = 2\Lambda_{A\beta}h_T \left[\frac{h_T\tanh(h_T\delta_\beta) + \text{Bi} - 1}{h_T + (\text{Bi} - 1)\tanh(h_T\delta_\beta)} \right] + \frac{2\Lambda_{A\beta}}{1 - \delta_\beta}.$$

The first contribution contains the intertwined effects of transport and bioreaction in the biofilm and aqueous phases. With increasing biofilm thickness, this term increases or decreases monotonically up or down to the asymptotic value of $2\Lambda_{A\beta}h_T$, depending on the relative importance of the internal (biofilm) and external (aqueous) resistances to mass transport. If the internal resistance to mass transport is lower than the external resistance (i.e., sufficiently low Biot and high Thiele numbers), then the first contribution increases up to the asymptotic value as the biofilm thickness increases. In the opposite case of higher internal resistance (i.e., sufficiently high Biot and low Thiele numbers), the first contribution decreases down to the asymptotic value with increasing biofilm thickness.

The second term in the formula for the overall Sherwood represents the effect of curvature on the concentration gradient that is evaluated at the core surface and increases monotonically with increasing biofilm thickness. In particular, each point on the oily core surface is projected to a surface element of finite area on the outer surface of the compound particle. As the distance between the two concentric spherical surfaces increases, the degree of geometric expansion also increases and causes a dilution in the solute concentration at the outer surface which, in turn, results in a higher concentration gradient. This geometric effect does not exist for a flat surface configuration. The interplay between the two contributions produces the pattern shown in Figure 3b.

The critical Biot number, above which the biofilm shell acts as a *diffusive barrier* and hinders the transport of oil compounds from the surface of the core to the surrounding aqueous phase, can be determined by the following condition

$$\left[\frac{d\text{Sh}_{\lambda/\beta}}{d\delta_\beta} \right]_{\delta_\beta=0} = 0, \tag{57}$$

With reference to Figure 3b, the above condition states that the constant Biot curve for the critical Biot value is normal to the $\text{Sh}_{\lambda/\beta}$ axis with abscissa $\delta_\beta = 0$. The first derivative of the Sherwood number with respect to the dimless biofilm thickness is determined from Equation (34) and, after some operations, obtains the following form

$$\frac{d\text{Sh}_{\lambda/\beta}}{d\delta_\beta} = \frac{2\Lambda_{A\beta}h_T^2[h_T^2 - (\text{Bi} - 1)^2]}{[h_T\cos h(h_T\delta_\beta) + (\text{Bi} - 1)\sinh(h_T\delta_\beta)]^2} + \frac{2\Lambda_{A\beta}}{(1 - \delta_\beta)^2}. \tag{58}$$

Substitution of the above expression in Equation (57), gives the following expression for the critical Biot

$$\text{Bi}_{crit} = 1 + \sqrt{h_T^2 + 1}. \tag{59}$$

The critical biofilm thickness, below which the biofilm hinders mass transport, corresponds to the abscissa of the minimum in the constant Biot curves (for $\text{Bi} > \text{Bi}_{crit}$). Therefore, for given Biot and Thiele numbers, it can be determined as the root of the nonlinear algebraic equation that is obtained by setting the first derivative given in Equation (58) equal to zero. Another straightforward way to obtain an estimate for the critical biofilm thickness is to consider that at this value the first contribution in the Sherwood expression is *sufficiently close* to the asymptotic value of $2\Lambda_{A\beta}h_T$. Therefore, by demanding that $\tanh(h_T\delta_{\beta,crit}) = 0.99$, the following simple estimate

$$\delta_{\beta,crit} \approx \frac{2.65}{h_T}, \tag{60}$$

is obtained. The critical biofilm thickness depends also on the Biot number but, as can be seen in Figure 3b, the dependence is weak and, thus, the above estimate is sufficient for practical purposes.

3.2. Relative Importance of the Bioreaction and Dissolution Processes

Part of the oil that dissolves at the oil–biofilm interface is biodegraded within the biofilm and the rest is released into the water column; where it might, or might not, be further degraded by suspended microbes. A key issue concerns the bioreactive effectiveness of the biofilm shell. The amount of dissolved oil that ends up either biodegraded or released can be determined by the overall mass balances (Section 2.4).

The overall dissolution rate $\widetilde{W}_{A,\lambda/\beta}$ of oil at the surface of the oily core is given in Equation (32). The oil dissolution rate from the particle surface to the surrounding aqueous phase is given in Equation (26) and can be expressed in the following equivalent form

$$\widetilde{W}_{A,p/v} = \widetilde{k}_{p/v}\widetilde{S}_{\beta v}c_{A\beta}(1)H_{A,v/\beta}\widetilde{c}_{A,\lambda/\beta}, \tag{61}$$

using the relation $\widetilde{c}_{Av}\left(\widetilde{R}_P\right) = c_{A\beta}(1)\widetilde{c}_{A,\lambda/v} = c_{A\beta}(1)H_{A,v/\beta}\widetilde{c}_{A,\lambda/\beta}$, with the dimless concentration $c_{A\beta}(1)$ given in Equation (30). The rate of oil bioreaction within the biofilm can be expressed as

$$\widetilde{W}_{A,\beta} \equiv \int_{V_\beta} \widetilde{r}_{A\beta}d\widetilde{V} = \widetilde{k}_{1\beta}\widetilde{V}_\beta\langle\widetilde{c}_{A\beta}\rangle, \tag{62}$$

with the average concentration of oil in the biofilm shell given by Equation (36). The mass fractions of biodegraded oil in the biofilm and released oil in the water column are defined as

$$\Phi_{brn} \equiv \frac{\widetilde{W}_{A,\beta}}{\widetilde{W}_{A,\lambda/\beta}} = \frac{2\Lambda_{A\beta}J_C}{\left(1 - \delta_\beta\right)^2\mathrm{Sh}_{\lambda/\beta}}, \tag{63a}$$

$$\Phi_{dis} \equiv \frac{\widetilde{W}_{A,p/v}}{\widetilde{W}_{A,\lambda/\beta}} = \frac{2\Lambda_{A\beta}\mathrm{Bi}\,c_{A\beta}(1)}{\left(1 - \delta_\beta\right)^2\mathrm{Sh}_{\lambda/\beta}}, \tag{63b}$$

respectively, with $\Phi_{brn} + \Phi_{dis} = 1$. Figure 4 presents the effect of the Thiele modulus on the biodegraded and released oil fractions for different values of the Biot number and the thickness of the biofilm shell. It is observed that the mass fraction of biodegraded oil increases with increasing Thiele modulus, decreasing Biot number, and increasing biofilm thickness. As a consequence, biofilms composed of fast oil-degrading microbes and lipophilic extracellular matrix would be ideal for retaining and biodegrading oil compounds in practical applications.

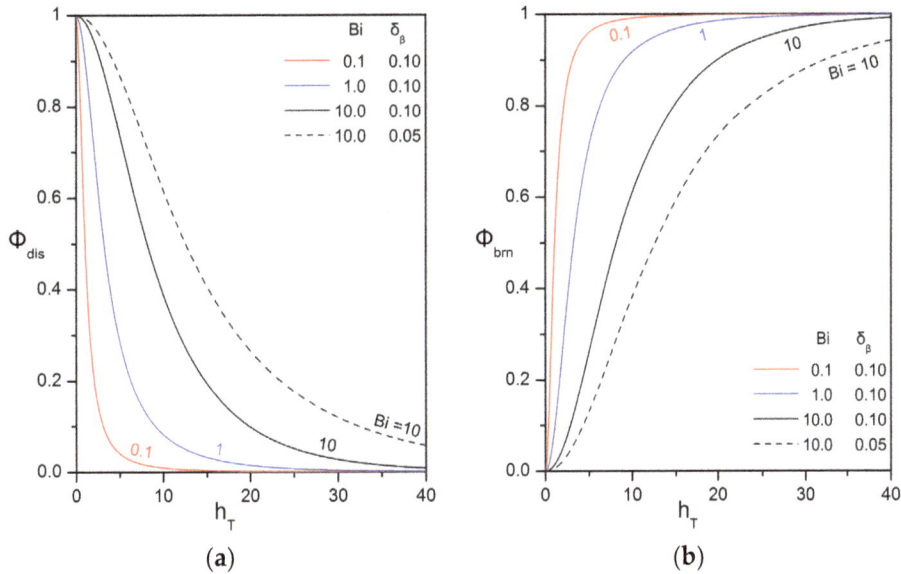

Figure 4. Impact of the Thiele and Biot numbers on: **(a)** the mass fraction of dissolved oil that is released into the aqueous phase; **(b)** the mass fraction of dissolved oil that is biodegraded within the biofilm. The values of the other parameters are: $\Lambda_{A\beta} = 1$; $H_{A,v/\beta} = 1$.

3.3. Impact of the Péclet and Thiele Numbers on the Particle Size Evolution

The evolution of the dimensions of the compound particle is determined by all the factors that affect the dissolution of oil into the surrounding phases, the direct uptake of oil at the surface of the oily core, and the volumetric growth of the biofilm phase. First, we examine the effects of the bioreaction in the biofilm (expressed by the Thiele number) and of the particle velocity (expressed by the Péclet number), while considering that the rates of direct uptake and biofilm growth are nil.

It is very convenient to use the dimless form of the coupled ODEs given in Equations (53a) and (53b), as it is only required to define certain dimless quantities without specifying the values of solubilities, kinetic and other system parameters. Figure 5 presents the strong effect of the initial Thiele number on the evolution of the particle dimensions, while keeping all other parameters constant. As expected, higher Thiele numbers result in higher shrinking rates and faster consumption of the oily core. An interesting feature is that the temporal change in the dimensions of the particle is non-linear.

For a given Thiele number, the diameter of the oily core decreases with an increasing shrinking rate (Figure 5a, concave function), whereas the diameter of the compound particle decreases with a decreasing shrinking rate (Figure 5b, convex function) until reaching an asymptotic value that corresponds to a residual particle that contains only biofilm.

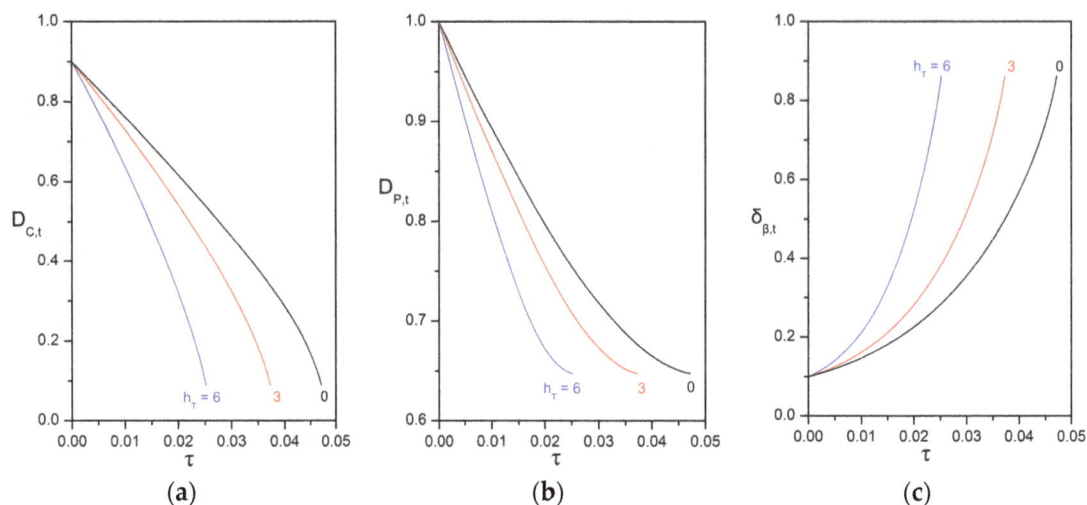

Figure 5. Impact of the initial Thiele modulus on the evolution of: (**a**) the dimless diameter of the oily core; (**b**) the dimless diameter of the compound particle; (**c**) the dimless thickness of the biofilm shell. The values of the other parameters are: $\delta_{\beta,0} = 0.1$; $Pe_{v,0} = 100$; $Da_v = 0$; $\Lambda_{A\beta} = 1$; $H_{A,v/\beta} = 1$; $Y_{\beta/A} = 0$; $k_{srn} = 0$; velocity mode = "const. U".

Figure 6 presents the effect of the initial Péclet number on the evolution of the particle dimensions, for the case of an inert biofilm shell (non-reactive, non-growing) and while keeping all other parameters constant. Upon the assumption that the diffusion coefficient and the initial particle size are held constant, the different Péclet numbers are associated with different characteristic velocities. Two cases are considered for the mode of change in the characteristic velocity as the particle shrinks. In the "free rise/fall" mode, the particle is considered to rise or fall in the water column under the action of gravity with the velocity given by Stokes' formula in Equation (51) and the Péclet number updated according to Equation (56).

In the "const. U" mode, the particle is considered to be carried by the aqueous stream at constant velocity. First of all, as expected, a higher Péclet number leads to faster dissolution and shrinkage of the oily core. Furthermore, for a given initial Péclet number, the velocity mode appears to have an appreciable effect on the shrinking rate. In particular, the "free rise/fall" mode exhibits an interesting dependence on the biofilm density (Figure 7).

Typically, the density of the oily phase is lower than that of the surrounding aqueous phase. Therefore, oil droplets without a biofilm shell tend to rise when released in a water column. The situation is different for compound droplets because the biofilm density shows significant variability as it depends on the content and type of cells and extracellular polymers. If the biofilm density is lower than or similar to the density of the aqueous phase, then the compound particle rises with decreasing velocity and Péclet number as the oily core is consumed over time (cases of $\rho_\beta \leq 1$ in Figure 7). On the other hand, if the biofilm density is higher than the density of the aqueous phase, then the compound particle rises until it becomes neutrally buoyant and, thereafter, begins to sink with increasing velocity as the oily core is shrinking (cases of $\rho_\beta > 1$ in Figure 7).

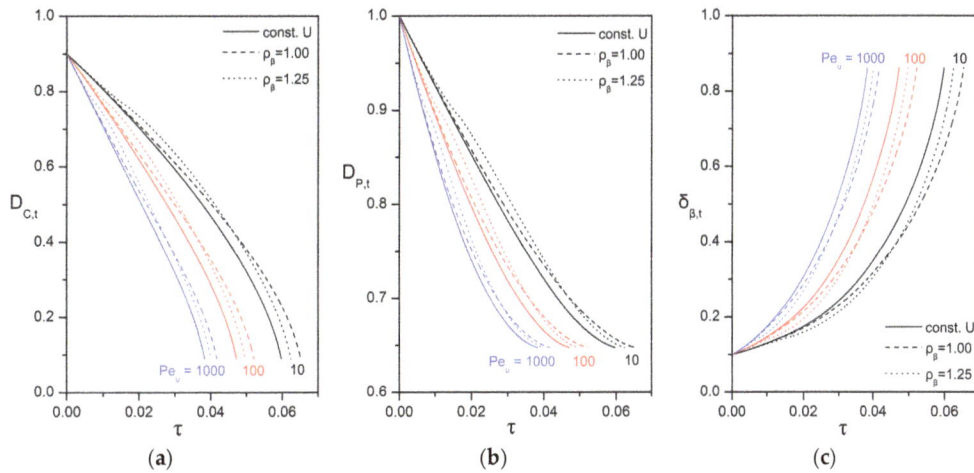

Figure 6. Impact of the velocity mode and the initial Péclet number on the evolution of: (**a**) the dimless diameter of the oily core; (**b**) the dimless diameter of the compound particle; (**c**) the dimless thickness of the biofilm shell. For each Péclet number, the continuous line corresponds to a compound particle that moves with constant velocity. For the free rise/fall mode: $\rho_\beta = 1.00$ (dashed line); $\rho_\beta = 1.25$ (dotted line). The values of the other parameters are: $\delta_{\beta,0} = 0.1$; $h_T = 0$; $Da_v = 0$; $\Lambda_{A\beta} = 1$; $H_{A,v/\beta} = 1$; $k_{srn} = 0$; $Y_{\beta/A} = 0$; $\rho_\lambda = 0.85$.

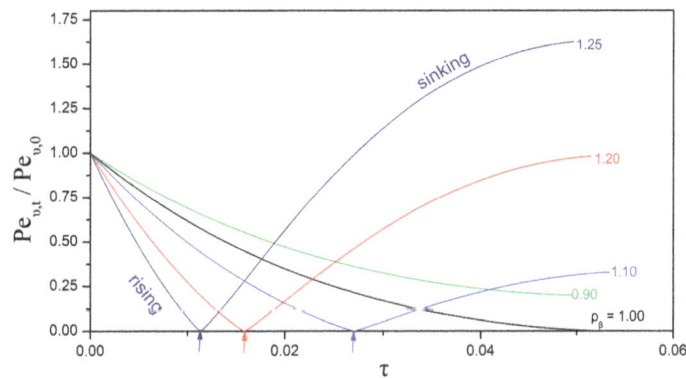

Figure 7. Impact of the biofilm density on the evolution of the Péclet number for a shrinking compound particle that freely rises/falls in a water column. The compound particle becomes neutrally buoyant at the time instants indicated by the arrows. The parameter values are: $Pe_{v,0} = 100$; $\delta_{\beta,0} = 0.1$; $h_T = 0$; $Da_v = 0$; $\Lambda_{A\beta} = 1$; $H_{A,v/\beta} = 1$; $k_{srn} = 0$; $Y_{\beta/A} = 0$; $\rho_\lambda = 0.85$.

3.4. Impact of Biofilm Growth and Direct Uptake on the Particle Size Evolution

The effect of a growing biofilm on the temporal evolution of the particle configuration is somewhat intricate as a thicker biofilm might not always enhance the droplet shrinking and oil biodegradation rates. If $H_{A,v/\beta}/\Lambda_{A\beta} < 1.2$, that is if the oil is more soluble and mobile in the biofilm than in the aqueous phase, then the internal resistance to mass transport is also lower than the external resistance ($Bi < Bi_{crit}$) and a net increase in the amount of biofilm due to growth results in higher rates of oil dissolution and droplet shrinking (Figure 8). Under such conditions, the enhancement of the biodegradation process is more profound with thicker biofilms, i.e., for higher values of the biofilm yield coefficient $Y_{\beta/A}$. In Figures 8 and 9, the value of $Y_{\beta/A} = 0$ corresponds to the case where the initial amount of biofilm remains constant over time and is just redistributed around the oily core as the latter shrinks. For $Y_{\beta/A} > 0$, additional volume of biofilm is produced.

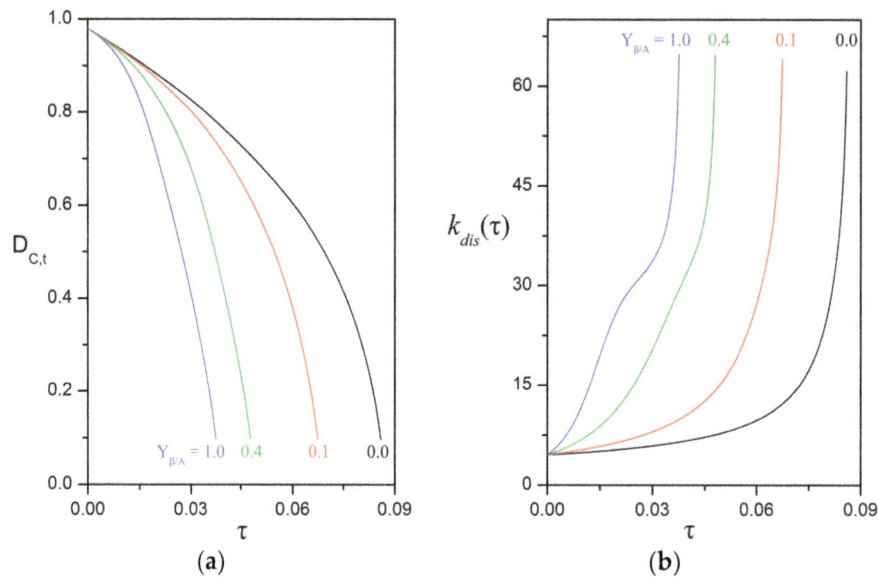

Figure 8. Impact of the biofilm yield coefficient $Y_{\beta/A}$ on the temporal evolution of: (**a**) the dimless diameter of the oily core; and (**b**) the dimless dissolution rate; for $H_{A,v/\beta} = 0.1$; $\Lambda_{A\beta} = 1$. The other parameters are: $Pe_{v,0} = 100$; $\delta_{\beta,0} = 0.02$; $h_T = 6$; $Da_v = 0$; $k_{srn} = 0$; velocity mode = "const. U".

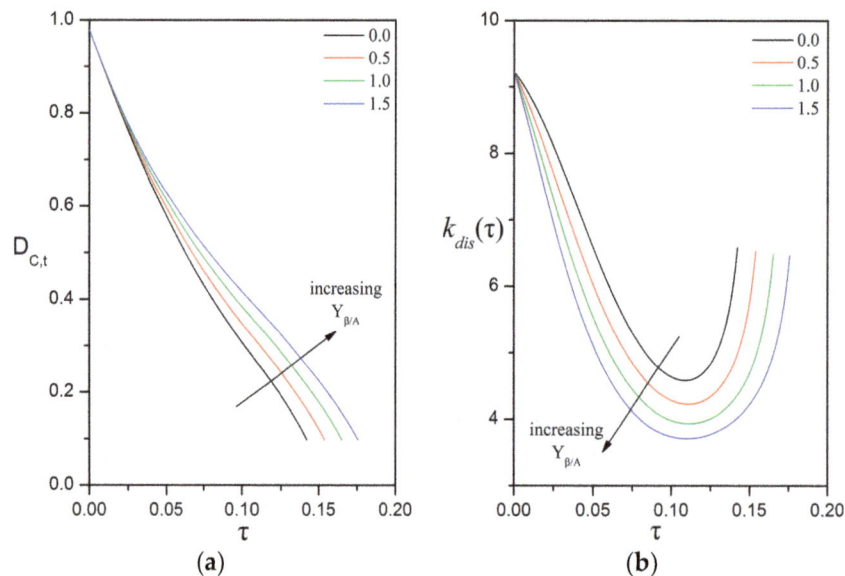

Figure 9. Impact of the biofilm yield coefficient $Y_{\beta/A}$ on the temporal evolution of: (**a**) the dimless diameter of the oily core; and (**b**) the dimless dissolution rate; for $H_{A,v/\beta} = 1.0$; $\Lambda_{A\beta} = 0.1$. The other parameters are: $Pe_{v,0} = 100$; $\delta_{\beta,0} = 0.02$; $h_T = 6$; $Da_v = 0$; $k_{srn} = 0$; velocity mode = "const. U".

On the other hand, if $H_{A,v/\beta}/\Lambda_{A\beta} > 1.2$, then the internal resistance to mass transport is higher than the external resistance (Bi > Bi_{crit}) and a net increase in the amount of biofilm due to growth results in lower rates of oil dissolution and droplet shrinking as compared to the non-growing case (Figure 9). The peculiar trend observed in Figure 9b for the dissolution rate is attributed to the role of biofilm as a diffusive barrier, which is discussed in detail in Section 3.1. In short, the dissolution rate is defined in Equation (54) as $k_{dis}(\tau) = 2Sh_{\lambda/\beta}/D_{P,t}$ and follows closely the dependence of the Sherwood number on the dimless biofilm thickness. As the oily core shrinks, the dimless biofilm thickness increases and, for Bi > Bi_{crit}, the Sherwood number decreases. This occurs until the biofilm exceeds the critical biofilm thickness, $\delta_\beta > \delta_{\beta,crit}$, so as the enhancement caused by the curvature effect to supersede the attenuation caused by the diffusive barrier. Thereafter, the Sherwood number increases with increasing biofilm thickness (see also Figure 3b and the discussion in Section 3.1).

Figure 10 illustrates the potential effect of the biodegradation and dissolution processes on the shrinking rate of the oily core. The parameter values are typical for the applications under consideration (see a detailed discussion in the next section), and have also been selected so as to exemplify that each mechanism might have a significant impact on the overall process. In real world applications, one or two or all mechanisms might act in parallel. For the direct uptake mechanism, the value of $k_{srn} = 10$ is obtained in Equation (54) by setting the direct uptake rate $\widetilde{k}_{srn} = 0.2$ nm/s (Table 1), the initial droplet diameter $\widetilde{D}_{P,0} = 100$ μm, the oil density $\widetilde{\rho}_\lambda = 0.85$ g/cm^3, the oil diffusivity in water $\widetilde{D}_{Av} = 10^{-6}$ cm^2/s, and the oil solubility in biofilm $\widetilde{c}_{A,\lambda/\beta} = 17$ μg/cm^3.

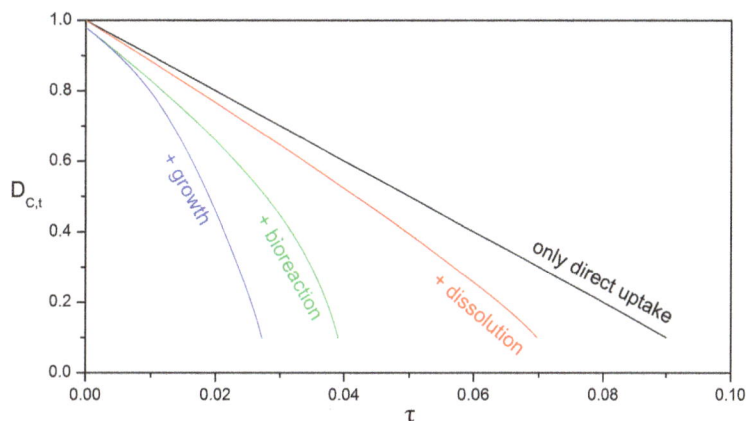

Figure 10. Impact of biodegradation and dissolution mechanisms on the temporal evolution of the dimless diameter of the oily core. For the scenario with only direct interfacial uptake (black line), the values of the parameters are: $k_{srn} = 10$; $Pe_{v,0} = 0$; $\delta_{\beta,0} = 0.0$; $h_T = 0$; $Y_{\beta/A} = 0$. For the scenario with added dissolution (red line), the parameters are: $k_{srn} = 10$; $Pe_{v,0} = 100$; $\delta_{\beta,0} = 0.0$; $h_T = 0$; $Y_{\beta/A} = 0$. For the scenario with added bioreaction in the biofilm (green line), the parameters are: $k_{srn} = 10$; $Pe_{v,0} = 100$; $\delta_{\beta,0} = 0.02$; $h_T = 6$; $Y_{\beta/A} = 0$. For the scenario with added biofilm growth (blue line), the parameters are: $k_{srn} = 10$; $Pe_{v,0} = 100$; $\delta_{\beta,0} = 0.02$; $h_T = 6$; $Y_{\beta/A} = 1$. In all scenarios, the other parameters are: $Da_v = 0$; $H_{A,v/\beta} = 0.1$; $\Lambda_{A\beta} = 1$; velocity mode = "const. U".

Table 1. Bioreaction constants ($\widetilde{k}_{1v}$; $\widetilde{k}_{1\beta}$; $\widetilde{k}_{srn}$) based on the kinetic parameters reported by *Vilcáez* et al. [32] for groups of hydrocarbons: ALK = alkanes; BTEX = monoaromatics (benzene, toluene, ethylbenzene, xylene); PAH = polyaromatics (naphthalene, fluorene, anthracene, etc.). For the calculation of the reaction constants, the cell concentrations are: $\widetilde{B}_v = 10^6$ cells/cm^3, $\widetilde{B}_\beta = 10^9$ cells/cm^3, $\widetilde{B}_{\lambda\beta} = 10^8$ cells/cm^2.

oil	$\widetilde{\mu}_m \left[h^{-1}\right]$	$\widetilde{K}_S \left[\frac{mg}{cm^3}\right]$	$Y_{C/A} \left[\frac{cells}{mg-oil}\right]$	$\widetilde{k}_{1\text{Æ}} \left[h^{-1}\right]$	$\widetilde{k}_{1\text{fi}} \left[h^{-1}\right]$	$\widetilde{k}_{srn} \left[\frac{\mu m}{s}\right]$
ALK	0.600	0.086	$1.25 \cdot 10^8$	0.056	55.8	0.00314
BTEX	0.320	0.129	$1.25 \cdot 10^8$	0.020	19.8	0.00168
PAH	0.053	0.028	$1.25 \cdot 10^8$	0.015	15.1	0.00028

3.5. Implications for the Biodegradation of Crude Oil Microdroplets in the Sea

At this point, a naturally arising question concerns the values of the characteristic dimless moduli and other system parameters for real world applications. First of all, as mentioned in the introduction, the microdroplet might be rising, sinking or drifting along underwater sea currents. For light crude oil microdroplets with a diameter in the range of $\widetilde{D}_{P,0} = 10 - 100$ μm and a density of $\widetilde{\rho}_\lambda = 0.85$ g/cm^3, freely *rising* due to buoyancy through an aqueous water column with density $\widetilde{\rho}_v = 1.02$ g/cm^3 and dynamic viscosity $\widetilde{\mu}_v = 0.01$ g/(cm·s), the Stokes velocity given by Equation (51) is in the range of $\widetilde{U}_{S,0} \cong 9 \cdot 10^{-4} - 0.09$ cm/s and the corresponding radius-based Reynolds number is $\text{Re}_v \approx 5 \cdot 10^{-5} - 5 \cdot 10^{-2}$. The density of the biofilm is expected to be similar to or larger than the density of the aqueous phase, depending on the type and volume fraction of cells and extracellular biopolymers within the biofilm. For instance, the density of marine snow particles collected from the Gulf of Mexico after the Deepwater Horizon incident exhibited great variability with values of *excess density* ranging from 0.07 g/cm^3 to 0.36 g/cm^3 [49]. Therefore, for compound particles with a diameter in the range of $10 - 100$ μm and density $\widetilde{\rho}_P = 1.25$ g/cm^3, the *settling* Stokes velocity is also on the order of $10^{-3} - 10^{-1}$ cm/s. In the case of (almost) neutrally buoyant particles, the characteristic velocity is determined by the velocity of the underwater sea current that carries the particles. The velocity of sea currents varies over several orders with a magnitude of a few mm/s for the vertical velocity [50], several cm/s for the horizontal velocity in deep sea (depth larger than 300 m) [6,50–52], and from tenths of cm/s up to a few m/s for the horizontal velocity near the sea surface [50–52]. For example, in the Gulf of Mexico, values of $2 - 3$ mm/s have been measured for the vertical velocity [50] and an average value of 7.8 cm/s has been reported for the horizontal velocity at a depth of 1100 m [6]. Thus, for compound particles drifting along an underwater current with a velocity in the range of $0.1 - 10$ cm/s, the Reynolds number is $\text{Re}_v \approx 5 \cdot 10^{-3} - 5$.

At atmospheric conditions, the interfacial tension between crude oil and water is in the range of $\widetilde{\gamma}_{\lambda v} = 10 - 30$ dyn/cm, depending on the detailed composition of the two phases [53,54]. An even higher tension might be expected between a compound particle and water, especially if the biofilm that covers the oily core is strongly hydrophobic. For instance, it has been recently found that microbes of the species *Bacillus subtilis* secrete a hydrophobic protein called BslA, which accumulates in the outer layer of the biofilm and results in strong repellence of aqueous drops (contact angle > 90°) [48,55]. Nonetheless, to the best of our knowledge, specific values of the interfacial tension for biofilm–water systems have not been reported yet. For a characteristic velocity in the range of $10^{-3} - 10$ cm/s and an interfacial tension of 20 dyn/cm the capillary number is $\text{Ca} = 5 \cdot 10^{-7} - 5 \cdot 10^{-3}$. Furthermore, for compound particles with a diameter in the range of $10 - 100$ μm and density of 1.25 g/cm^3 the Bond number is $\text{Bo} = 3 \cdot 10^{-6} - 3 \cdot 10^{-4}$. Finally, for a characteristic velocity in the range of $10^{-3} - 10$ cm/s, and for dissolved oil components with a diffusion coefficient on the order of $\widetilde{D}_{Av} = 10^{-6}$ cm^2/s, the Péclet number is $\text{Pe}_v = 0.5 - 5 \cdot 10^4$. In view of the above data, it is concluded that the hypotheses set out in Section 2.2 for the hydrodynamics and solute transport problems are reasonable for most of the cases considered in this work. Future extension of the CPM formulation to low-but-finite Reynolds numbers and retention of all terms in the solute mass balance is expected to improve substantially the domain of validity of the proposed model.

With regard to the bioreaction kinetic parameters, a major issue is raised. A theoretical model, focused on a specific spatial scale, requires the input of reaction rates and parameters which precisely correspond to the scale of focus. Typically, in oil biodegradation experiments, the physical state of the oil is not taken into explicit account and an *apparent* biodegradation rate is determined in terms of an average oil concentration that lumps together all forms of oil, i.e., dissolved, micellar, and/or dispersed in droplets. This lumped concentration and its spatial and temporal derivatives differ from the concentration field that is detected by the microbes in their microenvironment [56]; in the present context, from the concentrated oil detected by flatlanders at the oil–water interface, the concentration $\widetilde{c}_{Av}$ detected by drifters in the bulk aqueous phase, and the concentration $\widetilde{c}_{A\beta}$ detected by biofilm formers within the biofilm. Here, the term "apparent" is used to denote a reaction rate that pertains to a representative elementary volume with dimensions much larger than the size of a single microdroplet or a single microbial cell. For systems with microscale heterogeneity, such as porous media and multiphase dispersions, the apparent reaction rate incorporates the effects of the microscale structure and transport mechanisms and is, therefore, lower than the intrinsic (transport-free) reaction rate [57]. This important issue has recently attracted the attention of researchers working on the biodegradation of crude oil [58–64]. For example, it has been nicely demonstrated that the apparent biodegradation rate increases with decreasing droplet size, while keeping all other system parameters constant [30,58,61]. However, existing kinetic data for apparent reaction rates of oily compounds are not consistent with the CPM formulation, and can only be used to obtain *lower bounds* for the Damköhler and Thiele numbers.

Recent studies on the biodegradation of crude oil that was released by the Macondo well in the Gulf of Mexico after the Deepwater Horizon blowout, report that the apparent *half-life* of many biodegradable hydrocarbons is in the range of $\widetilde{T}_{1/2} = 0.1 - 10$ d (at ~5 °C) [7,30,62,63]. The first-order reaction constant is related to the half-life as $\widetilde{k}_{1,\alpha} = \ln 2 / \widetilde{T}_{1/2}$ and obtains values in the range of $\widetilde{k}_{1,\alpha} \approx 0.7 - 7 d^{-1}$. Therefore, for microdroplets of diameter $\widetilde{D}_{P,0} = 100$ μm and water diffusivity $\widetilde{D}_{Av} = 10^{-6}$ cm^2/s, the Damköhler number for the aqueous phase obtains values in the range of $Da_v = 2 \cdot 10^{-5} - 2 \cdot 10^{-3}$. The diffusion coefficient of oil compounds in the interstitial space of biofilms is expected to be lower, by one or more orders of magnitude, than the water diffusivity [46,47] and, thus, the Thiele number for the biofilm phase is in the range of $h_T = 0.0045 - 0.045$ (assuming $\Lambda_{A\beta} = 0.1$). According to the CPM formulation, Equation (4), the first-order reaction constant depends not only on the kinetic parameters ($\widetilde{\mu}_{m,\alpha}$, $\widetilde{K}_{S,\alpha}$, $Y_{C/A,\alpha}$), but also on the concentration $\widetilde{B}_\alpha$ of active microbial cells. In natural ecosystems, the concentration of cells residing in biofilms might be several orders higher than the concentration of suspended cells. For example, after the Deepwater Horizon incident, the concentration of marine microbes in the underwater oil plume was $\widetilde{B}_v = 10^4 - 10^6$ cells/cm^3 [7], whereas the cell concentration might reach values of $\widetilde{B}_\beta = 10^8 - 10^{10}$ cells/cm^3 within marine snow and biofilms, depending on the size of individual cells and the cell volume fraction. *Vilcáez* et al. [32] developed a theoretical model for the biodegradation of droplet populations on the basis of a shrinking particle model that accounts only for the direct uptake mode. They also analyzed kinetic data from the literature and reported lumped parameters for three groups of hydrocarbons, namely alkanes, monoaromatics (BTEX), and polyaromatics (PAHs). For 100 μm-sized droplets, the data of *Vilcáez* et al. (Table 1) give values in the range of $Da_v = 0.0001 - 0.0003$ for the Damköhler number in the aqueous phase and $h_T = 1.02 - 1.96$ for the Thiele number in the biofilm phase. Consequently, based on the available data, bioreaction is expected to be of considerable importance within the biofilm phase, whereas it is dominated by advection and diffusion in the aqueous phase. Nonetheless, because of the significant uncertainty with regard to the consistency between the CPM formulation and existing kinetic data for oily substrates, this discussion must be extended once data from microscale experiments become available.

4. Conclusions

In this paper, a compound particle model of the core-shell type is developed for the microbial degradation of solitary oil microdroplets and takes into account three fundamental biodegradation modes, namely the direct interfacial uptake at the oil surface, the bioreaction in the bulk aqueous phase, and the bioreaction in a biofilm formed around the droplet. Previous relevant models account only for the direct uptake mode. The major results of the theoretical analysis include an expression for the overall dissolution rate for a given particle configuration and two coupled ordinary differential equations for the evolution of the dimensions of the compound particle. An interesting finding is that biofilms consisting of a high concentration of fast oil-degrading microbes and lipophilic biopolymers (corresponding to a low-Biot and high-Thiele regime) are expected to be ideal for oil biodegradation applications because they retain the dissolved oil until complete degradation, instead of releasing it into the water column. The model is based on a large set of simplifying, yet justifiable, hypotheses; most of which rely on the consideration of microsized droplets with an immobilized interface by the presence of microbes and biopolymers. One of the most important hypotheses in the model is that the compound particle moves like a rigid sphere. This hypothesis negates the need to define the biofilm mechanics, which might range from a fluid-like to a solid-like behavior, depending on the composition of the biofilm and the applied stresses [65–68]. Another very important hypothesis is that the oily phase is treated as a single component, whereas crude oil and most natural or artificial oils are multi-component mixtures. Selective or faster biodegradation of certain components (e.g., alkanes) will result in concentration gradients within the oil droplet. The effects of these gradients on the droplet shrinking rate is expected to be minimal only if intra-droplet diffusion is much faster than diffusion in the surrounding biofilm phase (i.e., $\widetilde{D}_{A\lambda}/\widetilde{D}_{A\beta} \gg 1$). Besides the development of concentration gradients, selective biodegradation will also change the mass fraction of each oil compound within the droplet. In turn, the density of the droplet, ρ_λ, will also vary with time. For instance, if light alkanes are consumed faster than heavier compounds like PAHs, then the velocity of a compound particle undergoing free rise will decrease faster and the impact on the temporal evolution of the Péclet number might be appreciably stronger than that shown in Figure 7. These hypotheses will be relaxed in future work by adding more physical and mathematical complexity into the model formulation.

Acknowledgments: This work was completed in the context of the H2020-MSCA-IF project "OILY MICROCOSM" and received funding from the European Union's Horizon 2020 research and innovation programme under the Marie Sklodowska-Curie grant agreement No 741799.

Author Contributions: All authors have made substantial contribution to the development of the model, the analysis of the results, and the writing of the manuscript.

Conflicts of Interest: The authors declare no conflict of interest.

Nomenclature

$\widetilde{B}_\alpha$	concentration of active cells in the αth phase, $[\mathbf{cells \cdot cm^{-3}}]$;
$\widetilde{B}_{\lambda\beta}$	interfacial concentration of active cells, $[\mathbf{cells \cdot cm^{-2}}]$;
Bi	Biot number, $\mathbf{Bi} = H_{A,v/\beta}\widetilde{R}_P\widetilde{k}_{p/v}/\widetilde{D}_{A\beta} = \left(H_{A,v/\beta}/\Lambda_{A\beta}\right)\mathbf{Sh}_{p/v}/2$;
$\widetilde{c}_{A\alpha}$	mass concentration of oil in the αth phase, $[\mathbf{g \cdot cm^{-3}}]$, dimless $c_{A\alpha} = \widetilde{c}_{A\alpha}/\widetilde{c}_{A,\lambda/\alpha}$;
$\widetilde{c}_{A,\lambda/\alpha}$	solubility of oil in the αth phase, $[\mathbf{g \cdot cm^{-3}}]$;
$\left\langle \widetilde{c}_{A\beta}\right\rangle$	volume averaged concentration of oil in the β-phase, Equation (35), $[\mathbf{g \cdot cm^{-3}}]$;
$\widetilde{D}_{A\alpha}$	diffusion coefficient of the A solute in the αth phase, $[\mathbf{cm^2 \cdot s^{-1}}]$;
$\mathbf{Da}_\alpha$	Damköhler number in the αth phase, $\mathbf{Da}_\alpha = \widetilde{k}_{1\alpha}\widetilde{R}_P^2/\widetilde{D}_{A\alpha}$;
$\widetilde{D}_{c,t}$	diameter of the oily core, $[\mathbf{cm}]$, dimless $D_{c,t} = \widetilde{D}_{c,t}/\widetilde{D}_{P,0}$;
$\widetilde{D}_{P,t}$	diameter of the compound particle, $[\mathbf{cm}]$, dimless $D_{P,t} = \widetilde{D}_{P,t}/\widetilde{D}_{P,0}$;
$\widetilde{g}$	gravitational acceleration, $[\mathbf{cm \cdot s^{-2}}]$;
$H_{A,v/\beta}$	solubility ratio, $H_{A,v/\beta} = \widetilde{c}_{A,\lambda/v}/\widetilde{c}_{A,\lambda/\beta}$;
Ha	Hatta modulus, $\mathbf{Ha} = 2\sqrt{\mathbf{Da}_v}/\mathbf{Sh}^0_{p/v}$;
h_T	Thiele number in the biofilm shell, $h_T = \widetilde{R}_P\sqrt{\widetilde{k}_{1\beta}/\widetilde{D}_{A\beta}} = \sqrt{\mathbf{Da}_\beta}$;
$\widetilde{J}_{A\alpha}$	mass flux of oil in the αth phase, $[\mathbf{g \cdot cm^{-2} \cdot s^{-1}}]$, dimless $J_{A\alpha} = \widetilde{J}_{A\alpha}/(\widetilde{D}_{A\alpha}\widetilde{c}_{A,\lambda/\alpha}/\widetilde{R}_P)$;
$\widetilde{k}_{1\alpha}$	first-order reaction rate constant in the αth phase, Equation (4), $[\mathbf{s^{-1}}]$;
$\widetilde{k}^0_{p/v}$	mass transfer coefficient for external mass transfer with $\mathbf{Da}_v = \mathbf{0}$, Equation (21), $[\mathbf{cm \cdot s^{-1}}]$;
$\widetilde{k}_{p/v}$	mass transfer coefficient for external mass transfer, Equation (24), $[\mathbf{cm \cdot s^{-1}}]$;
$\widetilde{k}_{\lambda/\beta}$	mass transfer coefficient for the dissolution of the oily core, Equation (33), $[\mathbf{cm \cdot s^{-1}}]$;
$\widetilde{k}_{dis}$	droplet shrinking rate caused by dissolution, Equation (50b), $[\mathbf{cm \cdot s^{-1}}]$, dimless Equation (54);
$\widetilde{k}_{grt}$	biofilm expansion rate due to growth, Equation (50c), $[\mathbf{cm \cdot s^{-1}}]$, dimless Equation (54);
$\widetilde{k}_{srn}$	droplet shrinking rate caused by direct uptake, Equation (50a), $[\mathbf{cm \cdot s^{-1}}]$, dimless Equation (54);
$\widetilde{K}_{S,\alpha}$	half-saturation constant for the A solute in the αth phase, $[\mathbf{g \cdot cm^{-3}}]$;
$\mathbf{n}_{\alpha\omega}$	unit normal vector on the $\alpha\omega$-interface pointing from the α-phase to the ω-phase;
$\widetilde{r}$	radial coordinate, $[\mathbf{cm}]$, dimless $r = \widetilde{r}/\widetilde{R}_P$;
$\widetilde{r}_{A,\alpha}$	oil consumption rate in the αth phase, $[\mathbf{g \cdot cm^{-3} \cdot s^{-1}}]$;
$\widetilde{r}_{c,\alpha}$	microbial cell proliferation rate in the αth phase, $[\mathbf{cells \cdot cm^{-3} \cdot s^{-1}}]$;
$\widetilde{r}_\beta$	biofilm production rate, $\widetilde{r}_\beta = \widetilde{r}_{c,\beta}/Y_{c/\beta}$, $[\mathbf{g \cdot cm^{-3} \cdot s^{-1}}]$;
$\mathbf{Pe}_\alpha$	Péclet number in the αth phase, $\mathbf{Pe}_\alpha = \widetilde{R}_P\widetilde{U}/\widetilde{D}_{A\alpha}$;
$\widetilde{R}_c$	radius of the oily core, $[\mathbf{cm}]$, dimless $R_c = \widetilde{R}_c/\widetilde{R}_P$;
$\widetilde{R}_P$	radius of the compound particle, $[\mathbf{cm}]$, dimless $R_P = \widetilde{R}_P/\widetilde{R}_P = \mathbf{1}$;
$\widetilde{S}_{\beta v}$	area of the compound particle surface, $\widetilde{S}_{\beta v} = 4\pi\widetilde{R}_P^2$, $[\mathbf{cm^2}]$;
$\widetilde{S}_{\lambda\beta}$	area of the oily core surface, $\widetilde{S}_{\lambda\beta} = 4\pi\widetilde{R}_c^2$, $[\mathbf{cm^2}]$;
$\mathbf{Sh}^0_{p/v}$	Sherwood number for external mass transfer with $\mathbf{Da}_v = \mathbf{0}$, Equation (22);
$\mathbf{Sh}_{p/v}$	Sherwood number for external mass transfer, Equation (24);
$\mathbf{Sh}_{\lambda/\beta}$	overall Sherwood number for the dissolution of the oily core, Equation (34);
$\widetilde{U}$	undisturbed velocity of the approaching fluid, $[\mathbf{cm \cdot s^{-1}}]$;

$\widetilde{V}_\beta$	volume of the biofilm shell, $\widetilde{V}_\beta = \pi(\widetilde{D}_P^3 - \widetilde{D}_c^3)/6$, $[\text{cm}^3]$;
$\widetilde{v}_\upsilon$	velocity of the aqueous fluid, $[\text{cm}\cdot\text{s}^{-1}]$, dimless $\mathbf{v}_\upsilon = \widetilde{\mathbf{v}}_\upsilon/\widetilde{U}$;
$Y_{c/\beta}$	number of active cells per unit biofilm mass, [cells/mg-biofilm];
$Y_{c/A,\alpha}$	yield coefficient of cells in the αth phase, [cells/mg-oil];
$Y_{\beta/A}$	biofilm yield coefficient, $Y_{\beta/A} = Y_{c/A,\beta}/Y_{c/\beta}$, [mg-biofilm/mg-oil];
$\widetilde{W}_{A,p/\upsilon}$	external mass transfer rate: from the particle surface to the υ-phase, Equation (26), $[\mathbf{g}\cdot\mathbf{s}^{-1}]$;
$\widetilde{W}_{A,\lambda/\beta}$	overall dissolution rate at the surface of the oily core, Equation (32), $[\mathbf{g}\cdot\mathbf{s}^{-1}]$;

Greek letters

$\widetilde{\delta}_\beta$	thickness of the biofilm shell, $\widetilde{\delta}_\beta = \widetilde{R}_P - \widetilde{R}_c$, [cm];
δ_β	dimless thickness of the biofilm shell, $\delta_{\beta,t} = \widetilde{\delta}_{\beta,t}/\widetilde{R}_{P,t} = 1 - \widetilde{R}_{c,t}/\widetilde{R}_{P,t}$;
$\Delta\widetilde{\rho}$	excess density, $\Delta\widetilde{\rho} = |\widetilde{\rho}_\upsilon - \widetilde{\rho}_P|$, $[\mathbf{g}\cdot\mathbf{cm}^{-3}]$;
$\widetilde{\mu}_{m,\alpha}$	maximum specific growth rate of active cells, $[\mathbf{s}^{-1}]$;
$\widetilde{\mu}_\upsilon$	viscosity of the υ-phase, $[\mathbf{g}\cdot\mathbf{cm}^{-1}\cdot\mathbf{s}^{-1}]$;
$\Lambda_{A\beta}$	diffusivity ratio, $\Lambda_{A\beta} = \widetilde{D}_{A\beta}/\widetilde{D}_{A\upsilon}$;
$\widetilde{\rho}_\alpha$	density of the αth phase, $[\mathbf{g}\cdot\mathbf{cm}^{-3}]$, dimless $\varrho_\alpha = \widetilde{\rho}_\alpha/\widetilde{\rho}_\upsilon$;
$\widetilde{\tau}_D$	scaled characteristic diffusion time, Equation (52), [s];
τ	dimless time, $\tau = \widetilde{t}/\widetilde{\tau}_D$;
Φ_{brn}	mass fraction of oil biodegraded within the biofilm shell, Equation (63a);
Φ_{dis}	mass fraction of oil released into the aqueous phase, Equation (63b);

References

1. Li, C.; Miller, J.; Wang, J.; Koley, S.S.; Katz, J. Size distribution and dispersion of droplets generated by impingement of breaking waves on oil slicks. *J. Geophys. Res. Oceans* **2017**, *122*, 7938–7957. [CrossRef]

2. Nissanka, I.D.; Yapa, P.D. Oil slicks on water surface: Breakup, coalescence, and droplet formation under breaking waves. *Mar. Pollut. Bull.* **2017**, *114*, 480–493. [CrossRef] [PubMed]

3. Johansen, Ø.; Rye, H.; Cooper, C. DeepSpill—Field study of a simulated oil and gas blowout in deep water. *Spill Sci. Technol. Bull.* **2003**, *8*, 433–443. [CrossRef]

4. Brandvik, P.J.; Johansen, Ø.; Leirvik, F.; Farooq, U.; Daling, P.S. Droplet breakup in subsurface oil releases—Part 1: Experimental study of droplet breakup and effectiveness of dispersant injection. *Mar. Pollut. Bull.* **2013**, *73*, 319–326. [CrossRef] [PubMed]

5. Zhao, L.; Boufadel, M.C.; King, T.; Robinson, B.; Gao, F.; Socolofsky, S.A.; Lee, K. Droplet and bubble formation of combined oil and gas releases in subsea blowouts. *Mar. Pollut. Bull.* **2017**, *120*, 203–216. [CrossRef] [PubMed]

6. Camilli, R.; Reddy, C.M.; Yoerger, D.R.; Van Mooy, B.A.S.; Jakuba, M.V.; Kinsey, J.C.; McIntyre, C.P.; Sylva, S.P.; Maloney, J.V. Tracking hydrocarbon plume transport and biodegradation at Deepwater Horizon. *Science* **2010**, *330*, 201–204. [CrossRef] [PubMed]

7. Hazen, T.C.; Dubinsky, E.A.; DeSantis, T.Z.; Andersen, G.L.; Piceno, Y.M.; Singh, N.; Jansson, J.K.; Probst, A.; Borglin, S.E.; Fortney, J.L.; et al. Deep-sea oil plume enriches indigenous oil-degrading bacteria. *Science* **2010**, *330*, 204–208. [CrossRef] [PubMed]

8. Atlas, R.M.; Hazen, T.C. Oil biodegradation and bioremediation: A tale of the two worst spills in U.S. history. *Environ. Sci. Technol.* **2011**, *45*, 6709–6715. [CrossRef] [PubMed]

9. McGenity, T.J.; Folwell, B.D.; McKew, B.A.; Sanni, G.O. Marine crude-oil biodegradation: A central role for interspecies interactions. *Aquat. Biosyst.* **2012**, *8*, 10. [CrossRef] [PubMed]

10. Hazen, T.C.; Prince, R.C.; Mahmoudi, N. Marine oil biodegradation. *Environ. Sci. Technol.* **2016**, *50*, 2121–2129. [CrossRef] [PubMed]

11. Almeda, R.; Hyatt, C.; Buskey, E.J. Toxicity of dispersant Corexit 9500A and crude oil to marine microzooplankton. *Ecotoxicol. Environ. Saf.* **2014**, *106*, 76–85. [CrossRef] [PubMed]

12. Carroll, J.; Vikebø, F.; Howell, D.; Broch, O.J.; Nepstad, R.; Augustine, S.; Skeie, G.M.; Bast, R.; Juselius, J. Assessing impacts of simulated oil spills on the Northeast Arctic cod fishery. *Mar. Pollut. Bull.* **2018**, *126*, 63–73. [CrossRef] [PubMed]

13. Van Eenennaam, J.S.; Rahsepar, S.; Radović, J.R.; Oldenburg, T.B.P.; Wonink, J.; Langenhoff, A.A.M.; Murk, A.J.; Foekema, E.M. Marine snow increases the adverse effects of oil on benthic invertebrates. *Mar. Pollut. Bull.* **2018**, *126*, 339–348. [CrossRef] [PubMed]

14. Buskey, E.J.; White, H.K.; Esbaugh, A.J. Impact of oil spills on marine life in the Gulf of Mexico. *Oceanography* **2016**, *29*, 174–181. [CrossRef]

15. Lindo-Atichati, D.; Paris, C.B.; Le Hénaff, M.; Schedler, M.; Valladares Juárez, A.G.; Müller, R. Simulating the effects of droplet size, high-pressure biodegradation, and variable flow rate on the subsea evolution of deep plumes from the Macondo blowout. *Deep Sea Res. Part II Top. Stud. Oceanogr.* **2016**, *129*, 301–310. [CrossRef]

16. Leal, L.G. *Advanced Transport Phenomena: Fluid Mechanics and Convective Transport Processes*; Cambridge University Press: Cambridge, UK, 2007; ISBN 978-0-521-84910-4.

17. Daly, K.L.; Passow, U.; Chanton, J.; Hollander, D. Assessing the impacts of oil-associated marine snow formation and sedimentation during and after the Deepwater Horizon oil spill. *Anthropocene* **2016**, *13*, 18–33. [CrossRef]

18. Romero, I.C.; Toro-Farmer, G.; Diercks, A.-R.; Schwing, P.; Muller-Karger, F.; Murawski, S.; Hollander, D.J. Large-scale deposition of weathered oil in the Gulf of Mexico following a deep-water oil spill. *Environ. Pollut.* **2017**, *228*, 179–189. [CrossRef] [PubMed]

19. Law, A.M.J.; Aitken, M.D. Bacterial chemotaxis to naphthalene desorbing from a nonaqueous liquid. *Appl. Environ. Microbiol.* **2003**, *69*, 5968–5973. [CrossRef] [PubMed]

20. Wang, X.; Lanning, L.M.; Ford, R.M. Enhanced retention of chemotactic bacteria in a pore network with residual NAPL contamination. *Environ. Sci. Technol.* **2016**, *50*, 165–172. [CrossRef] [PubMed]

21. Lambert, R.A.; Variano, E.A. Collision of oil droplets with marine aggregates: Effect of droplet size. *J. Geophys. Res. Oceans* **2016**, *121*, 3250–3260. [CrossRef]

22. Levich, V.G. *Physicochemical Hydrodynamics*; Prentice-Hall: Englewood Cliffs, NJ, USA, 1962; ISBN 0136744400.

23. Miller, R.M.; Bartha, R. Evidence from liposome encapsulation for transport-limited microbial metabolism of solid alkanes. *Appl. Environ. Microbiol.* **1989**, *55*, 269–274. [PubMed]

24. Banat, I.M. Biosurfactants production and possible uses in microbial enhanced oil recovery and oil pollution remediation: A review. *Bioresour. Technol.* **1995**, *51*, 1–12. [CrossRef]

25. Brown, D.G. Relationship between micellar and hemi-micellar processes and the bioavailability of surfactant-solubilized hydrophobic organic compounds. *Environ. Sci. Technol.* **2007**, *41*, 1194–1199. [CrossRef] [PubMed]

26. Li, J.-L.; Chen, B.-H. Surfactant-mediated biodegradation of polycyclic aromatic hydrocarbons. *Materials* **2009**, *2*, 76–94. [CrossRef]

27. Antoniou, E.; Fodelianakis, S.; Korkakaki, E.; Kalogerakis, N. Biosurfactant production from marine hydrocarbon-degrading consortia and pure bacterial strains using crude oil as carbon source. *Front. Microbiol.* **2015**, *6*, 274. [CrossRef] [PubMed]

28. Hua, F.; Wang, H.Q. Uptake and transmembrane transport of petroleum hydrocarbons by microorganisms. *Biotechnol. Biotechnol. Equip.* **2014**, *28*, 165–175. [CrossRef] [PubMed]

29. Singh, A.K.; Sherry, A.; Gray, N.D.; Jones, D.M.; Bowler, B.F.J.; Head, I.M. Kinetic parameters for nutrient enhanced crude oil biodegradation in intertidal marine sediments. *Front. Microbiol.* **2014**, *5*, 160. [CrossRef] [PubMed]

30. Wang, J.; Sandoval, K.; Ding, Y.; Stoeckel, D.; Minard-Smith, A.; Andersen, G.; Dubinsky, E.A.; Atlas, R.; Gardinali, P. Biodegradation of dispersed Macondo crude oil by indigenous Gulf of Mexico microbial communities. *Sci. Total Environ.* **2016**, *557–558*, 453–468. [CrossRef] [PubMed]

31. Kapellos, G.E. Microbial strategies for oil biodegradation. In *Modeling of Microscale Transport in Biological Processes*; Becker, S.M., Ed.; Academic Press: Cambridge, MA, USA, 2017; pp. 19–39.

32. Vilcáez, J.; Li, L.; Hubbard, S.S. A new model for the biodegradation kinetics of oil droplets: Application to the Deepwater Horizon oil spill in the Gulf of Mexico. *Geochem. Trans.* **2013**, *14*, 4. [CrossRef] [PubMed]

33. Denis, B.; Pérez, O.A.; Lizardi-Jiménez, M.A.; Dutta, A. Numerical evaluation of direct interfacial uptake by a microbial consortium in an airlift bioreactor. *Int. Biodeterior. Biodegrad.* **2017**, *119*, 542–551. [CrossRef]

34. North, E.W.; Adams, E.E.; Thessen, A.E.; Schlag, Z.; He, R.; Socolofsky, S.A.; Masutani, S.M.; Peckham, S.D. The influence of droplet size and biodegradation on the transport of subsurface oil droplets during the Deepwater Horizon spill: A model sensitivity study. *Environ. Res. Lett.* **2015**, *10*, 024016. [CrossRef]

35. MacLeod, C.T.; Daugulis, A.J. Interfacial effects in a two-phase partitioning bioreactor: Degradation of polycyclic aromatic hydrocarbons (PAHs) by a hydrophobic *Mycobacterium*. *Process Biochem.* **2005**, *40*, 1799–1805. [CrossRef]

36. Stone, H.A. Dynamics of drop deformation and breakup in viscous fluids. *Annu. Rev. Fluid Mech.* **1994**, *26*, 65–102. [CrossRef]

37. Clift, R.; Grace, J.R.; Weber, M.E. *Bubbles, Drops, and Particles*; Academic Press: New York, NY, USA, 1978; ISBN 0-12-176950-X.

38. Michaelides, E.E. Hydrodynamic force and heat/mass transfer from particles, bubbles, and drops—The Freeman Scholar lecture. *J. Fluids Eng.* **2003**, *125*, 209–238. [CrossRef]

39. Deen, W.M. *Analysis of Transport Phenomena*; Oxford University Press: New York, NY, USA, 1998; ISBN 0195084942.

40. Johnson, A.I.; Akehata, T. Reaction accompanied mass transfer from fluid and solid spheres at low Reynolds numbers. *Can. J. Chem. Eng.* **1965**, *43*, 10–15. [CrossRef]

41. Goddard, J.D.; Acrivos, A. An analysis of laminar forced-convection mass transfer with homogeneous chemical reaction. *Q. J. Mech. Appl. Math.* **1967**, *20*, 471–497. [CrossRef]

42. Chen, W.C.; Pfeffer, R. Local and overall mass transfer rates around solid spheres with first-order homogeneous chemical reactions. *Ind. Eng. Chem. Fundam.* **1970**, *9*, 101–107. [CrossRef]

43. Ruckenstein, E.; Dang, V.-D.; Gill, W.N. Mass transfer with chemical reaction from spherical one or two component bubbles or drops. *Chem. Eng. Sci.* **1971**, *26*, 647–668. [CrossRef]

44. Hashimoto, H.; Kawano, S. Mass transfer around a moving encapsulated liquid drop. *Int. J. Multiph. Flow* **1993**, *19*, 213–228. [CrossRef]

45. Juncu, G. The influence of the Henry number on the conjugate mass transfer from a sphere: II—Mass transfer accompanied by a first-order chemical reaction. *Heat Mass Transf.* **2002**, *38*, 523–534. [CrossRef]

46. Stewart, P.S. A review of experimental measurements of effective diffusive permeabilities and effective diffusion coefficients in biofilms. *Biotechnol. Bioeng.* **1998**, *59*, 261–272. [CrossRef]

47. Kapellos, G.E.; Alexiou, T.S.; Payatakes, A.C. A multiscale theoretical model for diffusive mass transfer in cellular biological media. *Math. Biosci.* **2007**, *210*, 177–237. [CrossRef] [PubMed]

48. Arnaouteli, S.; MacPhee, C.E.; Stanley-Wall, N.R. Just in case it rains: Building a hydrophobic biofilm the Bacillus subtilis way. *Curr. Opin. Microbiol.* **2016**, *34*, 7–12. [CrossRef] [PubMed]

49. Passow, U.; Ziervogel, K.; Asper, V.; Diercks, A. Marine snow formation in the aftermath of the Deepwater Horizon oil spill in the Gulf of Mexico. *Environ. Res. Lett.* **2012**, *7*, 035301. [CrossRef]

50. Rivas, D.; Badan, A.; Sheinbaum, J.; Ochoa, J.; Candela, J. Vertical velocity and vertical heat flux observed within loop current eddies in the central Gulf of Mexico. *J. Phys. Oceanogr.* **2008**, *38*, 2461–2481. [CrossRef]

51. Chiswell, S.M. Mean velocity decomposition and vertical eddy diffusivity of the Pacific ocean from surface GDP drifters and 1000-m Argo floats. *J. Phys. Oceanogr.* **2016**, *46*, 1751–1768. [CrossRef]

52. North, E.W.; Adams, E.E.; Schlag, Z.; Sherwood, C.R.; He, R.; Hyun, K.H.; Socolofsky, S.A. Simulating oil droplet dispersal from the Deepwater Horizon spill with a Lagrangian approach. In *Monitoring and Modeling the Deepwater Horizon Oil Spill: A Record-Breaking Enterprise (Geophysical Monograph Series 195)*; Liu, Y., Macfadyen, A., Ji, Z.-G., Weisberg, R.H., Eds.; American Geophysical Union: Washington, DC, USA, 2011; pp. 217–226. [CrossRef]

53. Buckley, J.S.; Fan, T. Crude oil/brine interfacial tensions. *Petrophysics* **2007**, *48*, A1.

54. Moeini, F.; Hemmati-Sarapardeh, A.; Ghazanfari, M.-H.; Masihi, M.; Ayatollahi, S. Toward mechanistic understanding of heavy crude oil/brine interfacial tension: The roles of salinity, temperature and pressure. *Fluid Phase Equilibria* **2014**, *375*, 191–200. [CrossRef]

55. Epstein, A.K.; Pokroy, B.; Seminara, A.; Aizenberg, J. Bacterial biofilm shows persistent resistance to liquid wetting and gas penetration. *Proc. Natl. Acad. Sci. USA* **2011**, *108*, 995–1000. [CrossRef] [PubMed]

56. Stocker, R. Marine microbes see a sea of gradients. *Science* **2012**, *338*, 628–633. [CrossRef] [PubMed]

57. Meile, C.; Tuncay, K. Scale dependence of reaction rates in porous media. *Adv. Water Resour.* **2006**, *29*, 62–71. [CrossRef]

58. Medina-Moreno, S.A.; Jiménez-González, A.; Gutiérrez-Rojas, M.; Lizardi-Jiménez, M.A. Hexadecane aqueous emulsion characterization and uptake by an oil-degrading microbial consortium. *Int. Biodeterior. Biodegrad.* **2013**, *84*, 1–7. [CrossRef]

59. Nikolopoulou, M.; Pasadakis, N.; Kalogerakis, N. Evaluation of autochthonous bioaugmentation and biostimulation during microcosm-simulated oil spills. *Mar. Pollut. Bull.* **2013**, *72*, 165–173. [CrossRef] [PubMed]

60. Torlapati, J.; Boufadel, M.C. Evaluation of the biodegradation of Alaska North Slope oil in microcosms using the biodegradation model BIOB. *Front. Microbiol.* **2014**, *5*, 212. [CrossRef] [PubMed]

61. Brakstad, O.G.; Nordtug, T.; Throne-Holst, M. Biodegradation of dispersed Macondo oil in seawater at low temperature and different oil droplet sizes. *Mar. Pollut. Bull.* **2015**, *93*, 144–152. [CrossRef] [PubMed]

62.   Prince, R.C.; Butler, J.D.; Redman, A.D. The rate of crude oil biodegradation in the sea. *Environ. Sci. Technol.*
       2017, *51*, 1278–1284. [CrossRef] [PubMed]
63.   Thessen, A.E.; North, E.W. Calculating *in situ* degradation rates of hydrocarbon compounds in deep waters
       of the Gulf of Mexico. *Mar. Pollut. Bull.* **2017**, *122*, 77–84. [CrossRef] [PubMed]
64.   Rahsepar, S.; Langenhoff, A.A.M.; Smit, M.P.J.; van Eenennaam, J.S.; Murk, A.J.; Rijnaarts, H.H.M.
       Oil biodegradation: Interactions of artificial marine snow, clay particles, oil and Corexit. *Mar. Pollut. Bull.*
       2017, *125*, 186–191. [CrossRef] [PubMed]
65.   Kapellos, G.E.; Alexiou, T.S. Modeling momentum and mass transport in cellular biological media: From the
       molecular to the tissue scale. In *Transport in Biological Media*; Becker, S.M., Kuznetsov, A.V., Eds.; Academic
       Press: Waltham, MA, USA, 2013; pp. 1–40. [CrossRef]
66.   Billings, N.; Birjiniuk, A.; Samad, T.S.; Doyle, P.S.; Ribbeck, K. Material properties of biofilms—A review
       of methods for understanding permeability and mechanics. *Rep. Prog. Phys.* **2015**, *78*, 036601. [CrossRef]
       [PubMed]
67.   Sudarsan, R.; Ghosh, S.; Stockie, J.; Eberl, H.J. Simulating biofilm deformation and detachment with the
       immersed boundary method. *Commun. Comput. Phys.* **2016**, *19*, 682–732. [CrossRef]
68.   Hollenbeck, E.C.; Douarche, C.; Allain, J.-M.; Roger, P.; Regeard, C.; Cegelski, L.; Fuller, G.G.; Raspaud, E.
       Mechanical Behavior of a Bacillus subtilis Pellicle. *J. Phys. Chem. B* **2016**, *120*, 6080–6088. [CrossRef]
       [PubMed]

7

Comminution of Dry Lignocellulosic Biomass, a Review: From Fundamental Mechanisms to Milling Behaviour

Claire Mayer-Laigle *⬤, Nicolas Blanc, Rova Karine Rajaonarivony and Xavier Rouau

UMR Ingénierie des Agropolymères et des Technologies Emergentes (IATE), University of Montpellier, CIRAD, INRA, Montpellier SupAgro, Montpellier, France; nicolas.blanc@inra.fr (N.B.); karine.rajaonarivony@supagro.fr (R.K.R.); xavier.rouau@inra.fr (X.R.)
* Correspondence: claire.mayer@inra.fr

Abstract: The comminution of lignocellulosic biomass is a key operation for many applications as bio-based materials, bio-energy or green chemistry. The grinder used can have a significant impact on the properties of the ground powders, of those of the end-products and on the energy consumption. Since several years, the milling of lignocellulosic biomass has been the subject of numerous studies most often focused on specific materials and/or applications but there is still a lack of generic knowledge about the relation between the histological structure of the raw materials, the milling technologies and the physical and chemical properties of the powders. This review aims to point out the main process parameters and plant raw material properties that influence the milling operation and their consequences on the properties of ground powders and on the energy consumption during the comminution.

Keywords: plant materials; mechanical stresses; energy consumption; grinding law; grinding

1. Introduction

The use of biomass as a source of renewable energy, bio-based materials or green chemistry reagents has increased in recent years. In particular, the interest for lignocellulosic biomass has been growing as it presents the advantage for having a short production cycle while sequestering carbon. Lignocellulosic biomass produced annually by photosynthesis, represents an enormous feedstock estimated at approximately 79 GTOE/year (gigatons of oil equivalent, equivalent to about 320 Gtons/year) [1] while the annual world energy demand is about 12 GTOE/year (+2.5%/year). Lignocellulosic feedstocks occur as different types of raw materials which differ in compositions, properties, accessibility, processability. They can be categorized in different families: the agriculture and food processing by-products (straws, canes, stems, cobs, husks, hulls, brans, etc.), the dedicated energy crops (e.g., switchgrass, miscanthus, sorgho, etc.), the short and very short rotation coppices (e.g., willow, poplar, eucalyptus, etc.), the wood chain by-products (chips, sawdust, residues from felling and from second processing, etc.), the industrial and municipal wastes (residues of pallets, casing and packaging, green wastes, etc.) [2]. Whatever its origin, the input biomass needs to be calibrated to the transformation processes. The targeted size is function of the intended applications and can vary from several centimeters to a few micrometers. For example, in energetic applications, coarse particles in the range of few millimeters are targeted for anaerobic digestion processes [3] particles around 100 to 500 μm for biofuel production [4], and below 100 μm for gasification process [5] or direct combustion in engines [6,7]. Most of the time, in green chemistry applications, it is preferable to reduce the particle size below the cellular scale (~100 μm) in order to facilitate the extraction of molecules of interest [8,9]. When biomass is used as filler in bio-based materials, different particles properties could be sought by the comminution step. In case of composites intended for automotive equipment application, a significant aspect ratio is often sought to guarantee a good load transfer between the fiber and the polymeric matrix [10]. However in some applications of food packaging, it is preferable to decrease significantly the particle size (below 20 µm) in order to increase the proportion of filler content and reinforce the permeability properties [11]. In many cases, milling is still today seen as a common operation of particle size reduction for the pre-treatment of the biomass, and it is poorly studied in the scientific literature [12]. However, milling and especially fine milling can significantly influence the chemical and physical properties of the biomass and can be a lever to reveal some specific potentialities of the raw materials.

Based on a review of research works, this paper aims to highlight the major mechanisms occurring during the comminution of the biomass materials at the particle and process scales, their consequences on the properties of the ground powders and on the energy consumption.

2. Lignocellulosic Biomass Structure and Composition

Whole plants or their parts and primary products are constituted of different organs (trunks, stems, branches, leaves, roots …) ranging from several meters to few millimeters. At the histological scale (a few hundreds of µm) are the plant tissues made of clusters of cells with common natures and functions (vessels, parenchymae, supporting tissues, epidermis …). The cell sizes generally range from 100 to 10 µm depending on species and tissues. Plant cells are surrounded by a 0.1-to-several-µm-thick cell-wall and are generally devoid of cell content in most common lignocellulosic feedstocks. Lignocellulosic biomass comes mostly from secondary plant cell-walls [13]. The plant cell-wall is typically composed of cellulose (~20–60% dry weight basis, depending on biomasses), hemicelluloses (~10–40%) and lignin (~10–40%), with some associated minor constituents such as organic extractable compounds and mineral inclusions, for example silica deposits in rice husks [14].

The main component of the cell-walls, the cellulose, is a strictly water-insoluble assembly of linear polymers of beta-1-4-linked-D-gucopyranose units arranged in microfibrils containing both organized crystalline domains and amorphous regions. The degree of crystallinity refers to the ratio of crystalline to amorphous parts of cellulose [15]. Microfibrils are associated in cell-walls to form fibrils that can reach several µm length [16]. Hemicelluloses are a family of branched polysaccharides which form an amorphous matrix surrounding cellulose fibrils. They exhibit a great deal of compositional and structural variability depending on taxonomic classes, species and parts [17]. In the more common biomass feedstocks, main hemicellulose families are of glucurono-arabinoxylan type in grass plants and of galacto-glucomannantype and acetylated/methylated glucuronoxylan type in woods. Lignin is a tridimensional array of phenolic polymers resulting from the radical polymerisation of phenylpropenoid units. Its network impregnates secondarily the polysaccharide components of cell-walls and establish covalent linkages with hemicelluloses [18]. Lignin confers improved properties to the plant organs such as mechanical resistance, hydrophobicity, microbial protection, etc.

As an illustration, the composition of some biomasses is reported in Table 1 to give an idea of the variation range. The composition can vary significantly with the geographical area, the species and the analytical method thus, the data given here are indicative. Wheat straw and wood (pine), miscanthus and eucalyptus have been chosen to illustrate agriculture by-products, wood by-products, energy crops and short rotation coppices, respectively. The composition of industrial and agricultural wastes has not been reported as they highly depend on the origin and the formulation of the blend. Alfa is a perennial Mediterranean plant used for paper manufacturing, thanks to its high cellulose content. It is relatively close to wood pine, except for the lignin content which remains lower (12% against 25–30% for pine wood) and almost similar to miscanthus.

In addition, tissues from different parts of the plant can show different compositions. It is the case for example of flax and hemp fibers which are the external tissue of stem and present higher cellulose content (close to 80%) than flax shives and hemps hurds (core of the stalk). For full detailed compositions and other biomasses, reader can refer to the work of Vassilev et al. [19].

Table 1. Composition of different lignocellulosic biomasses, on a dry basis.

Biomass	Cellulose (%)	Hemicellulose (%)	Lignin (%)	Water–Soluble (%)	Ash (%)	References
Alfa	44–47	22–30	12–20	≈4	≈2	[20,21]
Eucalyptus	53–58	17–20	19–22	1–5	<1	[22]
Miscanthus	43–50	24–34	9–12	1–2	2–4	[23]
Wheat Straw	33–40	21–26	11–23	4–10	7–10	[24,25]
Wood (Pine)	45–50	20–30	25–30	2–10	<1	[26]
Flax fiber	78–80	6–13	2–5	2–4	1–2	[27,28]
Flax shives	32–53	13–21	23–25	1–2	2–3	[28–30]
Hemp fiber	67–76	8–12	2–5	2–16	<1	[27,31,32]
Hemp Hurds	39–49	16–23	16–23	0–2	2–4	[33–35]

From a mechanical point of view, the structural organization of plant materials results in a fibrous material with an anisotropic orientation that exhibits different mechanical properties (Young's modulus, tensile strength, Poisson coefficient, etc.) in the longitudinal and transverse directions [36].

As example, Figure 1 gives the range of the longitudinal Young's modulus (a) and tensile strength (b) for different biomasses. The Young's modulus or elastic modulus, is a measure of the stiffness of a material. It describes the elastic properties and the deformation behavior of the material. The tensile strength is the maximum stress that can be applied to a material before it breaks.

In Figure 1, we can note that these two properties are related partly to the cellulose content. Biomass with higher cellulose content and with cellulose microfibrils more aligned in the fiber direction (flax and hemp fibers) present higher mechanical performances in this direction [37]. However, the mechanical properties are also strongly influenced by the lignin content which depends on the growing and harvesting conditions, the plant maturity and the position in the stalk. In particular, Zhang et al. highlighted that a higher lignin content increases the Young's modulus and decreases the tensile strength and elongation of single wood fibers [38].

Some fragilities could also be induced by heat or hydric stresses, failure during growing or water resorption during drying process. As biomass is a living material, it evolves with time under the action of humidity, temperature and decomposition modifying its mechanical properties [39]. As a consequence, a same biomass specie could present a high variability in properties which explains the large range for the mechanical properties in Figure 1.

In many studies, the mechanical properties of the biomass are determined in tensile tests. However, to fully characterize a biomass in view of its comminution, traction data in both directions, compression and shear are necessary. Indeed, the fibrous nature of materials, requires most of the time the use of shear during the comminution to reduce the length of the fibers. Several authors have adapted three-points bending tests for wheat straw and corn stalk pith in order to determine their mechanical properties under shearing stress [40,41]. In many case, these data are not available in the literature and are difficult to measure on biomass samples due to the high variability of the materials. In addition, in a grinder or in a mill, the mechanical stresses are applied randomly in the longitudinal and transverse directions and it is still very difficult to predict accurately the behavior of a specific biomass in the machines in relation to its mechanical properties.

a) Young's modulus

Flax fiber	27-80
Hemp fiber	58-70
Alfa	18-25
Wheat Straw	2-8
Wood (Pine)	8-12

Longitudinal Young's modulus (GPa)

0   20   40   60   80   100

b) Tensile strength

Flax fiber	345-1850
Hemp fiber	550-1100
Alfa	188-308
Wheat Straw	58-146
Wood (Pine)	73-95

Tensile strength(MPa)

0   500   1000   1500   2000

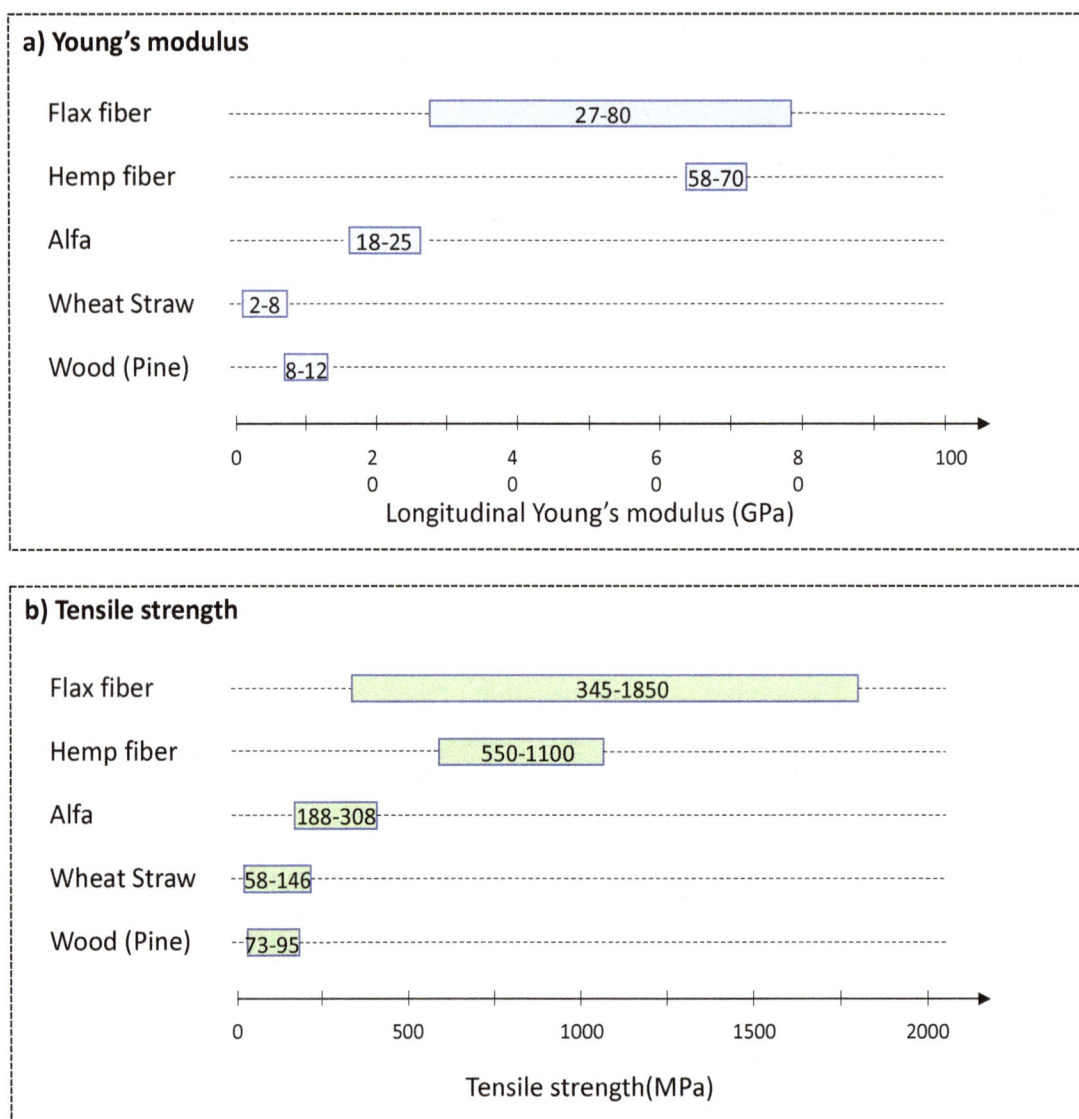

Figure 1. Main mechanical properties for different biomasses: (**a**) longitudinal Young's modulus and (**b**) tensile strength according to [37,42].

3. Milling Processes

Size reduction could be defined on the basis of an absolute scale associated to a metric size or an histological scale related to the structure of the plant materials as illustrated in Figure 2 [43]. Due to the heterogeneous structure of lignocellulosic biomass, the comminution involves generally several steps. The first one is generally coarse milling which leads to organ dissociation (from meter to centimeter) with cutting or crushing processes, then intermediate comminution (from cm to 1 mm) at the tissue scale, with shearing or impact processes, then fine milling at the tissue scale between 50 and 500 μm and finally ultra-fine milling, below the cell scale (<20 μm). Currently, it is not possible to reach particle size below 1 μm by dry milling processes. Thus the dissociation at the scale of the lignocellulosic matrix can be reached only by wet milling which is not in the scope of this review.

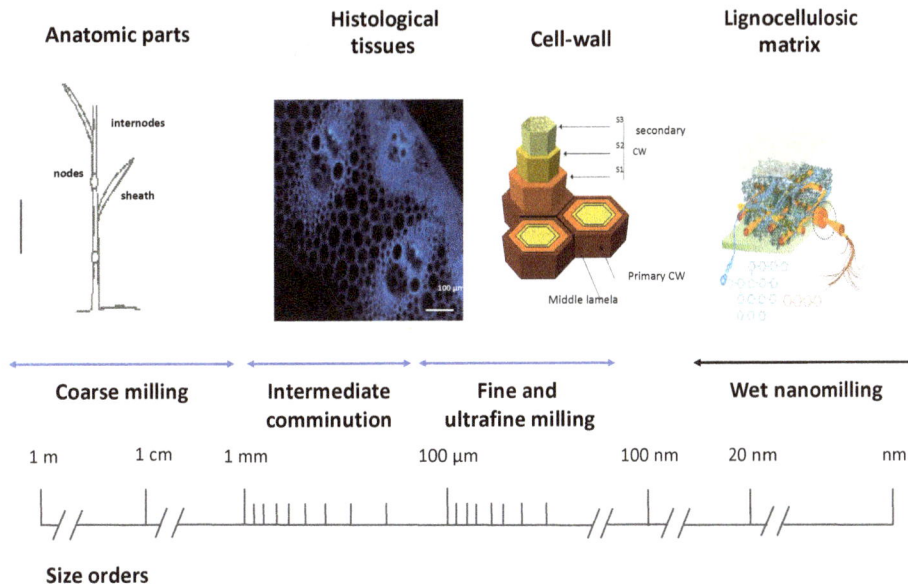

Figure 2. The metric and histological scales of a plant material in relation to the different milling steps.

In practice, at the process scale, the different milling devices can be compared on the basis of the main mechanical stress they generate. Among them we can distinguish: compression, impact, shearing and abrasion/attrition, illustrated schematically in Figure 3.

The two first mechanical stresses (impact and compression) are similar in the force applied to the materials but in the first case, the energy is given to the materials in a quite instantaneous manner whereas in the second case the energy is transferred during a longer contact time. Impact and compression can be generated by projection of milling media (balls, bars in ball-mills) on particles, particles against one another (as in jet-mills for example) or against a wall or a milling tool (as in hammer-mills).

Shear and attrition act both also in quite a similar way on the materials. However, attrition is a surface mechanism, comparable to erosion of the materials and is generated by friction against walls, other particles or beads. Shear acts rather in the bulk of the materials and is often created by rolling mechanisms, and/or differential velocities between two milling tools.

4. Interactions between Plant Material and Mechanical Stress during the Comminution

The mechanical constraints (stresses) generated by the grinder are applied to the plant materials during the comminution process. The constraints need to be sufficient in order to overcome the failure strength (σ) and lead to the rupture of the materials. Indeed, if the applied force is weaker (under the yield stress), the material will deform under the load and will recover its initial shape when the load will be removed (this is known as elastic deformation). A greater load but under the failure strength will conduct to irreversible deformation (plastic deformation) of the material but without failure. Finally a load greater than σ will conduct to failure which will propagate inside the material according to its histological structure and the presence of internal defects. Three loading modes (see Figure 4) that conduct to different failure behavior can be broadly distinguished, one in tension (mode I) and two in shear (mode II and III). In the mode I, a tensile stress is applied and conduct to tension failure. The mode II and mode III lead to in-plane (forward) and out of plane (transverse) shear failures, respectively [44].

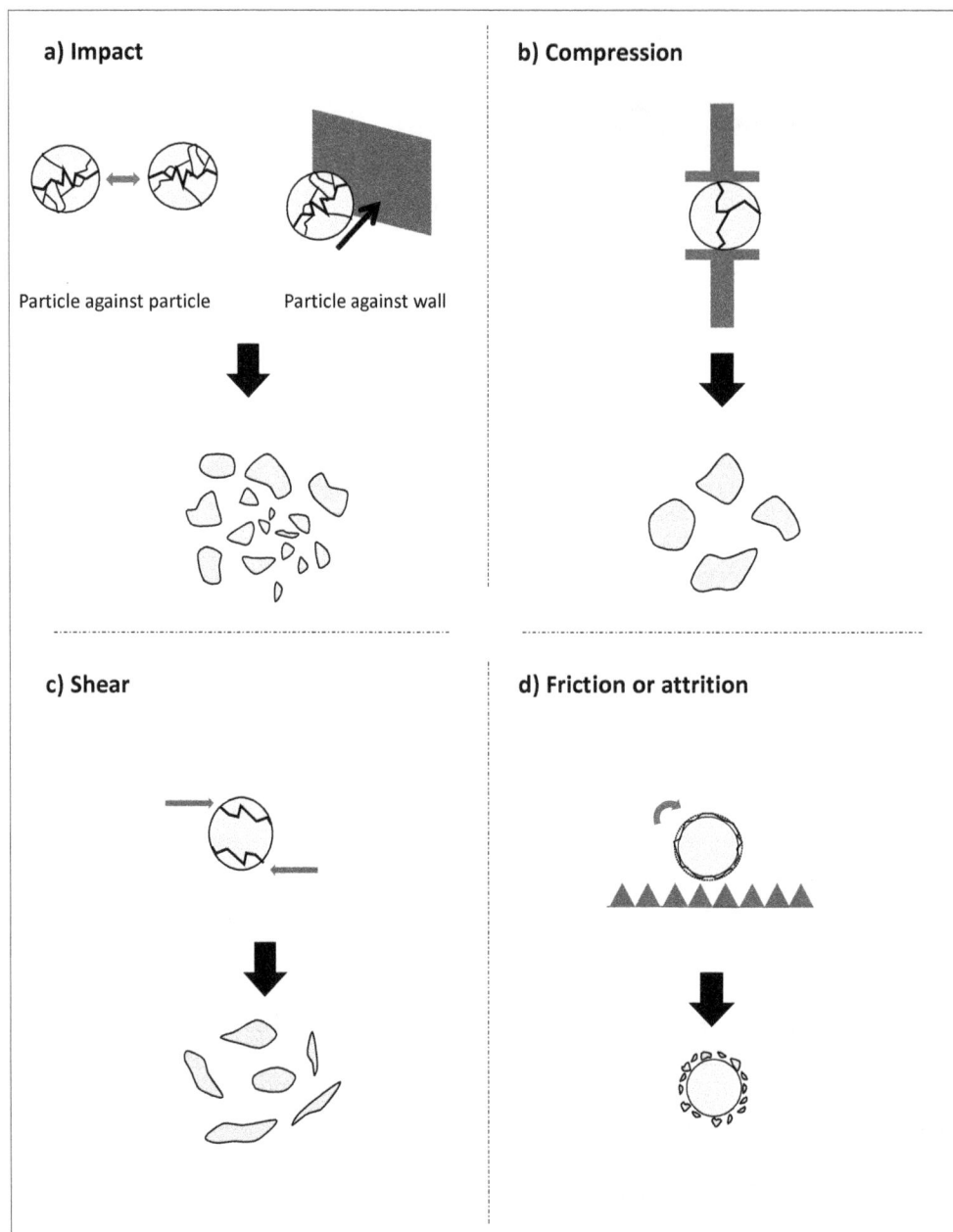

Figure 3. The main mechanical stresses in a grinder: (**a**) impact, (**b**) compression, (**c**) shear, and (**d**) friction and the particles that can be expected in case of comminution of an homogeneous materials according to Kaya et al. [45].

In a simplified approach, impact and compression could be associated to the loading mode I (traction) whereas shear and attrition to the loading mode II and III. In reality, at the scale of the particles, the mechanical loading is very complex. Indeed, in practice, the mechanical loading is applied to a more or less dense pack of particles and the local constraint on one particle is practically impossible to estimate. In addition, several mechanisms occur at the same time in the grinder but with various ratio of intensity according to the device and the process parameters [46]. For example, a rotation of the impactor during a compression leads to shearing action. Similarly, during shearing or attrition, the histological structure of the plant materials and the presence of internal defects could enable fracture in traction (mode I).

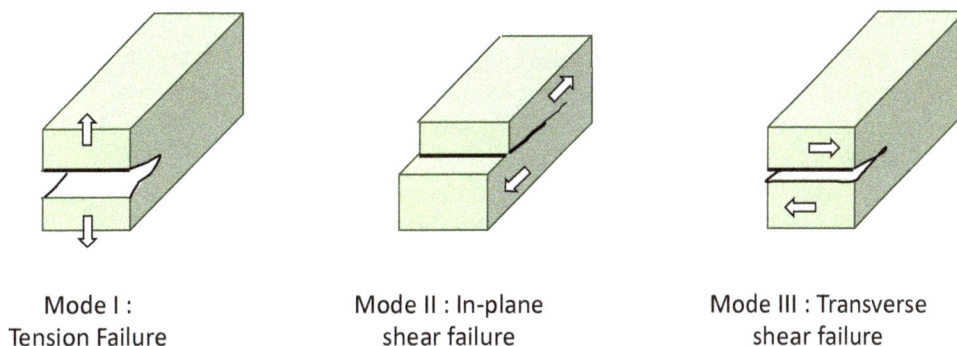

Mode I :
Tension Failure

Mode II : In-plane
shear failure

Mode III : Transverse
shear failure

Figure 4. The different loading modes applied at the micro-scale leading to the failure of the materials according to Karinkanta [44].

Whatever the way the breakage is initiated at the micro-scale, the fracture paths can propagate in different ways through the biomass. Its propagation depends on various process parameters (milling energy, mechanical stress, mechanical loading at the scale of the particle, etc.), properties of plant materials (histological structure, defects in the microstructure, mechanical properties, etc.) and environmental conditions (temperature, relative humidity, etc.) [47].

Two types of failures schematically represented in Figure 5 can be observed: (i) intercellular failure when the fracture path propagates between the cells leading to separation of the cells from each other (peeling) or (ii) intracellular failure when the breakage occurs across cells, disrupting secondary cell walls or trans-wall failure when the fracture path intersects the cell wall [48,49]. As an example, Motte et al. (2017) have shown that shear and compression generated by a millstone during the milling of cork induce intracellular failure (Figure 5b) whereas impact tends to generate intercellular failure (Figure 5a) [50].

a)

Intercellular failure

b)

Intracellular failure

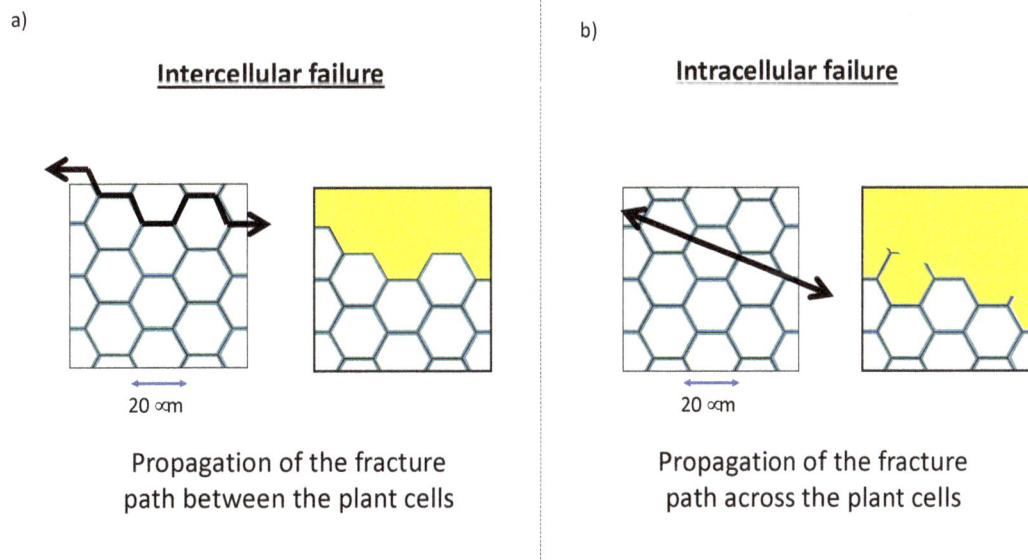

20 ∞m

Propagation of the fracture
path between the plant cells

20 ∞m

Propagation of the fracture
path across the plant cells

Figure 5. Intercellular failure (**a**) and Intracellular failure (**b**) observed during the milling of cork with impact and shear as main mechanical stresses, respectively [50]. (Tissue scale).

In addition, certain mechanical stresses can be more efficient than others according to mechanical properties of the plant materials. A study on milling of Douglas fir bark in view of histological dissociation has highlighted that a visco-elastic plant tissue, such as the suber, deforms under impact and compression (in a ball-mill) but recovers its initial shape immediately after. For this tissue, this kind of loading engenders a high energy expenditure for a weak particle size reduction. In contrast, the phloem which is a more brittle tissue results in very fine particles under the same mechanical stress [9]. In a same way, a study on olive-pomace milling (composed of pulp and fruit kernel) has shown that attrition regime in a ball-mill allows a very fast particle size reduction of the pulp (friable) but has only a very low effect on the kernel which is a harder material [51].

5. Influence of Milling on the Properties of Particles (Size, Shape, Surface Area)

The main property of ground biomass powders affected by grinding processes is the particle size distribution. Several factors can be used to summarize the whole distribution but the most commonly used are the median size ($d50$), the percentiles 10th ($d10$) and 90th ($d90$) related to the proportion of fine and coarse particles in the powder, and the span which reflects the width of the distribution.

$$SPAN = \frac{d(90) - d(10)}{d(50)}$$

This size distribution is influenced by the grinder type and the process parameters (residence time, velocity of the grinder tools, etc.) [12]. Depending on the mechanical stress, particles will break in different ways as illustrated in Figure 3 [45,52]. Abrasion has tendency to create two types of populations by erosion of the raw first particles: very fine particles and coarser particles resulting in a median average size close to the original particle size. Compression tends to result in particles of similar diameters with a tighten particle size distribution. Impact, by explosion due to the transmission of a high energy in a very short time, will conduct to a wide range of particle sizes.

The structure of the plant materials plays also a key role on the resulting particle size distribution obtained from ground biomass. Most of the time, the different populations obtained are related to the histology and their relative importance are function of the mechanical stresses as illustrated in the work of Mayer-Laigle et al. (2017) on the milling of oat bran in different devices. The authors have shown that four populations (≤ 1 μm, ≈ 20 μm, ≈ 100 μm, ≈ 500 μm) appear during the milling steps. However, shear generated by the high shear mill and the pin-mill is more efficient to grind the material at the cellular scale, as these devices favor the particle population around ≈ 20 μm [47].

The morphology of the particles may also change during the process and can have an effect on the properties of the end-product as for the combustion of densified biomasses [12]. Although there are still few data on particle shape evolution during grinding, largely due to the difficulties to obtain enough representative and quantitative measurements especially for fine particles, several authors have shown that the shape of the particles obtained after milling is related both to the milling technology, the type of biomass and its history and moisture content [12,53]. The shape seems also related to the size of the ground particles. For rather homogeneous materials like minerals, it has been demonstrated that the smaller the particles, the more they tends to a regular shape that can be assimilated to a sphere or a parallelepiped rectangle [16,45,54]). The same tendency has been also observed, to a lesser extent for wheat straw [16] and Douglas fir wood [55] although the heterogeneity of the raw materials is greater. This can be related to the decrease in anisotropy of the biomass and the diminution of the number of defects as the size of the particles decreases.

The modification of the particle size and shape has a direct influence on the bulk density and the rheological properties of the powder [55,56]. Bulk density takes into account the pores and the inter-particular voids. A densification of the powder is observed during the milling both by the suppression of porosity and the reduction of inter-particular gap due to the creation of fine particles which fill the voids between coarse particles [57]. However, an inverse evolution of the bulk density can be observed in case of ultrafine milling. Indeed, new interparticular voids can be created by bridges between particles due to the Van der Waals forces [58]. The particle size reduction leads to an increase of the balance between the interparticular forces and the gravity force, intensifying the cohesion and decreasing the flowability. This has a direct impact on the handling and processing of fine powders [59]. This is particularly true for biomass powders which have a low density and are highly sensitive to humidity, in contrast with mineral powders. The particle shapes also highly affect the flowability as illustrated by Tannous et al. on ground Douglas fir wood powder. In their study, they show that particle shapes have a more significant effect on flow properties than particle sizes [55]. In a similar way, Chen and Qu and Zhao et al. have shown, by measuring the angle of repose on red rice and ginger powders that superfine powders have a lower angle of repose than coarser ones, which can be related to better flowability [60,61]. They explain this observation by the formation of aggregates with regular shapes in the fine powders.

The specific surface area, related to the particle size distribution, increases during grinding processes [62]. In addition, ultrafine grinding, by opening cell-walls modify the surface aspect of the particles making accessible some constituents that were trapped inside the cell. This leads to an increase of extraction yield for components of interest as illustrated by Hemery et al. on wheat bran [8]. In addition, comminution allows the rupture of the polysaccharide chains bonds and may increase their exposure at the surface of the particles. The result is an improvement of the enzymatic hydrolysis and an increase of the conversion rate of the biomass into bioenergy in bioethanol production [4,63]. This increase is also due to the reduction of cellulose crystallinity that has been observed by several authors [64–66]. The amorphization of the cellulose structure is more pronounced in ball-milling in comparison to jet-milling [4], due to repeated impacts of balls on cellulose chain assemblies in crystals [67].

An increase of the water holding capacity (WHC) has been also reported attributed to the increase in specific surface area which leads to the exposure of more hydrophilic groups from cellulose and hemicelluloses [68,69]. However WHC could also decrease with the amorphization of the cellulose with reduction of porosity and it is difficult to predict this evolution in case of ultrafine biomass grinding [70].

6. Energy Consumption during Comminution

Several parameters are responsible for the energy consumption during the comminution such as the mechanical properties of the material, the size of the particles before and after milling, the moisture content, the feed rate and the mill characteristics.

Theoretically, in the case of a perfect single elastic sphere during a Brazilian test, it is possible to predict the energy of fracture depending of its tensile strength, Young modulus and Poisson's ratio [71]. This tensile strength needs to be determined experimentally and is varying with the composition of the material but also with its microstructure which is changing along with downsizing. As Weibull [72] and Epstein [73] have already explained, the number and size of defects in the material decrease with the size of the particles which leads to an increase of its resistance. This has been observed for single particle fracture with different rocks and minerals [74,75]. This effect can be particularly strong for lignocellulosic materials as they exhibit a complex and heterogeneous structure.

The resistance of a particle is also changing with the nature of the load and its application rate. In single particle fracture, a slow compression or a single or double impact configuration do not need the same amount of energy to break the particle. A higher stressing rate can either decrease the strength of some materials by the apparition of interferences between elastic waves leading to an early crack propagation in particular in case of visco-plastic materials like lignocellulosic materials [76,77].

In a bed of particles, particle breakage is controlled by the energy intensity but also by the relative position of particles compared to the direction of the force and by the number of contact between particles [78]. The number of contacts between fine and coarse particles increase with the proportion of fine particles. Some coarse particles can then be in an isostatic stress field and do not break under a compression load. If the energy still increases, it is the finest particles that break. As a consequence, a wide particle size distribution can slow down the comminution and can decrease the milling efficiency.

The attraction forces between particles (Van der Waals, electrostatic, capillary forces) also play a major role in energy consumption as they counteract the breaking process in promoting sometimes agglomeration phenomena. These forces are increasing as the particle sizes decrease [29]. Moreover, as the specific surface area is increasing, the friction between particles also increases leading to more energy loss by heat dissipation [79].

Moisture also strongly affect the energy consumption of biomass grinding as can be seen in several studies on various material milling (wood chips, pellets, miscanthus, switchgrass, various seeds, etc.) [69,80–83]. Increasing the moisture content increases the energy needed to mill the product. This effect can be explained by the fact that moisture increases the shear resistance of the material [56]. In addition, a dried material is leaving voids and cavities reducing the strength of the material facilitating thereby its milling [84].

There are also other sources of energy consumption related to the type and the size of the mill friction of the mechanical parts with the product, the milling media and the air, air flows, instrumentation, over-dimensioned device, etc.

The comparison of milling equipment on the basis of the energy efficiency is difficult to realize as there is no consensus on the definition of the energy efficiency. It is often computed as the ratio of the energy needed to reach a targeted size and the energy effectively used by the mill, which can be defined or measured in different ways. If authors are interested by the economic cost, the electrical consumption of the mill is considered. When a better understanding of the mechanical process is sought, the net energy (the energy consumed by the mill during grinding operation minus the void energy consumption when the mill is empty) measured as close as possible to the grinding chamber (measured by a torque meter mounted on the shaft of the mill for example) is often preferred. As example, in the case where the energy needed to create new surfaces is calculated from the energy of surface of the materials, the energy losses will represent a tremendous amount that can reach 96% to 99% of the total energy used by the mill [74,85]. But if it is computed from the energy needed for breaking single particles, greater efficiencies depending of the mill considered are observed (up to 80% for roller crushers [79]).

7. Conclusions

The milling and especially ultra-fine milling of biomass raises a growing interest in many sectors as it can influence the downstream operations and the yield of conversion in bioenergy. Through this literature review, we seek to underline that a good knowledge of the plant raw materials and process parameters can allow to modulate the powder properties. Such work opens the door to a real reverse engineering applied to the production of biomass particles.

We also underline that the factors influencing the energy consumption during the milling are numerous and difficult to predict from the mechanical properties of the raw materials. However, it is clearly stated that the energy consumption increases significantly as particle size decreases. In a perspective of sustainable technologies, it is important to adjust the target particle size to the intended application. For some applications with high-added value as building blocks for polymers and materials or molecule extraction for green chemistry, a higher energy expenditure for milling can be afforded whereas for biomass conversion into energy, it must be kept as low as possible to remain cost-effective.

Thus, the choice a grinder over another remains difficult as it has a direct impact on the ground powder and on energy consumption. Therefore, it must be based on a good knowledge of the technology and of the raw materials, and must integrate numerous constraints. In the following part of this review, we will seek to highlight the main elements to keep in mind for selecting a milling technology and evoke some approaches of process improvement through thermal pretreatment of starting raw materials.

Author Contributions: This paper is a collaborative work. Each author wrote specific parts of the article and all authors have contributed to define the plan of the article and to review in-depth the whole article.

Conflicts of Interest: The authors declare no conflicts of interest

References

1. IEA. *World Energy Outlook 2017*; International Energy Agency: Paris, France, November 2017.

2. Hadar, Y. Sources for lignocellulosic raw materials for the production of ethanol. In *Lignocellulose Conversion: Enzymatic and Microbial Tools for Bioethanol Production*; Faraco, V., Ed.; Springer: Berlin/Heidelberg, Germany, 2013; pp. 21–38.

3. Motte, J.C.; Escudié, R.; Bernet, N.; Delgenes, J.P.; Steyer, J.P.; Dumas, C. Dynamic effect of total solid content, low substrate/inoculum ratio and particle size on solid-state anaerobic digestion. *Bioresour. Technol.* **2013**, *144*, 141–148. [CrossRef] [PubMed]

4. Silva, G.G.D.; Couturier, M.; Berrin, J.-G.; Buléon, A.; Rouau, X. Effects of grinding processes on enzymatic degradation of wheat straw. *Bioresour. Technol.* **2012**, *103*, 192–200. [CrossRef] [PubMed]

5. Guizani, C.; Escudero Sanz, F.J.; Salvador, S. Influence of temperature and particle size on the single and mixed atmosphere gasification of biomass char with H_2O and CO_2. *Fuel Process. Technol.* **2015**, *134*, 175–188. [CrossRef]

6. McKnight, J. Powdered Fuels, Dispersions Thereof, and Combustion Devices Related Thereto. WO2008063549A2, 29 May 2008.

7. Piriou, B.; Vaitilingom, G.; Veyssière, B.; Cuq, B.; Rouau, X. Potential direct use of solid biomass in internal combustion engines. *Prog. Energy Combust. Sci.* **2013**, *39*, 169–188. [CrossRef]

8. Hemery, Y.; Chaurand, M.; Holopainen, U.; Lampi, A.-M.; Lehtinen, P.; Piironen, V.; Sadoudi, A.; Rouau, X. Potential of dry fractionation of wheat bran for the development of food ingredients, part i: Influence of ultra-fine grinding. *J. Cereal Sci.* **2011**, *53*, 1–8. [CrossRef]

9. Trivelato, P.; Mayer, C.; Barakat, A.; Fulcrand, H.; Aouf, C. Douglas bark dry fractionation for polyphenols isolation: From forestry waste to added value products. *Ind. Crops Prod.* **2016**, *86*, 12–15. [CrossRef]

10. Maya, J.J.; Sabu, T. Biofibres and biocomposites. *Carbohydr. Polym.* **2008**, *71*, 343–364.

11. Berthet, M.A.; Angellier-Coussy, H.; Chea, V.; Guillard, V.; Gastaldi, E.; Gontard, N. Sustainable food packaging: Valorising wheat straw fibres for tuning phbv-based composites properties. *Compos. Part A Appl. Sci. Manuf.* **2015**, *72*, 139–147. [CrossRef]

12. Womac, R.A.; Igathinathane, C.; Bitra, P.; Miu, P.; Yang, T.; Sokhansanj, S.; Narayan, S. Biomass pre-processing size reduction with instrumented mills. In *2007 ASAE Annual Meeting*; ASABE: St. Joseph, MI, USA, 2007.

13. Buchanan, B.B.; Gruissem, W.; Jones, R.L. *Biochemistry and Molecular Biology of Plants*, 2nd ed.; Wiley & sons Publisher: Somerset, NJ, USA, 2015.

14. Jauberthie, R.; Rendell, F.; Tamba, S.; Cisse, I. Origin of the pozzolanic effect of rice husks. *Constr. Build. Mater.* **2000**, *14*, 419–423. [CrossRef]

15. Fengel, D.; Wegener, G. *Wood: Chemistry, Ultrastructure, Reactions*; Fengel, D., Wegener, G., Eds.; Verlag Kessel: Remagen, Germany, 1984; pp. 227–239.

16. da Silva, G.G.D. Fractionnement par Voie Sèche de la Biomasse ligno-Cellulosique: Broyage Poussé de la Paille de Blé et Effets sur ses Bioconversions. Ph.D. Thesis, Supagro, Montpellier, France, 2011.

17. Monties, B. *Les Polymères Végétaux*; Gauthier-Villars: Paris, France, 1980.

18. Monties, B. *Plant Phenolics*; Elsevier Publisher: Amsterdam, The Netherlands, 1989; Volume 1.

19. Vassilev, S.V.; Baxter, D.; Andersen, L.K.; Vassileva, C.G. An overview of the chemical composition of biomass. *Fuel* **2010**, *89*, 913–933. [CrossRef]

20. Mallek-Fakhfakh, H.; Fakhfakh, J.; Walha, K.; Hassairi, H.; Gargouri, A.; Belghith, H. Enzymatic hydrolysis of pretreated alfa fibers (*Stipa tenacissima*) using β-D-glucosidase and xylanase of talaromyces thermophilus from solid-state fermentation. *Int. J. Biol. Macromol.* **2017**, *103*, 543–553. [CrossRef] [PubMed]

21. Trache, D.; Donnot, A.; Khimeche, K.; Benelmir, R.; Brosse, N. Physico-chemical properties and thermal stability of microcrystalline cellulose isolated from alfa fibres. *Carbohydr. Polym.* **2014**, *104*, 223–230. [CrossRef] [PubMed]

22. Pereira, H. Variability in the chemical composition of plantation eucalypts (*Eucalyptus globulus* labill.). *Wood Fiber Sci.* **1988**, *11*, 52–57.

23. Brosse, N.; Dufour, A.; Meng, X.; Sun, Q.; Ragauskas, A. Miscanthus: A fast-growing crop for biofuels and chemicals production. *Biofuels Bioprod. Biorefin.* **2012**, *6*, 580–598. [CrossRef]

24. Khan, T.; Mubeen, U. Wheat straw: A pragmatic overview. *Curr. Res. J. Biol. Sci.* **2012**, *4*, 673–675.

25. Waliszewska, B.; PrąDzyński, W.; Zborowska, M.; Stachowiak-Wencek, A.; Waliszewska, H.; Spek-Dźwigała, A. The diversification of chemical composition of pine wood depending on the tree age. *For. Wood Technol.* **2015**, *91*, 182–187.

26. Wyman, C. *Handbook on Bioethanol: Production and Utilization*; Taylor & Francis: Oxford, UK, 1996.

27. Dorez, G.; Ferry, L.; Sonnier, R.; Taguet, A.; Lopez-Cuesta, J.M. Effect of cellulose, hemicellulose and lignin contents on pyrolysis and combustion of natural fibers. *J. Anal. Appl. Pyrolysis* **2014**, *107*, 323–331. [CrossRef]

28. Sain, M.; Fortier, D. Flax shives refining, chemical modification and hydrophobisation for paper production. *Ind. Crops Prod.* **2002**, *15*, 1–13. [CrossRef]

29. Soulié, F. Cohésion par capillarité et comportement mécanique de milieux granulaires. Ph.D. Thesis, Université de Montpellier II, Montpellier, France, 2005.

30. Buranov, A.U.; Mazza, G. Extraction and characterization of hemicelluloses from flax shives by different methods. *Carbohydr. Polym.* **2010**, *79*, 17–25. [CrossRef]

31. Kostic, M.; Pejic, B.; Skundric, P. Quality of chemically modified hemp fibers. *Bioresour. Technol.* **2008**, *99*, 94–99. [CrossRef] [PubMed]

32. Shahzad, A. Hemp fiber and its composites—A review. *J. Compos. Mater.* **2011**, *46*, 973–986. [CrossRef]

33. Stevulova, N.; Cigasova, J.; Estokova, A.; Terpakova, E.; Geffert, A.; Kacik, F.; Singovszka, E.; Holub, M. Properties characterization of chemically modified hemp hurds. *Materials* **2014**, *7*, 8131–8150. [CrossRef] [PubMed]

34. Gandolfi, S.; Ottolina, G.; Riva, S.; Fantoni, G.P.; Patel, I. Complete chemical analysis of carmagnola hemp hurds and structural features of its components. *BioResources* **2013**, *8*, 2641–2656. [CrossRef]

35. Monteil-Rivera, F.; Phuong, M.; Ye, M.; Halasz, A.; Hawari, J. Isolation and characterization of herbaceous lignins for applications in biomaterials. *Ind. Crops Prod.* **2013**, *41*, 356–364. [CrossRef]

36. Guo, Q.; Chen, X.; Liu, H. Experimental research on shape and size distribution of biomass particle. *Fuel* **2012**, *94*, 551–555. [CrossRef]

37. Pickering, K.L.; Efendy, M.G.A.; Le, T.M. A review of recent developments in natural fibre composites and their mechanical performance. *Compos. Part A Appl. Sci. Manuf.* **2016**, *83*, 98–112. [CrossRef]

38. Zhang, S.-Y.; Fei, B.-H.; Yu, Y.; Cheng, H.-T.; Wang, C.-G. Effect of the amount of lignin on tensile properties of single wood fibers. *For. Sci. Pract.* **2013**, *15*, 56–60.

39. Annoussamy, M.; Richard, G.; Recous, S.; Guérif, J. Change in mechanical properties of wheat straw due to decomposition and moisture. *Appl. Eng. Agric.* **2000**, *16*, 657. [CrossRef]

40. Leblicq, T.; Vanmaercke, S.; Ramon, H.; Saeys, W. Mechanical analysis of the bending behaviour of plant stems. *Biosyst. Eng.* **2015**, *129*, 87–99. [CrossRef]

41. Chen, Z.; Qu, G. Shearing characteristics of corn stalk pith for separation. *BioResources* **2017**, *12*, 2296–2309. [CrossRef]

42. Green, D.W.; Winandy, J.E.; Kretschmann, D.E. *Wood Handbook: Wood as an engineering Material*; USDA Forest Service: Madison, WI, USA, 1999; pp. 1–45.

43. Barakat, A.; Mayer-Laigle, C.; Solhy, A.; Arancon, R.A.D.; de Vries, H.; Luque, R. Mechanical pretreatments of lignocellulosic biomass: Towards facile and environmentally sound technologies for biofuels production. *RSC Adv.* **2014**, *4*, 48109–48127. [CrossRef]

44. Karinkanta, P. *Dry Fine Grinding of Norway Spruce (Picea abies) Wood in Impact-Based Fine Grinding Mills*; Oulun Yliopiston Tutkijakoulu: Oulu, Finland, 2015.

45. Kaya, E.; Hogg, R.; Kumar, S. Particle shape modification in comminution. *KONA Powder Part. J.* **2002**, *20*, 188–195. [CrossRef]

46. Motte, J.C.; Delenne, J.Y.; Rouau, X.; Mayer-Laigle, C. Mineral–vegetal co-milling: An effective process to improve lignocellulosic biomass fine milling and to increase interweaving between mixed particles. *Bioresour. Technol.* **2015**, *192*, 703–710. [CrossRef] [PubMed]

47. Mayer-Laigle, C.; Barakat, A.; Barron, C.; Delenne, J.-Y.; Frank, X.; Mabille, F.; Rouau, X.; Sadoudi, A.; Samson, M.F.; Lullien, V. Dry biorefineries: Multiscale modeling studies and innovative processing. *Innov. Food Sci. Emerg. Technol.* **2017**. [CrossRef]

48. Conrad, M.P.C.; Smith, G.D.; Fernlund, G. Fracture of solid wood: A review of structure and properties at different length scales. *Wood Fiber Sci.* **2003**, *4*, 570–584.

49. Boatright, S.W.J.; Garrett, G.G. The effect of microstructure and stress state on the fracture behaviour of wood. *J. Mater. Sci.* **1983**, *18*, 2181–2199. [CrossRef]

50. Motte, J.-C.; Delenne, J.-Y.; Barron, C.; Dubreucq, E.; Mayer, C. Elastic properties of packing of granulated cork: Effect of particle size. *Ind. Crops Prod.* **2017**, *99*, 126–134. [CrossRef]

51. Lammi, S.; Barakat, A.; Mayer-Laigle, C.; Djenane, D.; Gontard, N.; Angellier-Coussy, H. Dry fractionation of olive pomace as a sustainable process to produce fillers for biocomposites. *Powder Technol.* **2018**, *326*, 44–53. [CrossRef]

52. Karinkanta, P.; Illikainen, M.; Niinimäki, J. Effect of different impact events in fine grinding mills on the development of the physical properties of dried norway spruce (*Picea abies*) wood in pulverisation. *Powder Technol.* **2014**, *253*, 352–359. [CrossRef]

53. Williams, O.; Newbolt, G.; Eastwick, C.; Kingman, S.; Giddings, D.; Lormor, S.; Lester, E. Influence of mill type on densified biomass comminution. *Appl. Energy* **2016**, *182*, 219–231. [CrossRef]

54. Domokos, G.; Kun, F.; Sipos, A.Á.; Szabó, T. Universality of fragment shapes. *Sci. Rep.* **2015**, *5*, 9147. [CrossRef] [PubMed]

55. Tannous, K.; Lam, P.S.; Sokhansanj, S.; Grace, J.R. Physical properties for flow characterization of ground biomass from douglas fir wood. *Part. Sci. Technol.* **2013**, *31*, 291–300. [CrossRef]

56. Mani, S.; Tabil, L.; Sokhansanj, S. *Mechanical Properties of Corn Stover Grind*; American Society of Agricultural and Biological Engineers: St. Joseph, MI, USA, 2004; Volume 47, pp. 1983–2013.

57. Chevanan, N.; Womac, A.R.; Bitra, V.S.; Sokhansanj, S. Effect of particle size distribution on loose-filled and tapped densities of selected biomass after knife mill size reduction. *Appl. Eng. Agric.* **2011**, *27*, 631. [CrossRef]

58. Li, Q.; Rudolph, V.; Weigl, B.; Earl, A. Interparticle van der waals force in powder flowability and compactibility. *Int. J. Pharm.* **2004**, *280*, 77–93. [CrossRef] [PubMed]

59. Tomas, J.; Kleinschmidt, S. Improvement of flowability of fine cohesive powders by flow additives. *Chem. Eng. Technol.* **2009**, *32*, 1470–1483. [CrossRef]

60. Zhao, X.; Yang, Z.; Gai, G.; Yang, Y. Effect of superfine grinding on properties of ginger powder. *J. Food Eng.* **2009**, *91*, 217–222. [CrossRef]

61. Chen, M.Q.; Fu, R.M.; Yue, L.F.; Cheng, Y.Y. Effect of superfine grinding on physicochemical properties, antioxidant activity and phenolic content of red rice (*Oryza sativa* l.). *Food Nutr. Sci.* **2015**, *6*, 1277–1284.

62. Gao, C.; Xiao, W.; Ji, G.; Zhang, Y.; Cao, Y.; Han, L. Regularity and mechanism of wheat straw properties change in ball milling process at cellular scale. *Bioresour. Technol.* **2017**, *241*, 214–219. [CrossRef] [PubMed]

63. Jiang, J.; Wang, J.; Zhang, X.; Wolcott, M. Evaluation of physical structural features on influencing enzymatic hydrolysis efficiency of micronized wood. *RSC Adv.* **2016**, *6*, 103026–103034. [CrossRef]

64. Khan, A.S.; Man, Z.; Bustam, M.A.; Kait, C.F.; Khan, M.I.; Muhammad, N.; Nasrullah, A.; Ullah, Z.; Ahmad, P. Impact of ball-milling pretreatment on pyrolysis behavior and kinetics of crystalline cellulose. *Waste Biomass Valorization* **2016**, *7*, 571–581. [CrossRef]

65. Stubičar, N.; Šmit, I.; Stubičar, M.; Tonejc, A.; Jánosi, A.; Schurz, J.; Zipper, P. An X-ray diffraction study of the crystalline to amorphous phase change in cellulose during high-energy dry ball milling. *Int. J. Biol. Chem. Phys. Technol. Wood* **1998**, *52*, 455–458. [CrossRef]

66. Yu, Y.; Wu, H. Effect of ball milling on the hydrolysis of microcrystalline cellulose in hot-compressed water. *AIChE J.* **2011**, *57*, 793–800. [CrossRef]

67. Avolio, R.; Bonadies, I.; Capitani, D.; Errico, M.E.; Gentile, G.; Avella, M. A multitechnique approach to assess the effect of ball milling on cellulose. *Carbohydr. Polym.* **2012**, *87*, 265–273. [CrossRef]

68. Tarafdar, J.C.; Meena, S.C.; Kathju, S. Influence of straw size on activity and biomass of soil microorganisms during decomposition. *Eur. J. Soil Biol.* **2001**, *37*, 157–160. [CrossRef]

69. Schell, D.J.; Harwood, C. Milling of lignocellulosic biomass. *Appl. Biochem. Biotechnol.* **1994**, *45*, 159–168. [CrossRef]

70. Raghavendra, S.N.; Ramachandra Swamy, S.R.; Rastogi, N.K.; Raghavarao, K.S.M.S.; Kumar, S.; Tharanathan, R.N. Grinding characteristics and hydration properties of coconut residue: A source of dietary fiber. *J. Food Eng.* **2006**, *72*, 281–286. [CrossRef]

71. Kanda, Y.; Kotake, N. Chapter 12 comminution energy and evaluation in fine grinding. In *Handbook of Powder Technology*; Salman, A.D., Ghadiri, M., Hounslow, M.J., Eds.; Elsevier Science B.V.: Amsterdam, The Netherlands, 2007; Volume 12, pp. 529–550.

72. Weibull, W. *A Statistical Theory of the Strength of Materials*; Generalstabens Litografiska Anstalts Förlag: Stockholm, Sweden, 1939.

73. Epstein, B. Statistical aspects of fracture problems. *J. Appl. Phys.* **1948**, *19*, 140–147. [CrossRef]
74. Tavares, L.M.; King, R.P. Single-particle fracture under impact loading. *Int. J. Miner. Process.* **1998**, *54*, 1–28. [CrossRef]
75. Yashima, S.; Kanda, Y.; Sano, S. Relationships between particle size and fracture energy or impact velocity required to fracture as estimated from single particle crushing. *Powder Technol.* **1987**, *51*, 277–282. [CrossRef]
76. Olsson, A.M.; Salmén, L. *Viscoelasticity of In Situ Lignin as Affected by Structure*; Swedish Pulp and Paper Research Institute: Stockholm, Sweden, 1992; pp. 133–143.
77. Sedan, D.; Pagnoux, C.; Smith, A.; Chotard, T. Mechanical properties of hemp fibre reinforced cement: Influence of the fibre/matrix interaction. *J. Eur. Ceram. Soc.* **2008**, *28*, 183–192. [CrossRef]
78. Gutsche, O.; Fuerstenau, D.W. Fracture kinetics of particle bed comminution—Ramifications for fines production and mill optimization. *Powder Technol.* **1999**, *105*, 113–118. [CrossRef]
79. Beke, B. *The Process of Fine Grinding*; Springer: Dordrecht, The Netherlands, 2012.
80. Temmerman, M.; Jensen, P.D.; Hébert, J. Von rittinger theory adapted to wood chip and pellet milling, in a laboratory scale hammermill. *Biomass Bioenergy* **2013**, *56*, 70–81. [CrossRef]
81. Miao, Z.; Grift, T.E.; Hansen, A.C.; Ting, K.C. Energy requirement for comminution of biomass in relation to particle physical properties. *Ind. Crops Prod.* **2011**, *33*, 504–513. [CrossRef]
82. Hess, J.; Wright, C.L.; Kenney, K. Cellulosic biomass feedstocks and logistics for ethanol production. *Biofuels Bioprod. Biorefin.* **2007**, *1*, 181–190. [CrossRef]
83. Mani, S.; Tabil, L.G.; Sokhansanj, S. Grinding performance and physical properties of wheat and barley straws, corn stover and switchgrass. *Biomass Bioenergy* **2004**, *27*, 339–352. [CrossRef]
84. Chand, N.; Hashmi, S.A.R. Mechanical properties of sisal fibre at elevated temperatures. *J. Mater. Sci.* **1993**, *28*, 6724–6728. [CrossRef]
85. Ghorbani, Z.; Masoumi, A.A.; Hemmat, A. Specific energy consumption for reducing the size of alfalfa chops using a hammer mill. *Biosyst. Eng.* **2010**, *105*, 34–40. [CrossRef]

8

Process Analytical Technology for Advanced Process Control in Biologics Manufacturing with the Aid of Macroscopic Kinetic Modeling

Martin Kornecki and Jochen Strube *

Institute for Separation and Process Technology, Clausthal University of Technology, Leibnizstr. 15, 38678 Clausthal-Zellerfeld, Germany; kornecki@itv.tu-clausthal.de
* Correspondence: strube@itv.tu-clausthal.de

Abstract: Productivity improvements of mammalian cell culture in the production of recombinant proteins have been made by optimizing cell lines, media, and process operation. This led to enhanced titers and process robustness without increasing the cost of the upstream processing (USP); however, a downstream bottleneck remains. In terms of process control improvement, the process analytical technology (PAT) initiative, initiated by the American Food and Drug Administration (FDA), aims to measure, analyze, monitor, and ultimately control all important attributes of a bioprocess. Especially, spectroscopic methods such as Raman or near-infrared spectroscopy enable one to meet these analytical requirements, preferably in-situ. In combination with chemometric techniques like partial least square (PLS) or principal component analysis (PCA), it is possible to generate soft sensors, which estimate process variables based on process and measurement models for the enhanced control of bioprocesses. Macroscopic kinetic models can be used to simulate cell metabolism. These models are able to enhance the process understanding by predicting the dynamic of cells during cultivation. In this article, in-situ turbidity (transmission, 880 nm) and ex-situ Raman spectroscopy (785 nm) measurements are combined with an offline macroscopic Monod kinetic model in order to predict substrate concentrations. Experimental data of Chinese hamster ovary cultivations in bioreactors show a sufficiently linear correlation ($R^2 \geq 0.97$) between turbidity and total cell concentration. PLS regression of Raman spectra generates a prediction model, which was validated via offline viable cell concentration measurement (RMSE ≤ 13.82, $R^2 \geq 0.92$). Based on these measurements, the macroscopic Monod model can be used to determine different process attributes, e.g., glucose concentration. In consequence, it is possible to approximately calculate ($R^2 \geq 0.96$) glucose concentration based on online cell concentration measurements using turbidity or Raman spectroscopy. Future approaches will use these online substrate concentration measurements with turbidity and Raman measurements, in combination with the kinetic model, in order to control the bioprocess in terms of feeding strategies, by employing an open platform communication (OPC) network—either in fed-batch or perfusion mode, integrated into a continuous operation of upstream and downstream.

Keywords: process analytical technology; macroscopic modeling; biologics manufacturing; upstream; downstream; turbidity; Raman; Chinese hamster ovary; process control

1. Introduction

1.1. Process Analytical Technology

The steadily increasing demand for process robustness and understanding has led to the introduction of the process analytical technology (PAT) initiative in the biotechnological, biopharmaceutical, and food industry by the Food and Drug Administration (FDA) in 2004 [1–3]. The PAT initiative aims to measure, analyze, monitor, and ultimately control all important attributes of a bioprocess in order to maintain or improve product quality [1,4–6]. These attributes comprise process parameters (e.g., pH, pO_2, gas flow) and variables (e.g., biomass/viability, substrate/metabolite/product concentration). The overall goal is to control critical process parameters that influence the cellular growth rate μ, production rates q_i of the product, host cell proteins, and metabolites ($q_{product}$ q_{HCP}, q_{lac}, q_{amm}), as well as product quality (e.g., structure, post-translational modifications, and efficacy) [7,8]. Hence, the real-time monitoring of bioreactors is crucial for an efficient, well controlled, robust bioprocess. Depending on the location of the analysis, bioreactor monitoring techniques distinguish between in-situ, which can be invasive, non-invasive, or placed in a sampling loop, or ex-situ [9]. In-situ or ex-situ spectroscopic methods are employed for the quantitative or qualitative description of process variables, due to their fast, sensitive, and reliable characteristics [9].

Exemplary methods are shown in Table 1. Spectroscopic methods aim to measure the cellular condition, as well as substrate, product, and metabolite concentration online. The common online determination of these variables using dedicated measurement devices (online glucose analyzer, cell imager) is either laborious, susceptible to maintenance efforts, or has the tendency to be used specifically for only one application. Spectroscopic probes eliminate those drawbacks by multiplexing, and process independent economic use and simple integration [4,10–12]. However, the analysis and interpretation of spectral data can be complex, and the correlation between responses and factors may not always be obvious [13].

Table 1. Exemplary overview of methods that are able to measure process variables of mammalian cell culture quantitatively or qualitatively, according to [4,13]. UV/Vis, ultraviolet and visible spectroscopy; NIR, near-infrared spectroscopy; MIR mid-infrared spectroscopy; VCD, viable cell density.

Method	Component	Quantitative or Qualitative
UV/Vis	Cell density	Quantitative
Fluorescence	VCD, titer, tyrosine, tryptophan	Quantitative
Raman	Glc, Lac, Gln, Glu, Amm, VCD	Quantitative
NIR	Glc, Lac, Biomass, Gln, AMM, titer, viscosity	Quantitative and qualitative
MIR	Glc, EtOH, organic acids	Quantitative

Despite this, the analysis of complex spectral data, e.g., Raman, NIR, MIR, and fluorescence, can be conducted using multivariate chemometric techniques. Chemometrics comprises mathematical and statistical methods in order to design or select optimal measurement procedures and to obtain maximal information by analyzing chemical data [14]. Techniques such as Principal Component Analysis (PCA), Principal Components Regression (PCR), and Partial Least Squares Regression (PLS) are used in order to describe the correlation between responses and factors, especially in spectral data [4,9,14,15]. The application of these methods is mainly due to the non-selectivity of the spectroscopic measurements and the high collinearity of the variables [9].

The high resolution data acquisition of target variables such as biomass or substrates is crucial for process control and gaining deeper process understanding, due to the correlation between process parameters, the process state, and target attributes. The online, offline, and soft sensor/model determination of these variables can be seen in Figure 1. Process variables such as stirrer rate, air flow, or base addition are measured, monitored, and controlled using dedicated control units (i.e., DCU, digital control unit) or OPC (open platform communication) platforms. Variables, which are in-situ not directly accessible (e.g., biomass, substrate concentration), can be monitored employing spectroscopic techniques, for example.

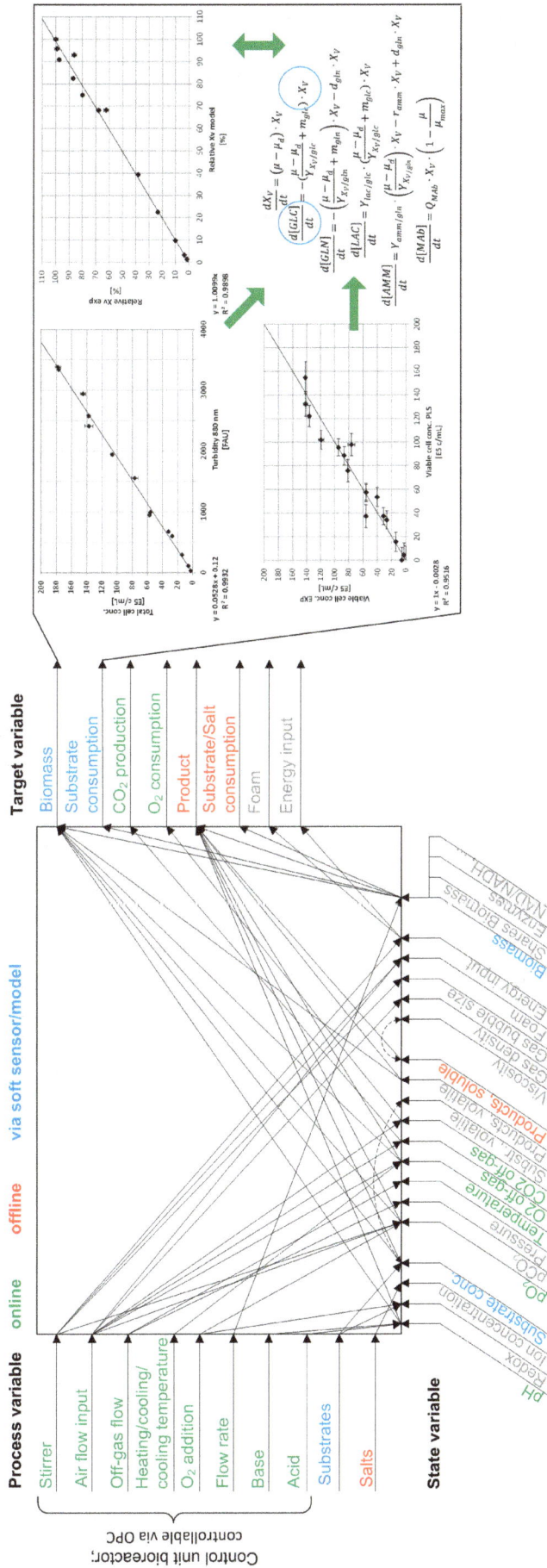

Figure 1. Main correlations between process, state, and target variables according to [22]. In this experimental set-up green, red, and blue variables can be measured/controlled online, offline, or via a soft sensor/kinetic model, respectively (**left**). The target variable biomass can be measured indirectly via turbidity (linear correlation) or Raman spectroscopy (PLS). The substrate consumption, e.g., glucose, can be calculated due to the relationship between cell and substrate concentration in a macroscopic model based on Monod equations (**right**).

Commercially available in-situ turbidity sensors, which measure the optical density (OD) of suspensions, can be implemented for the determination of total cell concentrations of mammalian cell cultures [2,16]. Inline turbidity measurement techniques distinguish between transmission, absorption, reflection, or light scattering [2,17]. The most significant advantages of turbidity probes are their measurement resolution, simple use, and implementation. However, depending on the cellular viability, the optical density can considerably differ from the viable cell concentration.

Nevertheless, turbidity probes are able to directly collect online data regarding total cell concentration and can be used for process control if the cellular viability is not decreasing (i.e., viable cell concentration does not significantly differ from total cell concentration). This makes turbidity sensors a viable monitoring and controlling technique for bioprocesses [16,18] (especially for continuous bioprocessing, e.g., perfusion turbidostat).

Therefore, techniques that provide insight into the cellular viability are favorable and should be monitored in real-time according to the ICH Q8-R2 guidance [16,19]. Raman spectroscopy, in combination with chemometric analysis methods, can, for example, be used for the determination of the total or viable cell concentration, as well as for the substrate and metabolite concentration [4,20].

The combination of turbidity and/or Raman spectroscopy with a macroscopic model such as the soft sensor enables the estimation and/or prediction of process variables, which are not directly measureable [21].

1.2. Macroscopic Models

Macroscopic models that predict the dynamic state of a mammalian cell culture help to gather vast information about the cellular condition, including correlations between growth and substrate uptake, as well as metabolite and product accumulation [23–29]. According to [23], the relationship between input (e.g., substrate concentration) and output (e.g., cell concentration) variables can be macroscopically modeled by using methods such as statistics (e.g., PCA), empirical observations (e.g., yield coefficients), and metabolic networks (e.g., metabolic flux analysis). The kinetic model can be established by neural networks and logistic or Monod-type approaches. Initiating with experimental and literature data, the correlation between input and output variables using empirical observations (i.e., yield coefficients), as well as the establishment of the kinetic using a Monod-type approach, seems to be the most straightforward method for implementation. Metabolic networks, however, represent a more detailed image of the cell [30]. However, the construction of such models is more time consuming than empirical observations, if reaction and flux rates of metabolic pathways are experimentally determined using metabolomics and not adapted from literature [31].

1.3. Main Metabolism of Mammalian Cells

In terms of the metabolism of mammalian cells, Figure 2 shows a schematic overview of the main catabolic metabolism, which is divided into three stages [32–35]:

1. Nutrients such as polysaccharides, as well as proteins and lipids, are broken down into their components.
2. Components, derived in stage one, are converted into their common compounds, pyruvate, and acetyl-CoA.
3. Finally, acetyl-CoA is integrated into the citric acid cycle, which is accompanied by oxidative phosphorylation.

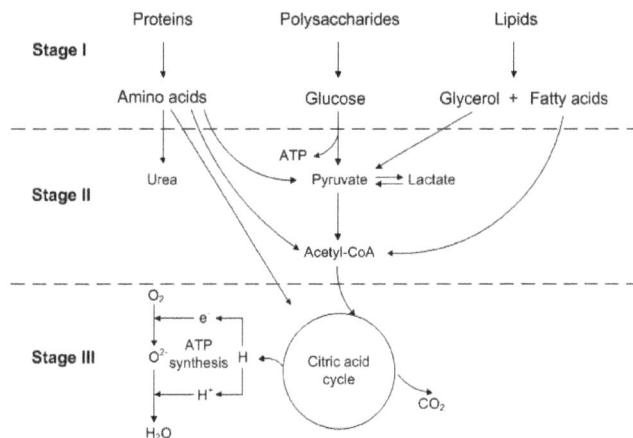

Figure 2. Schematic overview of the main catabolic metabolism divided into three stages, adapted from [32].

The main glucose metabolism that initiates with glycolysis alongside the pentose phosphate cycle (synthesis of ribose 5-phosphate, a precursor of nucleotides) is depicted in Figure 3. Previous research has shown that energy production from glucose alone is not sufficient [36]. Glucose cannot, however, be added in arbitrary quantities, as excess glucose leads to substrate inhibition. This means alternative energy sources, for example the 20 proteinogenic amino acids, are required to manage increased energy requirements [37]. Due to the efficient absorption of glutamine, it is added to most cultivation media as an alternative energy source. Glutamine plays a central role in the metabolism, since it serves as a starting material in numerous reactions and syntheses. These include, for example, peptide and protein synthesis, as well as the formation of amino sugars and nucleic acid synthesis [38].

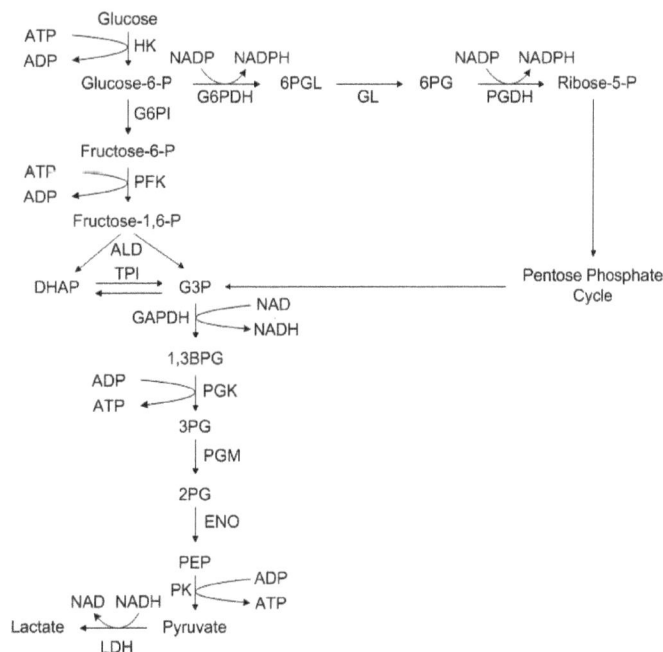

Figure 3. Production of lactate and pyruvate in glycolysis. HK, Hexokinase; G6PI, Glucose-6-phosphate isomerase; PFK, Phosphofructokinase; ALD, Aldolase; TPI, Triosephosphate isomerase; GAPDH, Glyceraldehyde-3-phosphate dehydrogenase; PGK, Phosphoglycerate kinase; PGM, Phosphoglycerate mutase; ENO, Enolase; PK, Pyruvate kinase; LDH, Lactate dehydrogenase; G6PDH, Glucose-6-phosphate dehydrogenase; GL, 6-Phosphogluconolactonase; PGDH, Phosphogluconate dehydrogenase; DHAP, Dihydroxyacetone phosphate; G3P, Glyceraldehyde 3-phosphate; 1,3BPG, 1,3-Bisphosphoglycerate; 3PG, 3-Phosphoglycerate; 2PG, 2-Phosphoglycerate; PEP, Phosphoenolpyruvate; 6PGL, 6-Phosphogluconolactone; 6PG, 6-Phosphogluconate.

Besides the main energy sources, the resulting metabolites also play an important role in cultivation. The main by-products of glucose and glutamine metabolism are lactate and ammonium, which have a strong influence on the life span of cells and the production of the antibody [39–41].

In addition to pyruvate, the crucial metabolite lactate is being produced in glycolysis during the exponential growth phase. The consumption of lactate is observed during the stationary phase upon depletion of glutamine and is desirable, since it seems to correlate to improved optimal process performance [42]. However, the reduction of pH, as well as the inhibition of cell growth and production of IgG, constitutes side-effects, resulting in the increase of lactate [37]. Nevertheless, variation of pH can be reduced by an appropriate controller during a bioprocess. However, the addition of ions leads to an increased osmotic pressure, which in turn has negative effects on cell culture. At most, non-controlled cultivations in shaking flasks may be exposed to pH variation. For these reasons, the lactate content should be kept as low as possible.

The also critical by-product ammonium (NH_4^+) is mainly produced via the metabolism of glutamine [38,39,43]. The degradation of glutamine to α-ketoglutarate in glutaminolysis can take place via two metabolic pathways: on the one hand, via the transamination reaction, and on the other hand, via the complete oxidation of glutamine catalyzed by the glutamine dehydrogenase (GDH). The amount of ammonium has a stronger impact on the culture than the lactate. However, it is difficult to determine an exact threshold for the ammonium concentration, since different critical values, between 2 and 10 mM, are given in the literature [39,40,44]. Glutamine is first converted into glutamate in glutaminolysis with ammonium splitting off in order to be introduced into the TCA as α-ketoglutarate, during which further ammonium is split off. Up to this point, 1 mol of glutamine has been converted into 2 mol of ammonium (see Figure 4).

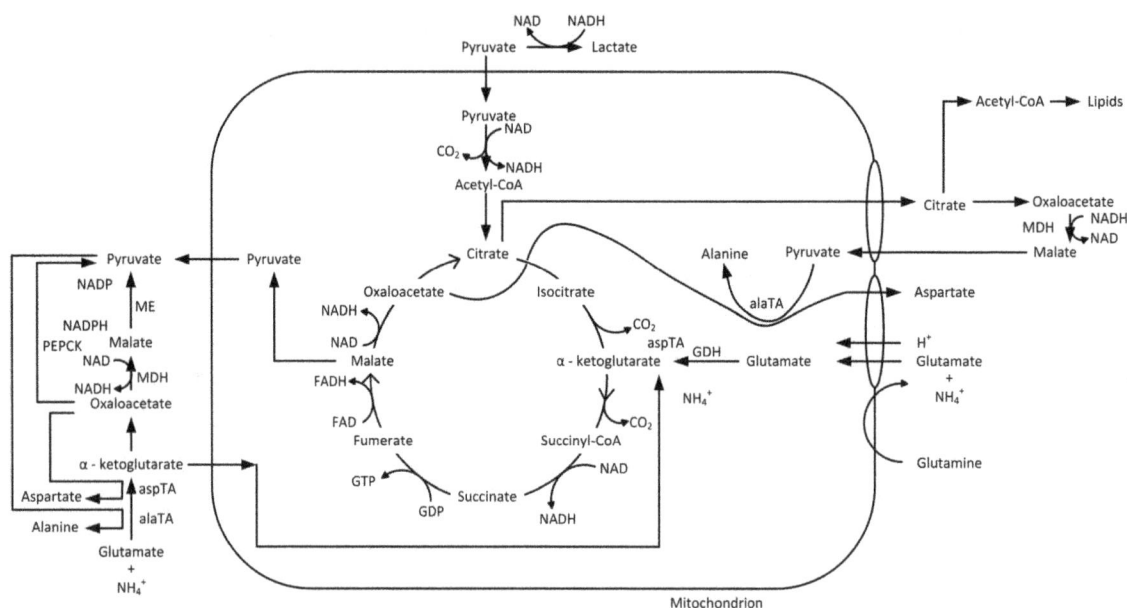

Figure 4. Citrate cycle, which represents the entry point into the energy metabolism, adapted from [37].

As ammonium dissociates, ammonia is present as an uncharged molecule, which can diffuse freely through the cell membrane. As soon as a proton is released into the culture medium, the pH value in the medium decreases. In the cell, on the other hand, the pH value is increased by the uptake of a proton to form ammonium, which results in a pH gradient between the intracellular and extracellular space. In order to compensate for this gradient, the cell has to re-establish the pH-value under energy consumption by means of ion pumps. As a result, the basic energy requirement for maintaining the organism increases significantly and can therefore no longer be used for the proliferation or production of the antibody [36]. As a consequence, ammonium has a negative effect on cell growth and viability [38–40,43].

The concentration of inhibiting metabolites, such as lactate and ammonium, has to be monitored and preferably controlled for the above-mentioned reasons.

In this approach, in-situ turbidity and ex-situ Raman spectroscopy measurements are combined with an offline macroscopic Monod kinetic model in order to estimate substrate and metabolite concentrations (i.e., glucose and its metabolite lactate) as proof of principle. Macroscopic models lead to a more detailed process understanding and, in combination with experimental data, these models can be employed for the online prediction of substrates or metabolites. The connection between online cell concentration measurement and reliable online substrate or metabolite prediction will be one important step towards the online monitoring and controlling of bioprocesses.

2. Materials and Methods

Chinese hamster ovary cells (CHO DG44) were used in bioreactor (Biostat® B, Sartorius Stedim Biotech GmbH, Göttingen, Germany) cultivations for the production of a monoclonal antibody in a serum-free medium. The cultivation conditions were 36.8 °C, 5.0% carbon dioxide, 433 rpm, pH 7.1 and 60.0% dissolved oxygen (pO_2). The in-situ turbidity probe (transmission, 880 nm, HiTec Zang GmbH, Herzogenrath, Germany) was connected to a LabManager® (HiTec Zang GmbH). Raman spectroscopy was performed using an ex-situ Raman probe (Diode laser, 785 nm, Ocean Optics B.V, Ostfildern, Germany) in order to correlate the spectral data and the viable cell concentration using the NIPALS algorithm derived from the software The Unscrambler® (CAMO Software AS., Oslo, Norway).

Offline viable and total cell concentration were repeatedly determined by using a Neubauer chamber (BRAND GMBH + CO KG, Wertheim, Germany), microscope (Motic BA 310, Motic Deutschland GmbH, Wetzlar, Germany), and trypan blue solution (0.4%, Sigma-Aldrich, St. Louis, MO, USA) as dye for the detection of dead cells. Glucose and lactate were repeatedly measured using a LaboTRACE compact (TRACE Analytics GmbH, Braunschweig, Germany). The monoclonal antibody was quantified by Protein A chromatography (PA ID Sensor Cartridge, Applied Biosystems, Bedford, MA, USA). Dulbecco's PBS buffer (Sigma-Aldrich, St. Louis, MO, USA) was used as loading buffer at pH 7.4 and as elution buffer at pH 2.6. The absorbance was monitored at 280 nm.

The macroscopic kinetic model was developed in Aspen Custom Modeler V8.4 (Aspen Technology, Inc., Bedford, MA, USA). Since the cell cultures were performed in fed-batch mode with daily bolus feed additions, the model equations were extended by feeding terms. Consequently, the bioreactor volume was considered as well.

3. Results

3.1. Online Cell Concentration Measurement

The basic principle of the correlation between process, state, and target variables is shown in Figure 1. This relationship, alongside the results from online turbidity and Raman measurements, as well as macroscopic modeling, is shown and discussed in detail in the following. The in-situ measurement of the turbidity of the cell culture suspension is one of the most applied techniques for biomass monitoring, due to the simplicity of this method [2]. Here, an online transmission probe using a wavelength of 880 nm was implemented in order to record the total cell concentration of two CHO DG44 bioreactor cultivations. The correlation between the turbidity measurements (formazine attenuation units, FAU) and total cell concentration (E5 cells/mL) is shown as an example for one bioreactor cultivation in Figure 5. The linear relation was found to be in good agreement for two bioreactor cultivations ($R^2 \geq 0.97 \pm 0.02$).

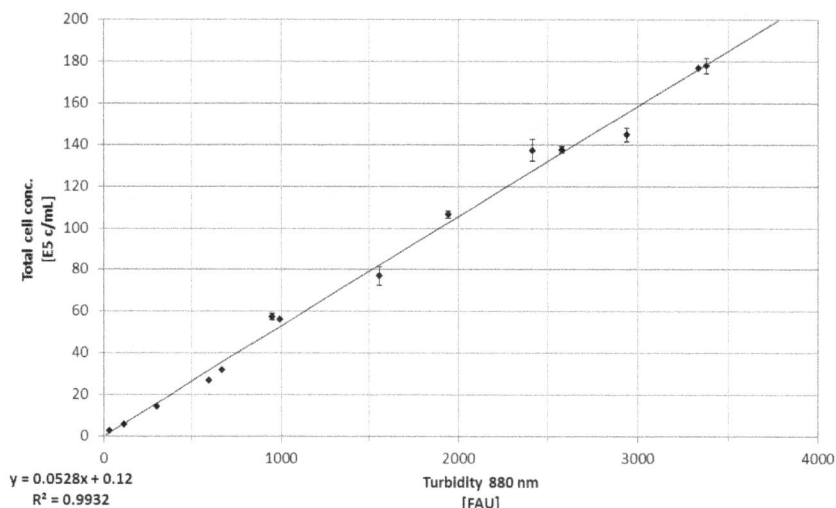

$$y = 0.0528x + 0.12$$
$$R^2 = 0.9932$$

Figure 5. Correlation between offline determination of total cell concentration (E5 cells/mL) and online turbidity measurements (880 nm, FAU). Error bars represent the double determination of cell concentration, as well as the error of the probe (0.75%). The coefficient of determination for this linear relation is >0.99.

Compared to the determination of the total cell concentration using turbidity probes, Raman spectroscopy combined with chemometric techniques can be used to predict substrate concentration, i.e., glutamine or glutamic acid, or viable and total cell concentration [2]. Using this approach, ex-situ Raman spectroscopy and PLS were used in order estimate the viable cell concentration offline. The PLS model was established using the NIPALS algorithm alongside cross-validation. Four factors are needed in order to explain 94% of the response variance. The resulting response prediction generates a sufficient model error (RMSE $\leq$ 13.82, $R^2 \geq$ 0.92) for estimating the experimental viable cell concentration.

The correlation between the predicted viable cell concentration that results from the NIPALS algorithm and the experimental viable cell concentration is shown in Figure 6. Considering the nature of ex-situ spectral data acquisition (e.g., possible long hold up, variations of process variables in sampling loop), this prediction method seems to be quite sufficient for determining the viable cell concentration in the bioreactor.

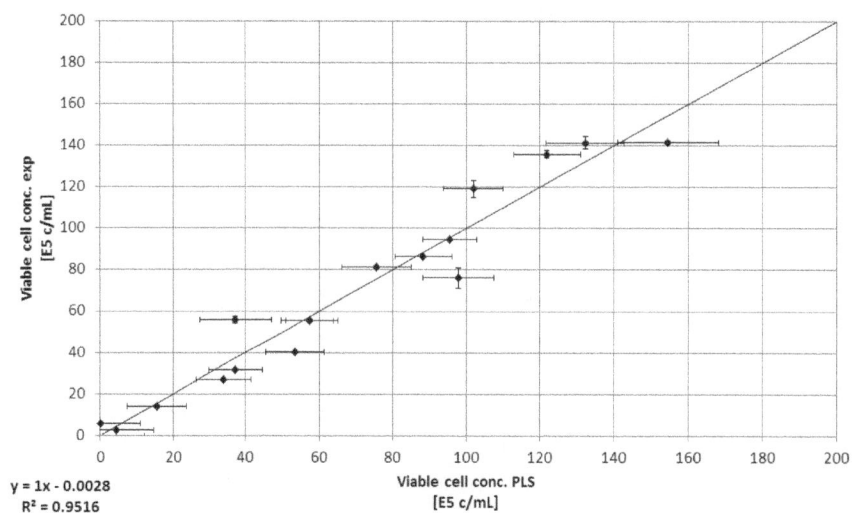

$$y = 1x - 0.0028$$
$$R^2 = 0.9516$$

Figure 6. Correlation between offline viable cell concentration (E5 cells/mL) determination and PLS (NIPALS)-based viable cell concentration (E5 cells/mL). Error bars represent the double determination of cell concentration, as well as the model error (RMSE $\leq$ 13.82). The coefficient of determination for this linear relation is >0.95.

Both techniques generate reproducible data and predictions. The acquisition of turbidity is more easily available and is simpler to integrate and interpret than Raman spectroscopy coupled to PLS. However, turbidity probes are only able to determine the total cell concentration based on the optical density of the culture suspension instead of the viable concentration and are affected by air bubbles and noncellular particles [2]. Regarding the productivity of secondary metabolites or recombinant proteins like monoclonal antibodies in the stationary phase of a (fed-)batch culture, the cellular state of the culture, and therefore its viability, is of utmost importance [45]. Hence, techniques that are able to determine the viable cell concentration are in favor. This prediction can efficiently be accomplished by Raman spectroscopy or, for example, near-infrared and impedance measurements [2]. The discrepancy between the results of Raman spectroscopy and turbidity can be seen in Figure 7.

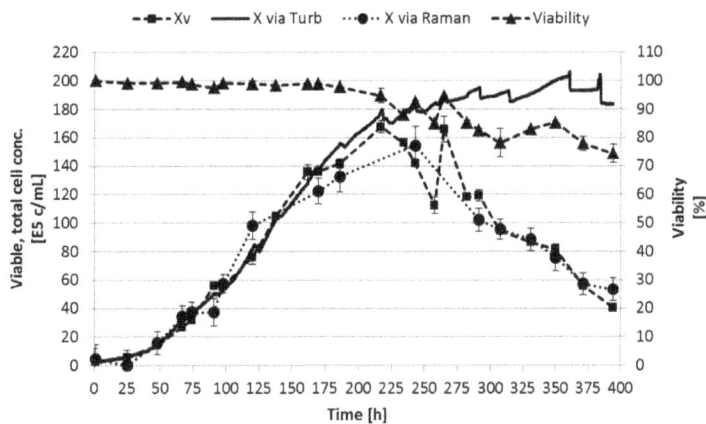

Figure 7. Viable and total cell concentration based on Raman spectroscopy coupled to PLS and turbidity, respectively, compared to experimental viable cell concentration. Error bars represent the double determination of cell concentration and viability, as well as the model error of PLS (RMSE $\leq$ 13.82).

The difference between the results of turbidity (total cell concentration) and Raman (viable cell concentration) measurements is expected. As can be seen in Figures 7 and 8, the turbidity is in good correlation to the viable cell concentration due to the high viability of the cell culture. However, after approximately 234 h culture time, the turbidity probe measurements do not correlate to the viable cell concentration any more. Here, the PLS model is able to predict the decreasing viable cell concentration. The slowly rising cell concentration based on turbidity after 234 h is mainly due to cell accumulation in the probe gap. After the probe positioning in the bioreactor was adjusted, this increase did not occur again.

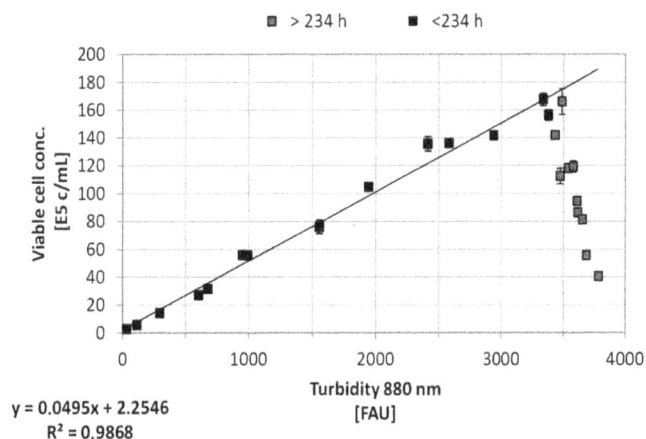

Figure 8. Correlation between offline viable cell concentration (E5 cells/mL) determination and online turbidity measurements (880 nm, FAU) prior to and after 234 h bioreactor cultivation. Error bars represent the double determination of cell concentration, as well as the deviation of the probe (0.75%). The coefficient of determination for this linear relation is >0.98 for $t <$ 234 h.

In conclusion, turbidity data can efficiently be used as process variable for process control if the cellular viability does not significantly differ ($\geq$10%, see Figure 7, $\geq$234 h) between viable and total cell concentration. Besides, according to the more complex interpretation of Raman spectra, this technique is able to predict the cellular state more reliable. Methods are available that can acquire spectra and integrate them into PLS models, resulting in the online determination of viable cell, substrate, and metabolite concentrations [46,47]. For this, PLS models have to be robust and validated using validation data sets. By increasing the range of these validation data sets it is possible to reduce the model error significantly [48].

3.2. Macroscopic Kinetic Modeling

Macroscopic kinetic models are being used in order to simulate the cellular state and productivity in-silico [23,24,29,49]. In this approach, a Monod type model is used for the offline prediction of glucose concentration. This methodology will be enhanced in future experimental set-ups to predict the substrate concentration online and implement a control strategy utilizing an OPC network. The model consists of Monod-type equations that describe the time-dependent state of the viable cell $[X_V]$, glucose [GLC], glutamine [GLN], lactate [LAC], ammonium [AMM], and monoclonal antibody [mAb] concentration. Model parameters such as Monod constants (K_{glc}, K_{gln}, K_{Ilac}, K_{Iamm}), μ_{max}, and yield coefficients ($Y_{X/glc}$, $Y_{X/glc}$, $Y_{X/glc}$) are calculated based on bioreactor cultivations or are adopted from literature. The Equations (1) to (9) are given in the following and are mostly adopted from Xing et al. [49]. The employed model parameters are given in the following Table 2.

$$\frac{d[X_V]}{dt} = (\mu - \mu_d) \cdot [X_V], \tag{1}$$

$$\mu = \mu_{max} \cdot \frac{[GLC]}{K_{glc} + [GLC]} \cdot \frac{[GLN]}{K_{gln} + [GLN]} \cdot \frac{K_{Ilac}}{K_{Ilac} + [LAC]} \cdot \frac{K_{Iamm}}{K_{Iamm} + [AMM]}, \tag{2}$$

$$\mu_d = k_d \cdot \frac{[LAC]}{K_{Dlac} + [LAC]} \cdot \frac{[AMM]}{K_{Damm} + [AMM]}, \tag{3}$$

$$\frac{d[GLC]}{dt} = -\left(\frac{\mu - \mu_d}{Y_{\frac{X}{GLC}}} + m_{glc}\right) \cdot [X_V], \tag{4}$$

$$\frac{d[GLN]}{dt} = -\left(\frac{\mu - \mu_d}{Y_{\frac{X}{GLN}}} + m_{gln}\right) \cdot [X_V], \tag{5}$$

$$m_{gln} = \frac{a_1 \cdot [GLN]}{a_2 + [GLN]}, \tag{6}$$

$$\frac{d[LAC]}{dt} = Y_{\frac{LAC}{GLC}} \cdot \left(\frac{\mu - \mu_d}{Y_{\frac{X}{GLC}}} + m_{glc}\right) \cdot [X_V], \tag{7}$$

$$\frac{d[AMM]}{dt} = Y_{\frac{AMM}{GLN}} \cdot \left(\frac{\mu - \mu_d}{Y_{\frac{X}{GLC}}}\right) \cdot [X_V] - r_{Amm} \cdot [X_V], \tag{8}$$

$$\frac{d[mAb]}{dt} = Q_{mAb} \cdot [X_V] \cdot \left(1 - \frac{\mu}{\mu_{max}}\right), \tag{9}$$

The cellular state, together with the product and glucose concentration, can be predicted sufficiently, as seen in Figure 9. The coefficient of determination for each prediction is greater than 0.92. In order to improve the prediction quality, the Markov Chain Monte Carlo method can be used for the prediction of parameters that are difficult to obtain such as m_{glc}, K_{gln}, or K_{glc} [49]. However, even parameters that can be derived from experimental data [23], such as μ_{max} or yield coefficients (e.g., $Y_{X/glc}$, $Y_{X/gln}$), exhibit a significant influence on model variables, especially $[X_v]$ and [GLC].

The model parameter μ_{max}, for example, varied according to the minimal and maximum values reported for mammalian cells, as well as $\pm 5\%$ and $\pm 10\%$ of the experimentally determined value in order to examine its influence on the course of the viable cell and glucose concentration (Figure 10).

Table 2. Model parameters used in the macroscopic Monod model for the simulation of CHO cultivations based on literature (Lit.) or experimental data (exp). Literature data were determined using fed-batch CHO cell culture and the markov chain monte carlo method for parameter estimation [49] or by the differential evolution technique [50].

Parameter	Description	Value	Unit	Source
μ_{max}	Maximum growth rate	0.039	h^{-1}	exp
k_d	Maximum death rate	0.004	h^{-1}	exp
K_{glc}	Monod constant glucose	1.00	mM	[50]
K_{gln}	Monod constant glutamine	0.047	mM	[49]
K_{Ilac}	Monod constant lactate for inhibition	43.00	mM	[49]
K_{Iamm}	Monod constant ammonium for inhibition	6.51	mM	[49]
K_{Dlac}	Monod constant lactate for death	45.8	mM	[49]
K_{Damm}	Monod constant ammonium for death	6.51	mM	[49]
$Y_{X/glc}$	Yield coefficient cell conc./glucose	0.357	E9 cells mmol^{-1}	exp
$Y_{X/gln}$	Yield coefficient cell conc./glutamine	0.974	E9 cells mmol^{-1}	[49]
$Y_{lac/glc}$	Yield coefficient lactate/glucose	0.70	mmol mmol^{-1}	exp
$Y_{amm/gln}$	Yield coefficient ammonium/glutamine	0.67	mmol mmol^{-1}	[49]
r_{amm}	Ammonium removal rate	6.3	E-12 mmol cell$^{-1}\cdot$h^{-1}	[49]
m_{glc}	Glucose maintenance coefficient	69.2	E-12 mmol cell$^{-1}\cdot$h^{-1}	[49]
a_1	Coefficient for m_{gln}	3.2	E-12 mmol cell$^{-1}\cdot$h^{-1}	[49]
a_2	Coefficient for m_{gln}	2.1	mM	[49]
Q_{mAb}	Specific production rate	1.51	E-12 g$\cdot$c$^{-1}\cdot$h^{-1}	exp

As can be seen in Equations (1)–(9) and Figure 10, the maximum growth rate μ_{max} may be the key parameter of this macroscopic model. An increase of the maximum growth rate has significant impact on growth and maximum cell concentration, and consequently on substrate concentration. Accompanied by the increased cell concentration, the glucose concentration decreases more rapidly (constant feed strategy assumed). This relation is due to the dependency of the substrate concentration on the cell concentration (Equation (4)), which therefore affects the temporal growth rate (Equation (1)) and hence the maximum growth rate μ_{max} (Equation (2)). This effect is not surprising, but clearly shows the importance of an exact parameter determination methodology for highly sensitive parameters. Here, μ_{max} exhibited the most significant influence on the course of model variables. For example, online measurements such as turbidity can efficiently be used for the determination of μ_{max}, due to their very high sample resolution. The improved data sets result in a significantly improved data consistency and in a more precise model parameter prediction.

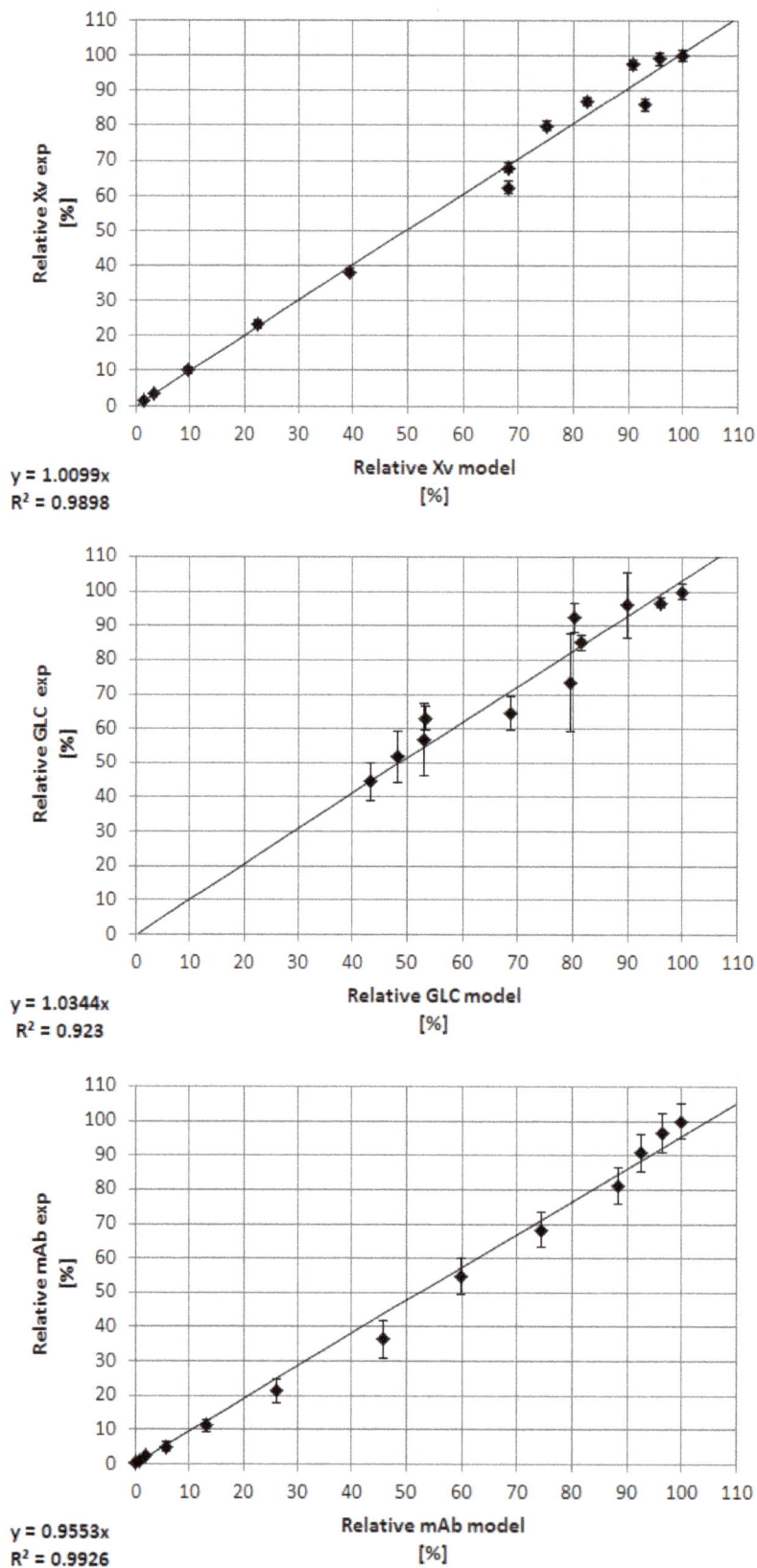

y = 1.0099x
R² = 0.9898

y = 1.0344x
R² = 0.923

y = 0.9553x
R² = 0.9926

Figure 9. Correlation between experimental and predicted values for the viable cell, product, and glucose concentration. The error bars represent the standard deviation of various bioreactor cultivations. The coefficient of determination is ≥0.92.

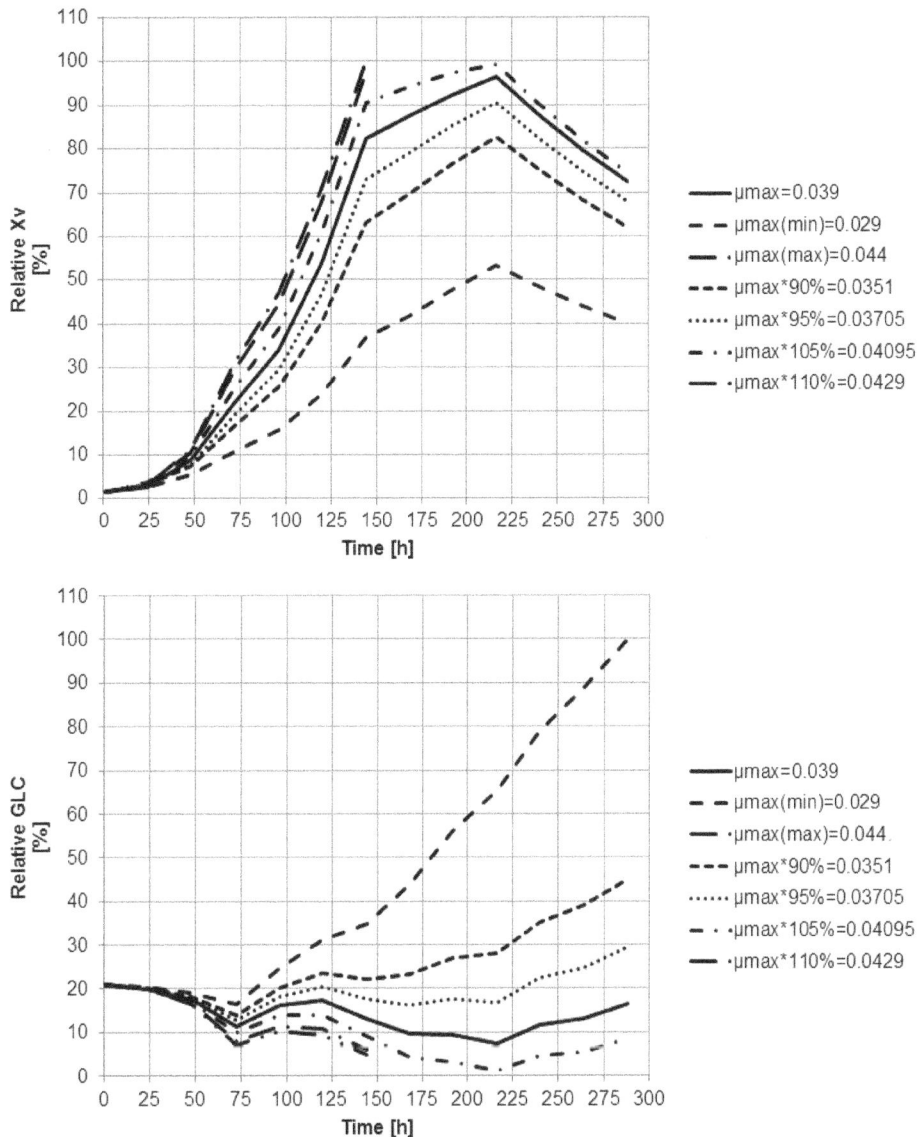

Figure 10. Influence of μ_{max} on the viable cell and glucose concentration. μ_{max} values represent data reported for mammalian cells (min, max), as well as ±5% and ±10% of the experimental determined value.

3.3. Combination of Experimental Data and Predictive Modeling

The prediction of the experimental viable cell and substrate concentration using a macroscopic kinetic model shows a sufficient agreement. In order to integrate this Monod type model into the process environment, online turbidity data, which represents the cell concentration, were connected to the model for the prediction of glucose concentration (Equation (4)). In future approaches, this shall lead to a simple method for the approximate determination of growth limiting substrates, such as glucose. The comparison between offline measurements and model-based prediction of the glucose concentration can be seen in the following Figure 11.

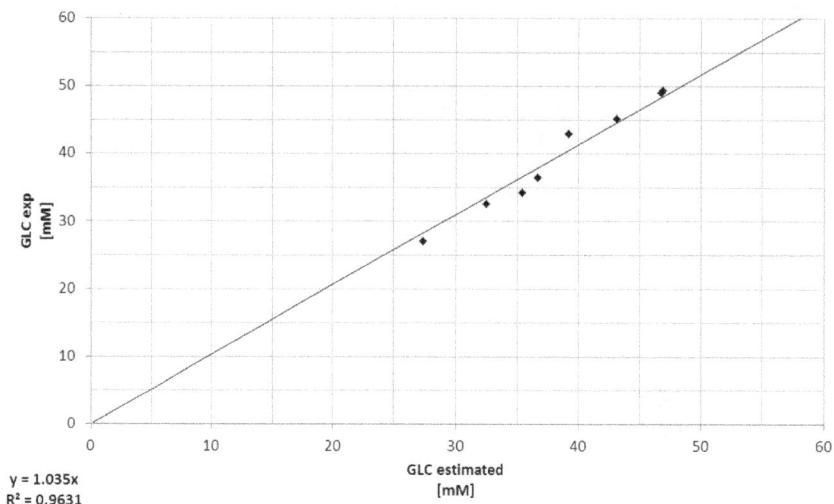

Figure 11. Comparison between offline measurements (exp) and model-based prediction of the glucose concentration. The deviation represents the offline double determination of the glucose concentration. The coefficient of determination for this relation is ≥ 0.96.

As it can be seen in Figure 11, the model based determination of the glucose concentration sufficiently ($R^2 \geq 0.96$) predicts the offline measured values, especially in the pre-feed phase ($\leq 67\,h$) of the cultivation. The prediction of the glucose concentration in the feed phase is significantly more challenging, since physiological effects of the feed media on the cellular state (e.g., substrate limitation/inhibition, change from cell growth to production, influence of feed media components) are not described by this model at this time. Furthermore, any process disturbances result in a non-ideality of the cellular state or substrate concentration and are difficult to predict using macroscopic models. In addition, the viable cell concentration determination based on turbidity measurements is dependent on a viable cell state (Viability $\geq 90\%$). The discrepancy between experimentally determined viable cell concentration and turbidity-based calculation (Figure 8) results in an additional deviation of the estimated glucose concentration from the available glucose in culture. Besides the need for a reliable estimation or even prediction of the substrate concentration, metabolites have to be predicted, because lactate or ammonium are responsible for inhibiting cell proliferation (see Figures 3 and 4). The comparison between offline measurements and model-based prediction of the lactate concentration can be seen in the following Figure 12.

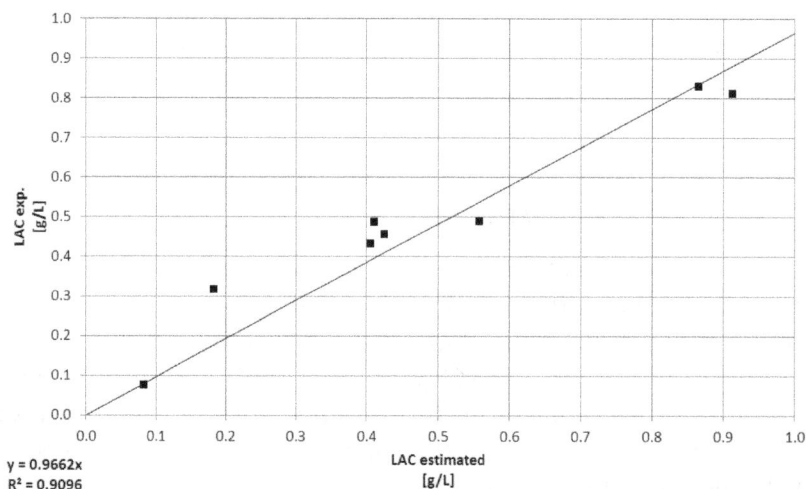

Figure 12. Comparison between offline measurements (exp) and model-based prediction of the lactate concentration. The deviation represents the offline double determination of the lactate concentration. The coefficient of determination for this relation is ≥ 0.90.

Nevertheless, turbidity measurements combined with a macroscopic kinetic model can be used for an online estimation of substrate concentration and, in future approaches, will be validated and extended to metabolites (e.g., ammonium), since they play a significant role in cell dynamics and protein production as described above (see Figures 3 and 4).

4. Conclusions

The presented approach of online turbidity and Raman spectroscopy depicts their advantages and disadvantages regarding their implementation in bioprocesses. Turbidity measurements seem to be more suited to a fast and simple technique for interpreting cellular state monitoring, despite the limitations on total cell concentration. Therefore, spectroscopy methods that display the cellular viability, such as Raman coupled to chemometric analysis (e.g., PLS), are in favor. However, the interpretation of the acquired spectra can be complex and are not always obvious. The correlation of total and viable cell concentration using online turbidity ($R^2 \geq 0.99$) and Raman spectroscopy ($R^2 \geq 0.95$), respectively, is sufficient and capable of approximately predicting the glucose ($R^2 \geq 0.96$) and lactate concentration ($R^2 \geq 0.90$) by employing a macroscopic Monod-type model.

In order to measure, analyze, monitor, and ultimately control all important attributes of a bioprocess, reliable measurement techniques are essential. In future approaches, the prediction of the glucose concentration will be further validated and implemented into a control strategy using turbidity or Raman spectroscopy, as well as an OPC platform. This strategy will merge online measurements, macroscopic models, and finally advanced process control for robust and reliable batch or continuous operation.

Those steps are also crucial for any innovation and optimization of the manufacturing technology of biologics such as, e.g., the transfer form batch to continuous processes. In a continuous operation, process robustness is even more crucial in order to prevent failures or instabilities, as the annual costs would be significantly higher than in a comparable batch operation. First studies on continuous biomanufacturing prove better reproducibility [51]. Appropriate process control strategies based on sound PAT tools, in combination with process modeling technologies, as shown here, will keep critical product quality attributes reliably within specification by controlling operation parameters within their design space.

Improved reliability in process operation will enable manufacturing optimization projects and even innovation in industrialization towards more economic manufacturing of innovative medicines for the patients' sake [52–57].

Acknowledgments: The authors would like to thank the ITVP lab team, especially Frank Steinhäuser, Volker Strohmeyer, and Thomas Knebel, as well as Christian Siemers (Institute of Electrical Information Technology, Clausthal University of Technology) for their efforts and support. Special thanks are also addressed to M.Sc. students Lara Lohmann for her excellent laboratory work and Tianqi Liu for his contributions to the employed macroscopic model.

Author Contributions: Martin Kornecki conceived, designed and performed the experiments as well as wrote the paper. All authors interpreted the data. Jochen Strube substantively revised the work and contributed the materials and analysis tools. Jochen Strube is responsible for conception and supervision.

Conflicts of Interest: The authors declare no conflict of interest.

References

1. Hinz, D.C. Process analytical technologies in the pharmaceutical industry: The FDA's PAT initiative. *Anal. Bioanal. Chem.* **2006**, *384*, 1036–1042. [CrossRef] [PubMed]
2. Biechele, P.; Busse, C.; Solle, D.; Scheper, T.; Reardon, K. Sensor systems for bioprocess monitoring. *Eng. Life Sci.* **2015**, *15*, 469–488. [CrossRef]
3. Food and Drug Administration. Guidance for Industry. PAT—A Framework for Innovative Pharmaceutical Development, Manufacturing, and Quality Assurance; 2004. Available online: https://www.fda.gov/downloads/drugs/guidances/ucm070305.pdf (accessed on 19 February 2018).

4. Musmann, C.; Joeris, K.; Markert, S.; Solle, D.; Scheper, T. Spectroscopic methods and their applicability for high-throughput characterization of mammalian cell cultures in automated cell culture systems. *Eng. Life Sci.* **2016**, *16*, 405–416. [CrossRef]

5. Zobel-Roos, S.; Mouellef, M.; Siemers, C.; Strube, J. Process Analytical Approach towards Quality Controlled Process Automation for the Downstream of Protein Mixtures by Inline Concentration Measurements Based on Ultraviolet/Visible Light (UV/VIS) Spectral Analysis. *Antibodies* **2017**, *6*, 24. [CrossRef]

6. Bechmann, J.; Rudolph, F.; Gebert, L.; Schaub, J.; Greulich, B.; Dieterle, M.; Bradl, H. Process parameters impacting product quality. *BMC Proc.* **2015**, *9*, O7. [CrossRef]

7. Alt, N.; Zhang, T.Y.; Motchnik, P.; Taticek, R.; Quarmby, V.; Schlothauer, T.; Beck, H.; Emrich, T.; Harris, R.J. Determination of critical quality attributes for monoclonal antibodies using quality by design principles. *Biologicals* **2016**, *44*, 1–15. [CrossRef] [PubMed]

8. Del Val, I.J.; Kontoravdi, C.; Nagy, J.M. Towards the implementation of quality by design to the production of therapeutic monoclonal antibodies with desired glycosylation patterns. *Biotechnol. Prog.* **2010**, *26*, 1505–1527. [CrossRef] [PubMed]

9. Lourenço, N.D.; Lopes, J.A.; Almeida, C.F.; Sarraguça, M.C.; Pinheiro, H.M. Bioreactor monitoring with spectroscopy and chemometrics: A review. *Anal. Bioanal. Chem.* **2012**, *404*, 1211–1237. [CrossRef] [PubMed]

10. Arnold, S.A.; Crowley, J.; Woods, N.; Harvey, L.M.; McNeil, B. In-situ near infrared spectroscopy to monitor key analytes in mammalian cell cultivation. *Biotechnol. Bioeng.* **2003**, *84*, 13–19. [CrossRef] [PubMed]

11. Rhiel, M.; Ducommun, P.; Bolzonella, I.; Marison, I.; von Stockar, U. Real-time in situ monitoring of freely suspended and immobilized cell cultures based on mid-infrared spectroscopic measurements. *Biotechnol. Bioeng.* **2002**, *77*, 174–185. [CrossRef] [PubMed]

12. Berry, B.; Moretto, J.; Matthews, T.; Smelko, J.; Wiltberger, K. Cross-scale predictive modeling of CHO cell culture growth and metabolites using Raman spectroscopy and multivariate analysis. *Biotechnol. Prog.* **2015**, *31*, 566–577. [CrossRef] [PubMed]

13. Teixeira, A.P.; Oliveira, R.; Alves, P.M.; Carrondo, M.J.T. Advances in on-line monitoring and control of mammalian cell cultures: Supporting the PAT initiative. *Biotechnol. Adv.* **2009**, *27*, 726–732. [CrossRef] [PubMed]

14. Otto, M. *Chemometrics. Statistics and Computer Application in Analytical Chemistry*, 3rd ed.; Wiley-VCH Verlag GmbH et Co. KGaA: Weinheim, Germany, 2017.

15. Faassen, S.M.; Hitzmann, B. Fluorescence spectroscopy and chemometric modeling for bioprocess monitoring. *Sensors* **2015**, *15*, 10271–10291. [CrossRef] [PubMed]

16. Kroll, P.; Stelzer, I.V.; Herwig, C. Soft sensor for monitoring biomass subpopulations in mammalian cell culture processes. *Biotechnol. Lett.* **2017**, *39*, 1667–1673. [CrossRef] [PubMed]

17. Hausmann, R.; Henkel, M.; Hecker, F.; Hitzmann, B. Present Status of Automation for Industrial Bioprocesses. In *Current Developments in Biotechnology and Bioengineering*; Elsevier: Amsterdam, The Netherlands, 2017; pp. 725–757.

18. Pörtner, R.; Platas Barradas, O.; Frahm, B.; Hass, V.C. Advanced Process and Control Strategies for Bioreactors. In *Current Developments in Biotechnology and Bioengineering*; Elsevier: Amsterdam, The Netherlands, 2017; pp. 463–493.

19. FDA; CDER; CBER; FDA; USDHHS. Pharmaceutical Development Q8(R2); 2009. Available online: https://www.ich.org/fileadmin/Public_Web_Site/ICH_Products/Guidelines/Quality/Q8_R1/Step4/Q8_R2_Guideline.pdf (accessed on 19 February 2018).

20. Esmonde-White, K.A.; Cuellar, M.; Uerpmann, C.; Lenain, B.; Lewis, I.R. Raman spectroscopy as a process analytical technology for pharmaceutical manufacturing and bioprocessing. *Anal. Bioanal. Chem.* **2017**, *409*, 637–649. [CrossRef] [PubMed]

21. Luttmann, R.; Bracewell, D.G.; Cornelissen, G.; Gernaey, K.V.; Glassey, J.; Hass, V.C.; Kaiser, C.; Preusse, C.; Striedner, G.; Mandenius, C.-F. Soft sensors in bioprocessing: A status report and recommendations. *Biotechnol. J.* **2012**, *7*, 1040–1048. [CrossRef] [PubMed]

22. Präve, P. (Ed.) *Handbuch der Biotechnologie. Mit 150 Tabellen, 26 Fließschemata sowie 80 Strukturformeln und Zahlreichen Weiteren Zusammenstellungen und Formeln*; Oldenbourg: München, Germany, 1994.

23. Yahia, B.B.; Malphettes, L.; Heinzle, E. Macroscopic modeling of mammalian cell growth and metabolism. *Appl. Microbiol. Biotechnol.* **2015**, *99*, 7009–7024. [CrossRef] [PubMed]

24. Yahia, B.B.; Gourevitch, B.; Malphettes, L.; Heinzle, E. Segmented linear modelling of CHO fed-batch culture and its application to large scale production. *Biotechnol. Bioeng.* **2016**, *9999*, 1–13. [CrossRef]

25. Goudar, C.T. Computer programs for modeling mammalian cell batch and fed-batch cultures using logistic equations. *Cytotechnology* **2012**, *64*, 465–475. [CrossRef] [PubMed]

26. Goudar, C.T.; Konstantinov, K.B.; Piret, J.M. Robust parameter estimation during logistic modeling of batch and fed-batch culture kinetics. *Biotechnol. Prog.* **2009**, *25*, 801–806. [CrossRef] [PubMed]

27. Goudar, C.T.; Joeris, K.; Konstantinov, K.B.; Piret, J.M. Logistic equations effectively model mammalian cell batch and fed-batch kinetics by logically constraining the fit. *Biotechnol. Prog.* **2005**, *21*, 1109–1118. [CrossRef] [PubMed]

28. Jang, J.D.; Barford, J.P. An unstructured kinetic model of macromolecular metabolism in batch and fed-batch cultures of hybridoma cells producing monoclonal antibody. *Biochem. Eng. J.* **2000**, *4*, 153–168. [CrossRef]

29. Sidoli, F.R.; Mantalaris, A.; Asprey, S.P. Modelling of Mammalian Cells and Cell Culture Processes. *Cytotechnology* **2004**, *44*, 27–46. [CrossRef] [PubMed]

30. Galleguillos, S.N.; Ruckerbauer, D.; Gerstl, M.P.; Borth, N.; Hanscho, M.; Zanghellini, J. What can mathematical modelling say about CHO metabolism and protein glycosylation? *Comput. Struct. Biotechnol. J.* **2017**, *15*, 212–221. [CrossRef] [PubMed]

31. Huang, Z.; Lee, D.-Y.; Yoon, S. Quantitative intracellular flux modeling and applications in biotherapeutic development and production using CHO cell cultures. *Biotechnol. Bioeng.* **2017**, *114*, 2717–2728. [CrossRef] [PubMed]

32. Christen, P.; Jaussi, R.; Benoit, R. *Biochemie und Molekularbiologie. Eine Einführung in 40 Lerneinheiten*; Springer: Berlin, Germany, 2016.

33. Schaub, J.; Clemens, C.; Schorn, P.; Hildebrandt, T.; Rust, W.; Mennerich, D.; Kaufmann, H.; Schulz, T.W. CHO gene expression profiling in biopharmaceutical process analysis and design. *Biotechnol. Bioeng.* **2010**, *105*, 431–438. [CrossRef] [PubMed]

34. Konakovsky, V.; Clemens, C.; Müller, M.M.; Bechmann, J.; Berger, M.; Schlatter, S.; Herwig, C. Metabolic Control in Mammalian Fed-Batch Cell Cultures for Reduced Lactic Acid Accumulation and Improved Process Robustness. *Bioengineering* **2016**, *3*. [CrossRef] [PubMed]

35. Schaub, J.; Clemens, C.; Kaufmann, H.; Schulz, T.W. Advancing biopharmaceutical process development by system-level data analysis and integration of omics data. *Adv. Biochem. Eng. Biotechnol.* **2012**, *127*, 133–163. [CrossRef] [PubMed]

36. Glacken, M.W. Catabolic Control of Mammalian Cell Culture. *Nat. Biotechnol.* **1988**, *6*, 1041–1050. [CrossRef]

37. Ozturk, S.S.; Hu, W.-S. *Cell Culture Technology for Pharmaceutical and Cell-Based Therapies*; Taylor & Francis: New York, NY, USA; London, UK, 2006.

38. Newsholme, P.; Procopio, J.; Lima, M.M.R.; Pithon-Curi, T.C.; Curi, R. Glutamine and glutamate—Their central role in cell metabolism and function. *Cell Biochem. Funct.* **2003**, *21*, 1–9. [CrossRef] [PubMed]

39. Andersen, D.C.; Goochee, C.F. The effect of ammonia on the O-linked glycosylation of granulocyte colony-stimulating factor produced by Chinese hamster ovary cells. *Biotechnol. Bioeng.* **1995**, *47*, 96–105. [CrossRef] [PubMed]

40. Ozturk, S.S.; Riley, M.R.; Palsson, B.O. Effects of ammonia and lactate on hybridoma growth, metabolism, and antibody production. *Biotechnol. Bioeng.* **1992**, *39*, 418–431. [CrossRef] [PubMed]

41. Zhou, M.; Crawford, Y.; Ng, D.; Tung, J.; Pynn, A.F.J.; Meier, A.; Yuk, I.H.; Vijayasankaran, N.; Leach, K.; Joly, J.; et al. Decreasing lactate level and increasing antibody production in Chinese Hamster Ovary cells (CHO) by reducing the expression of lactate dehydrogenase and pyruvate dehydrogenase kinases. *J. Biotechnol.* **2011**, *153*, 27–34. [CrossRef] [PubMed]

42. Zagari, F.; Jordan, M.; Stettler, M.; Broly, H.; Wurm, F.M. Lactate metabolism shift in CHO cell culture: The role of mitochondrial oxidative activity. *New Biotechnol.* **2013**, *30*, 238–245. [CrossRef] [PubMed]

43. Hong, J.K.; Cho, S.M.; Yoon, S.K. Substitution of glutamine by glutamate enhances production and galactosylation of recombinant IgG in Chinese hamster ovary cells. *Appl. Microbiol. Biotechnol.* **2010**, *88*, 869–876. [CrossRef] [PubMed]

44. Xing, Z.; Li, Z.; Chow, V.; Lee, S.S. Identifying inhibitory threshold values of repressing metabolites in CHO cell culture using multivariate analysis methods. *Biotechnol. Prog.* **2008**, *24*, 675–683. [CrossRef] [PubMed]

45. Klein, T.; Heinzel, N.; Kroll, P.; Brunner, M.; Herwig, C.; Neutsch, L. Quantification of cell lysis during CHO bioprocesses: Impact on cell count, growth kinetics and productivity. *J. Biotechnol.* **2015**, *207*, 67–76. [CrossRef] [PubMed]

46. Bhatia, H.; Mehdizadeh, H.; Drapeau, D.; Yoon, S. In-line monitoring of amino acids in mammalian cell cultures using raman spectroscopy and multivariate chemometrics models. *Eng. Life Sci.* **2018**, *18*, 55–61. [CrossRef]

47. Buckley, K.; Ryder, A.G. Applications of Raman Spectroscopy in Biopharmaceutical Manufacturing: A Short Review. *Appl. Spectrosc.* **2017**, *71*, 1085–1116. [CrossRef] [PubMed]

48. Zhao, N.; Wu, Z.-S.; Zhang, Q.; Shi, X.-Y.; Ma, Q.; Qiao, Y.-J. Optimization of Parameter Selection for Partial Least Squares Model Development. *Sci. Rep.* **2015**, *5*, 11647. [CrossRef] [PubMed]

49. Xing, Z.; Bishop, N.; Leister, K.; Li, Z.J. Modeling kinetics of a large-scale fed-batch CHO cell culture by markov chain monte carlo method. *Biotechnol. Prog.* **2010**, *26*, 208–219. [CrossRef] [PubMed]

50. Craven, S.; Shirsat, N.; Whelan, J.; Glennon, B. Process model comparison and transferability across bioreactor scales and modes of operation for a mammalian cell bioprocess. *Biotechnol. Prog.* **2013**, *29*, 186–196. [CrossRef] [PubMed]

51. Subramanian, G. *Continuous Biomanufacturing—Innovative Technologies and Methods*; Wiley-VCH Verlag GmbH & Co. KGaA: Weinheim, Germany, 2017.

52. Gronemeyer, P.; Thiess, H.; Zobel, S.; Ditz, R.; Strube, J. Integration of Upstream and Downstream in Continuous Biomanufacturing. In *Continuous Biomanufacturing*; Subramanian, G., Ed.; Wiley-VCH Verlag GmbH & Co. KGaA: Weinheim, Germany, 2017.

53. Gronemeyer, P.; Ditz, R.; Strube, J. Trends in Upstream and Downstream Process Development for Antibody Manufacturing. *Bioengineering* **2014**, *1*, 188–212. [CrossRef] [PubMed]

54. Kornecki, M.; Mestmäcker, F.; Zobel-Roos, S.; Heikaus de Figueiredo, L.; Schlüter, H.; Strube, J. Host Cell Proteins in Biologics Manufacturing: The Good, the Bad, and the Ugly. *Antibodies* **2017**, *6*, 13. [CrossRef]

55. Zobel, S.; Helling, C.; Ditz, R.; Strube, J. Design and operation of continuous countercurrent chromatography in biotechnological production. *Ind. Eng. Chem. Res.* **2014**, *53*, 9169–9185. [CrossRef]

56. Wurm, F.M. Production of recombinant protein therapeutics in cultivated mammalian cells. *Nat. Biotechnol.* **2004**, *22*, 1393–1398. [CrossRef] [PubMed]

57. Sommerfeld, S.; Strube, J. Challenges in biotechnology production—Generic processes and process optimization for monoclonal antibodies. *Chem. Eng. Process. Process Intensif.* **2005**, *44*, 1123–1137. [CrossRef]

9

Large Scale Production and Downstream Processing of Labyrinthopeptins from the Actinobacterium *Actinomadura namibiensis*

Zeljka Rupcic [1,2] [iD], Stephan Hüttel [1,2,*], Steffen Bernecker [1,2], Sae Kanaki [3] and Marc Stadler [1,2,*] [iD]

[1] Department Microbial Drugs, Helmholtz Centre for Infection Research GmbH, Inhoffenstraße 7, 38124 Braunschweig, Germany; zeljka.rupcic@helmholtz-hzi.de (Z.R.); steffen.bernecker@helmholtz-hzi.de (S.B.)

[2] German Centre for Infection Research (DZIF), partner site Hannover-Braunschweig, 38124 Braunschweig, Germany

[3] Toyama Prefectural University, 5180 Kurokawa Imizu-shi, Toyama 939-0398, Japan; t416010@st.pu-toyama.ac.jp

* Correspondence: stephan.huettel@helmholtz-hzi.de (S.H.); marc.stadler@helmholtz-hzi.de (M.S.)

Abstract: A method was established for the production of 1.2-fold and 4.2-fold increased amounts of the antiviral and central nervous system-active lantipeptides, labyrinthopeptins A1 and A2, respectively, isolated from the actinobacterium *Actinomadura namibiensis*, to enable production in gram scale. We then performed in vivo characterization of this promising compound class. The labyrinthopeptins A1 and A2 have similar chemical structures and physical properties but differ drastically in their bioactivities. Therefore, large quantities of highly pure material are required for pharmacological studies. An effective methodology was established for the first time for their production in bioreactors, their separation involving gel permeation chromatography on LH20 material, followed by reversed phase-high performance liquid chromatography. With an optimized methodology, 580 mg of labyrinthopeptin A1 and 510 mg of labyrinthopeptin A2 were quantitatively isolated with recovery rates of 72.5% and 42.3% from 7.5 L of culture broth, respectively. However, the fermentation that had already resulted in maximum yields of over 100 mg/L of both target molecules after 300 h in a 10-L scale bioreactor, still requires further optimisation.

Keywords: anti-viral agents; bioprocess; central nervous system; lantibiotics; optimization production; reversed phase-high performance liquid chromatography; scale-up

---

1. Introduction

Several groups of Gram-positive bacteria, such as actinobacteria, lactobacilli, and staphylococci, are able to ribosomally biosynthesize oligopeptides from 18–38 amino acids, called lantibiotics or lantipeptides. Their common structural feature is the presence of noncanonical thioether amino acids like lanthionine and/or methyllanthionine (MeLan) [1,2]. The food preservative nisin is the most important commercial representative of this compound family [3]. According to their biosynthetic pathways, lantibiotics are classified into three major types: Class I lantibiotics are modified by separate dehydratases (LanB) and cyclases (LanC), whereas class II lantibiotics are dehydrated and cyclized by a single LanM-type enzyme. Unlike classes I and II, class III lantibiotics have little or no antibiotic activity; rather, they provide alternative physiological functions [1]. An example of class I lantibiotics, aside from the aforementioned nisin, are the recently characterized pinensins, which represent the first antifungal lantibiotics isolated from a Gram-negative bacterium [4]. Pseudomycoicidin, an antibacterial representative of class II lantibiotics, was isolated from a Gram-positive bacterium [5]. From actinobacteria, only four peptides have previously been described and structurally characterized as type III lantipeptides, although comparative genomic studies have revealed that homologous gene clusters are abundant in other strains of this bacterial group [6]. An example of lantibiotics in phase II clinical trials is duramycin (moli1901), which is being developed for the treatment of cystic fibrosis as it increases chloride transport in the airway epithelium [7].

The current study was dedicated to the labyrinthopeptins, which were first isolated from the desert actinobacterium *Actinomadura namibiensis* within Aventis in 1988 [8], but their structure was only determined recently [9]. Electrospray Ionization Fourier Transform Ion Cyclotron Resonance (EIFTCR) mass spectrometry of its crude extract showed three labyrinthopeptin derivatives: labyrinthopeptin A1 (**1**) (Figure 1a), labyrinthopeptin A2 (**2**) (Figure 1b), and labyrinthopeptin A3 (**3**), a degradation product of **1** [10]. All three derivatives possess a post translationally modified triamino acid labionin (Figure 2) [10,11].

Because of their biological activities, labyrinthopeptins have the potential to become lead compounds for drug development in two different indications. For instance, **2** displayed an activity in a spared nerve injury mouse model of neuropathic pain, whereas **1** exhibited in vitro antiviral effects against Human Immunodeficiency Virus (HIV) and Herpes Simplex Virus (HSV) at sub-micromolar concentrations [10]. Compound **1** also showed synergistic effects with standard antiretroviral drugs and the absence of a PBMC inflammatory response. Since they did not interact with vaginal lactobacilli, labyrinthopeptin A1 (**1**) is an ideal candidate for the treatment of sexually transmitted viruses, as well as for development as a broad spectrum antiviral agent [12]. The latter hypothesis was confirmed in a study where both labyrinthopeptins showed activity against human respiratory syncytial virus (hRSV) subtype A and B [13]. The mechanism of action for this antiviral activity is still not completely understood, but the activity may be related to the blockage of viral entry by interacting with the viral envelope and preventing cell-to-cell transmission [12].

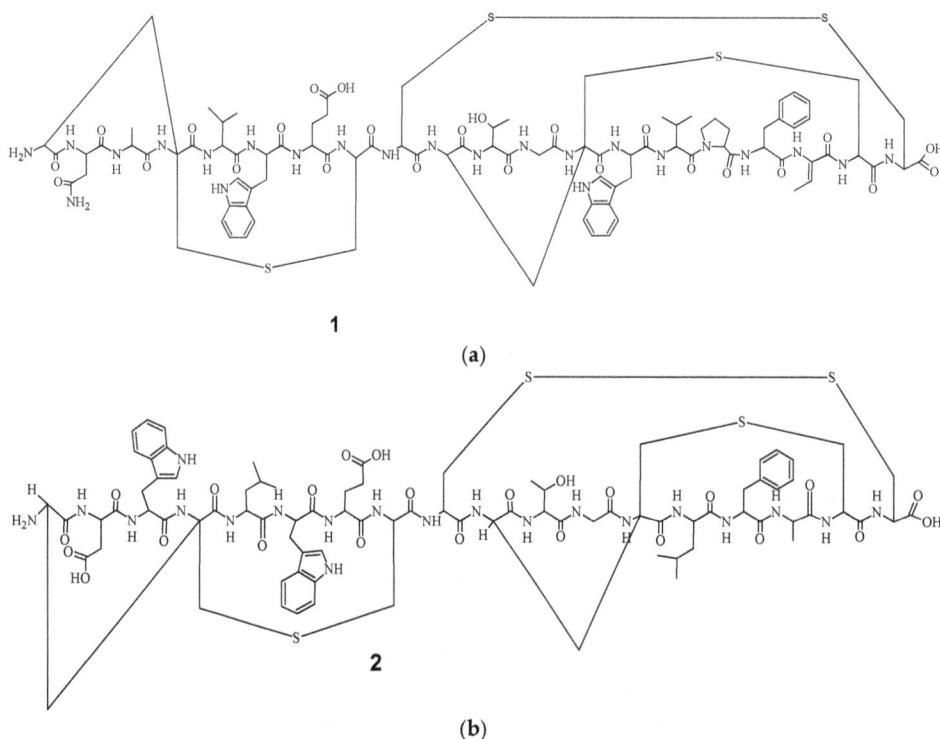

1

(a)

2

(b)

Figure 1. Chemical structures of (**a**) labyrinthopeptin A1 (**1**) and (**b**) labyrinthopeptin A2 (**2**).

Labyrinthopeptins were produced by biotechnological means yielding 6.2 mg/L of **2**, obtained from a shake flask batch cultivation in 10 L scale [10]. Another study by Krawczyk et al. [14] reported 90 mg/L for **1** and 39 mg/mL for **2** obtained from the wild type strain of *Actinomadura namibiensis*. However, details on the purity of the compounds were not provided in that study. In the current study, for the first time, we describe a scalable batch-process using a 10 L bioreactor and *Actinomadura namibiensis* strain DSM 6313 as the producer organism, which led to a 1.2-fold increase in the yield of **1**, and a 4.2-fold increase in the yield of **2** compared to the data in the literature [14]. Moreover, we outline an optimized downstream processing procedure that ensures a quantitative supply of **1** and **2** with sufficient purity for further pre-clinical studies.

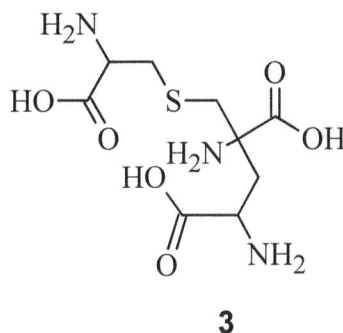

3

Figure 2. Chemical structure of labionin (**3**).

2. Materials and Methods

2.1. General Experimental Procedure

All HPLC analyses were performed on an Agilent 1200 Series instrument (Agilent, Santa Clara, CA, USA), equipped with degasser, binary Pump SL (Agilent Technologies 1260 Infinity, Agilent Technologies, Santa Clara, CA, USA), autosampler and a combined diode array detector/electron light scattering detector Corona Ultra RS (Dionex, Sannyvale, CA, USA), using the conditions described by Beckmann et al. [15].

High performance liquid chromatography-electrospray ionization mass spectrometry (HPLC-ESI-MS) spectra were recorded with an Agilent 1200 series HPLC system with Acquity UPLC BEH C18 column (2.1 × 50 mm, 1.7 µm) from Waters (Eschborn, Germany), coupled to an ion trap mass spectrometer, amZon™ (Bruker, Bremen, Germany) (scan range 100 to 2000 *m/z*, capillary voltage 4000 V, dry temperature 250 °C).

Ultra High Resolution-Time of Flight (UHR-TOF-MS) data were obtained using an Ultimate 3000RS Thermo Scientific™ Dionex™ (Waltham, MA, USA) instrument equipped with a Kinetex C18 column (150 × 2.1 mm, 1.7 µm) from Phenomenex (Torrance, CA, USA) as stationary phase. The HPLC system was coupled to a mass spectrometer (maXis HD™, Bruker, Bremen, Germany), with a scan range from 250 to 2500 *m/z*, a set collision energy of 8.0 eV, capillary voltage 4500 V, nebulizer 4.0 bar, and the dry heater set to 200 °C.

Chemicals and solvents were obtained from AppliChem GmbH (Darmstadt, Germany), Avantor Performance Materials (Deventor, The Netherlands), Carl Roth GmbH & Co. KG (Karlsruhe, Germany), and Merck KGaA (Darmstadt, Germany) in analytical and HPLC grade.

2.2. Fermentation in a 10 Liter Scale Bioreactor

A seed culture of *Actinomadura namibiensis* DSM 6313 was prepared in two steps. An aliquot of 1.8 mL from a cryo culture in 10% glycerol was inoculated in 100 mL of the production medium, starch-glucose-glycerol (SGG) consisting of 10 g/L starch (Cargill, Sas van Gent, The Netherlands), 2 g/L yeast extract (Ohly®GmbH, Hamburg, Germany), 10 g/L glucose (Cerestar, Neuilly-Sur-Seine,

France), 10 g/L glycerol (Roth, Karlsruhe, Germany), 2.5 g/L corn steep powder (Sigma-Aldrich, Darmstadt, Germany), 2 g/L peptone (Markor, Carlstadt, NJ, USA), 1 g/L sodium chloride (NaCl), 3 g/L calcium carbonate (CaCO$_3$), both from Roth in tap water at pH 7.0 before sterilization, in a 250 mL Erlenmeyer flask and incubated on a rotary shaker for 72 h at 30 °C and 160 rpm. For the preparation of a secondary seed culture, 20 mL of each of the first inoculum were added to 1000 mL sterile Erlenmeyer flasks with a 400 mL working volume of SGG medium, and then incubated for 48 h at 30 °C and 160 rpm.

The batch cultivation was performed in a 10 L steel reactor (xCUBIO in situ, bbi biotech, Berlin, Germany). The system was equipped with an aseptic sampling system probe, and analyzers for exhaust O$_2$ and CO$_2$, pH, pO$_2$ (clarc) and temperature, level and foam sensors, and three rushton impellers for agitation. The pH was adjusted prior to 7.2 fermentation by addition of H$_2$SO$_4$ or KOH, and was not maintained during the fermentation (Figure 3b). The gas flow rate was maintained at 0.25 vvm (2.5 nL/min), and the temperature was set to 30 °C. Stirring was set to a minimum of 100 rpm and automatically increased to maintain a pO$_2$ of 30%. Batch fermentation was started with a cultivation volume of 9.5 L of SGG medium. A 5% inoculum was added and the fermentation was continued for 672 h.

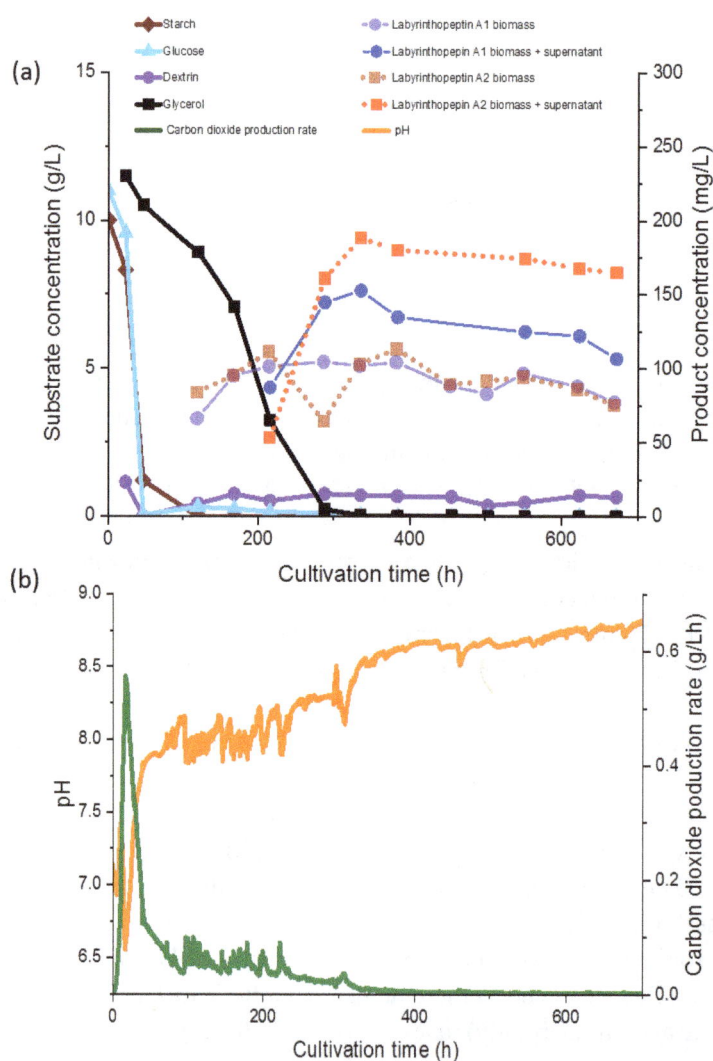

Figure 3. (**a**) Correlation between product concentration (mg/L) and substrate concentrations (g/L); (**b**) carbon dioxide production (CPR; g/Lh) and pH determined during regular time intervals in a 10 L batch fermentation. Product concentration is expressed as a cumulative value for the biomass and the supernatant together (dashed lines) and for the biomass only (solid lines).

2.3. Determination of Substrate and Product Concentrations during the Fermentation

Glucose, glycerol, and dextrin concentrations were determined as described previously [14] using an Agilent 1260 series HPLC and a Phenomenex (Aschaffenburg, Germany) REZEX ROA-Organic Acid H+ (8%) column (300 × 7.8 mm × 8 μm) at 65 °C with a refractive index detector. Separation occurred under isocratic conditions of 0.05 mM H_2SO_4 for 45 min (Figure 3a). The carbon dioxide production rate (CPR) was automatically monitored and the time course is shown in Figure 3b.

The product concentration was determined separately for the biomass and the supernatant. For this purpose, samples (2 × 10 mL) were obtained at regular intervals. To the first sample, 1% (v/v) Amberlite XAD™-16N adsorber resin (Sigma-Aldrich, St. Louis, MO, USA) was added and centrifuged for 20 min at 5000 rpm. Thereafter, the supernatant was discarded, 12 mL of a mixture of acetone, methanol (MeOH), and water (1:3:2) was added, sonicated for 30 min at 40 °C, and centrifuged for 20 min at 5000 rpm. The organic solvent was evaporated *in vacuo* and the aqueous residue was freeze-dried.

The second sample was treated as described above, but without the Amberlite XAD™-16N adsorber resin, to determine the product concentration in pellet for only. Both extracts were reconstituted in the MeOH and H_2O mixture (50:50, v/v), centrifuged for 5 min at 14,000 rpm and the supernatants were injected into a Dionex Ultimate 3000 (Thermo Scientific, Waltham, MA, USA) HPLC system, using a Nucleodur Phenyl Hexyl column (1.8 μm; 100 × 2 mm, Macherey Nagel, Düren, Germany) as stationary phase, equipped with a pre-column consisting of the same material (1.8 μm; 4 × 2 mm). The mobile phase was composed of solvent A, H_2O + 0.1% formic acid (FA), and solvent B, acetonitrile (ACN + 0.1% FA). A gradient was run from 30% B for 1 min, followed by a linear increase to 50% B over 10 min, which was afterward increased to 100% B in 1 min, and then maintained under isocratic conditions at 100% for 5 min, at a flow rate of 0.3 mL/min and ultraviolet (UV) detection at 280 nm. The product concentrations were determined via a calibration curve using five defined standard solutions, as displayed in Figure 3a in relation to the substrate concentration.

2.4. Downstream Processing and Isolation of Labyrinthopeptins

The culture broth (7.5 L) was centrifuged at 9000 rpm (Sorvall® RC5B centrifuge, DuPont Instruments, Wilmington, DE, USA) for 30 min to separate the biomass (975 mL) from the supernatant. The resulting biomass cake (529 g) was defatted with an equal amount of n-heptane, extracted two times with a mixture of 7.5 L acetone, MeOH, and H_2O (1:3:2) for 1 h in an ultraconic bath Sonorex Digital 10P (Bandelin, Berlin, Germany). The organic solvent was evaporated *in vacuo* and the remaining aqueous residue was freeze-dried to avoid losses due to the extensive foaming of the extract at lower pressure. We obtained 8.8 g of crude product.

The culture broth was incubated with 1% (v/v) of Amberlite XAD™-16N adsorber resin under stirring for 1 h. The resin was removed by filtration and subsequently extracted with 1 L of solvent mixture, according to the same protocol described above, to obtain 1.7 g of crude extract.

Both the mycelial and the supernatant extracts were dissolved in a mixture of MeOH and H_2O (50:50, v/v), centrifuged for 15 min at 5000 rpm, and purified via gel permeation chromatography (GPC) on Sephadex®LH 20 material (Pharmacia Fine Chemicals, Inc., New York, NY, USA). A medium pressure liquid chromatography (MPLC) device was used for the GPC, equipped with Minipuls S3 pump (Gilson, Inc.; Middleton, WI, USA), a B 687 mixing device (Büchi, Flawil, Switzerland), a 2138 UVICORDS UV detector set to 220 nm (LKB, Bromma, Stockholm, Sweden), and a REC 102 writer (Pharmacia Biotech Inc., Piscataway, NJ, USA).

The mobile phase was composed of MeOH:H_2O (50:50, v/v), the flow rate of 8 mL/min was applied, with an overall solvent consumption of 7.7 L to create two fractions at the retention times of 108 min and 124 min. The fractions were subsequently checked by HPLC-ESI-MS to determine peptide co-elution and therefore, both were pooled together. A total of 5 g and 0.8 g of intermediate product was obtained after GPC, which was further purified via the preparative HPLC. For this purpose, a Gilson GX270 Series HPLC system was used with a Nucleodur Phenyl hexyl column

(5 μm; 150 × 40 mm; Macherey-Nagel, Düren, Germany), solvent A with H_2O + 0.1% FA, and solvent B with ACN + 0.1% FA. Gradient elution was performed as follows: 5 min isocratic on 30% of B, followed by a linear increase to 80% of B over 30 min, then an increase of B to 100% over 5 min, and maintaining at 100% of B for the next 5 min. A flow rate of 30 mL/min was used and detection was performed at 254 nm. Fractions were collected according to the observed peaks and were submitted to UHR-TOF-MS. For chromatograms, see Supplementary Figures S1–S7. Finally, 420 mg and 160 mg of **1** were isolated from the biomass and supernatant, respectively, with 99.34% purity. From **2**, 380 mg were isolated from the biomass crude extract, whereas 140 mg were obtained from the supernatant crude extract with 99.13% purity. For their isolation at the preparative scale, 26 L of ACN and 22.7 L of H_2O were consumed, whereas 42 h were required for obtaining the yields. The mass balance for the major unit operations of the downstream processing is displayed in Tables 1 and 2 for the biomass and the supernatant, respectively, whereas recovery of the isolation process is provided in Table 3.

Table 1. Mass balance of the input and the output of major unit operations in the biomass downstream processing.

Process Step	Input (g)	Output A1		
		A1 (g)	Recovery A1 (%)	Loss A1 (%)
Crude extract	8.8	0.57	-	0
GPC (3 runs) [1]	5.0	0.45	78.9	21.1
Prep. HPLC (25 runs) [1]	0.8	0.42	93.3	6.7

Process Step	Input (g)	Output A2		
		A2 (g)	Recovery A2 (%)	Loss A2 (%)
Crude extract	8.8	0.55	-	0
GPC (3 runs) [1]	5.0	0.45	81.8	18.2
Prep. HPLC (25 runs) [1]	0.8	0.38	84.4	15.6

[1] For the complete downstream process of the biomass.

Table 2. Mass balance of the input and the output of major unit operations in the supernatant downstream processing.

Process Step	Input (g)	Output A1		
		A1 (g)	Recovery A1 (%)	Loss A1 (%)
Crude extract	1.7	0.23	-	0
GPC (1 run) [2]	0.8	0.18	78.3	21.7
Prep. HPLC (5 runs) [2]	0.3	0.16	88.9	11.1

Process Step	Input (g)	Output A2		
		A2 (g)	Recovery A2 (%)	Loss A2 (%)
Crude extract	1.7	0.68	-	0
GPC (1 run) [2]	0.8	0.28	41.2	58.8
Prep. HPLC (5 runs) [2]	0.3	0.14	42.4	57.6

[2] For the complete downstream process of the supernatant.

Table 3. Overall-recovery of labyrinthopeptins A1 and A2 from the isolation and separation process.

Compounds	Amount Estimated (g)	Amount Isolated (g)	Recovery (%)	Loss (%)
Labyrinthopeptin A1	0.80	0.58	72.5	27.5
Labyrinthopeptin A2	1.23	0.51	42.3	57.7

3. Results and Discussion

According to unpublished experiments, **1** and **2** differ drastically in their bioactivity. To enable further in vivo drug development studies, substantial amounts of the compounds are required as well as high purity peptides. Since cost-efficient synthesis is not possible, their production can only be attained by means of biotechnological production.

Conversely, heterologous production of these ribosomally synthesized and posttranslationally modified peptides was reported previously [14]. Since attempts to establish a suitable genetic system in the wild-type producer *Actinomadura namibiensis* failed, a heterologous *Streptomyces* host was used instead. A general obstacle to the heterologous expression in *S. lividans* is the undesired production of labyrinthopeptin variants with additional N-terminal amino acids. The authors reported using different constructs, pLab_SG6, for the exclusive generation of **2**, resulting in a yield of 14 mg/L, whereas **1** was generated by the pLab_SG6 construct (86 mg/L). The titers obtained were in the range of the wild-type *Actinomadura namibiensis*, at 90 and 39 mg/L for **1** and **2**, respectively [14]. However, this process was never scaled up to attain quantitative amounts of the desired products. These constructs may not provide a path forward to transfer production into pilot scale bioreactors, despite the fact that the considerably increased yields compared to those initially reported yields from shake flasks (6.2 mg/L for **2**) [10].

During fermentation, various observations were recorded regarding the correlations of the substrate consumption, which was monitored during the course of our study for the first time during production of the labyrinthopeptins, and metabolite production. All carbon sources used in the media, except for glycerol, were depleted within 48 to 96 h. The starch was metabolized at the initiation of the fermentation and consequently, after the starch was hydrolyzed, an increased concentration of dextrin was measured (<1.0 g/L). Free glucose was only detected at the beginning of the fermentation at a concentration of 0.05 g/L, until 48 h after inoculation, whereas glycerol decreased gradually until 336 h (Figure 3a). The lowest measured pH value over the fermentation time was 6.93 (Figure 3b), but the pH showed a growing tendency toward the end of the fermentation. The sampling started 120 h after the inoculation and 65 mg/L and 83 mg/L of labyrinthopeptins from A1 (**1**) and A2 (**2**) were calculated, respectively, although we assumed from preliminary shake flask experiments the production started shortly before the first measurement (data not shown). Nevertheless, preliminary experiments showed that the production occurred after sugars and their oligomers had been consumed and the strain entered a phase of limited respiratory activity as illustrated by the CPR where glycerol was then used as a carbon source (Figure 3a).

At the end of the fermentation, the concentrations of **1** and **2** were estimated to be approximately 106 and 165 mg/L, respectively. Interestingly, compound **1** was predominant in terms of biomass throughout the entire cultivation, whereas **2** was equally represented in both the supernatant and the biomass. After 672 h of fermentation, increased concentrations of **2** in the supernatant were observed (Figure 3a,b). An anomaly in the product concentration was observed after 500 h of fermentation, where a decreased concentration in the product obtained from the supernatant was measured. This measuring point was eliminated from the graph in Figure 3a. We depicted the fermentation from which the product was ultimately isolated, but since the product titers remained stable, terminating the fermentation much earlier in the future would be possible, after about 300 h. Several other options to further increase the titers and reduce the fermentation time, such as by increasing the inoculum and conduction of fed-batch experiments, are presently being planned, and those will commence in parallel to the transfer of the process to 70- and 250-L scales.

After preparative separation, only about 50% of the estimated value of compound **2** was recovered by chromatography, resulting in 510 mg of isolated compound. Nevertheless, this yield constitutes a 4.2-fold increase compared to the data found in the literature [14]. A titer of 106 mg/L was obtained for compound **1** with a recovery rate of 72.5%, resulting in the isolation of 580 mg of the compound (Table 3). As compared to the data in the literature, this yield constitutes a 1.2-fold increase [14]. However, this study did not include experiments on the quantitative isolation of **1** and no recovery rates are provided. In comparison to the titer for **2**, we obtained a considerably lower titer for **1**; however, larger amounts of **1** were quantitatively separated from the crude product (Tables 1 and 2).

Since the compounds have a high molecular weight and as large sample amounts had to be handled, the downstream processing was accomplished with the combination of a GPC, as a pre-purification step, with the Sephadex® LH 20 material, and finally by the purification by preparative HPLC. As expected, the labyrinthopeptins were eluted concurrently, but separating the majority of the co-metabolites and media constituents from the mixture of **1** and **2** was possible. For the final purification of the two target compounds, preparative chromatography had to be used.

Our work clearly demonstrates a substantial increase in the production of labyrinthopeptins and provides a straightforward approach to downstream processing, even for future attempts to further scale-up the process to the pilot scale. From our hindsight of the bioprocess development, different media and different ratios of media components are concurrently being tested to achieve even higher titers in the future, and the process is ready for transfer to the pilot scale.

4. Conclusions

A method for the biotechnological production, together with a new method for providing sustainable accessibility of the two structurally similar lantibiotics was established in the current study. By improving the isolation/separation conditions, labyrinthopeptins A1 and A2 were obtained at final recovery rates of 72.5% and 42.3%, respectively, and purities of over 99% as estimated by HPLC-UV were attained for both metabolites. To fulfil the requirements for future evaluation of the compound class, involving both, animal studies, and the formulation experiments, gram amounts of labyrinthopeptins are needed. The results presented here provide a path forward, since the compounds were obtained at very high purity, and both the GPC method and the final HPLC separation step can be scaled up in a straightforward manner. A concurrent extensive optimization of culture media and fermentation parameters, in order to further increase the yields and recovery rates will be necessary in any case, in order to attain favorable costs of goods and will be the subject of our further research.

Author Contributions: Z.R. planned and developed new approach in downstream processing. S.B. collected process data and helped with the technical issues during the fermentation. S.H. supervised the work during the fermentation. Z.R. performed small scale extractions to determine the kinetic profile of the fermentation, harvested bioreactor and with the help of an intern student. S.K. performed extraction GPC and preparative HPLCs. M.S. and S.H. supervised the progress of the project and helped finalizing the draft.

Funding: This research received no external funding.

Acknowledgments: We kindly want to thank Cäcilia Bergmann and Axel Schulz for their technical support during the fermentation and Esther Surges for her help during the downstream processing.

Conflicts of Interest: The authors declare no conflict of interest.

References

1.   Willey, J.M.; van der Donk, W.A. Lantibiotics: Peptides of diverse structure and function. *Ann. Rev. Microbiol.* **2007**, *61*, 477–511. [CrossRef] [PubMed]

2.   Jung, G. Lantibiotics-ribosomally synthesized biologically active polypeptides containing sulfide bridges and α,β-didehydroamino acids. *Angew. Chem. Int. Ed.* **1991**, *30*, 1051–1068. [CrossRef]

3.   Breukink, E.; Wiedemann, I.; van Kraaij, C.; Kuipers, O.P.; Sahl, H.G.; de Kruijff, B. Use of the cell wall precursor lipid II by a pore-forming peptide antibiotic. *Science* **1999**, *286*, 2361–2364. [CrossRef] [PubMed]

4.   Mohr, K.I.; Volz, C.; Jansen, R.; Wray, V.; Hoffmann, J.; Bernecker, S.; Wink, J.; Gerth, K.; Stadler, M.; Müller, R. Pinensins: The first antifungal lantibiotics. *Angew. Chem. Int. Ed.* **2015**, *54*, 11254–11258. [CrossRef] [PubMed]

5.   Basi-Chipalu, S.; Dischinger, J.; Josten, M.; Szekat, C.; Zweynert, A.; Sahl, H.G.; Bierbaum, G. Pseudomycoicidin, a class II lantibiotic from *Bacillus pseudomycoides*. *Appl. Environ. Microbiol.* **2015**, *81*, 3419–3429. [CrossRef] [PubMed]

6. Völler, G.; Krawczyk, J.; Pesic, A.; Krawczyk, B.; Nachtigall, J.; Süssmuth, R.D. Characterization of new class III lantibiotics—Erythreapeptin, avermipeptin and griseopeptin from Saccharopolyspora erythraea, Streptomyces avermitilis and Streptomyces griseus demonstrates stepwise n-terminal leader processing. *ChemBioChem* **2012**, *13*, 1174–1183. [CrossRef] [PubMed]

7. Grasemann, H.; Stehling, F.; Brunar, H.; Widmann, R.; Laliberte, T.W.; Molina, L.; Döring, G.; Ratjen, F. Inhalation of moli1901 in patients with cystic fibrosis. *Chest* **2007**, *131*, 1461–1466. [CrossRef] [PubMed]

8. Wink, J.; Kroppenstedt, R.M.; Seibert, G.; Stackebrandt, E. *Actinomadura namibiensis* sp. nov. *Int. J. Syst. Evol. Microbiol.* **2003**, *53*, 721–724. [CrossRef] [PubMed]

9. Meindl, K.; Schmiederer, T.; Schneider, K.; Reicke, A.; Butz, D.; Keller, S.; Güring, H.; Vértesy, L.; Wink, J.; Hoffmann, H.; et al. Labyrinthopeptins: A new class of carbacyclic lantibiotics. *Angew. Chem. Int. Ed.* **2010**, *49*, 1151–1154. [CrossRef] [PubMed]

10. Seibert, G.; Vértesy, L.; Wink, J.; Winkler, I.; Süssmuth, R.D.; Sheldrick, G.; Meindl, K.; Brönstrup, M.; Hoffmann, H.; Guehring, H.; et al. Antibacterial and Antiviral Peptides from *Actinomadura namibiensis*. International Application No. PCT/EP2007/008294, 10 April 2008.

11. Férir, G.; Petrova, M.I.; Andrei, G.; Huskens, D.; Hoorelbeke, B.; Snoeck, R.; Vanderleyden, J.; Balzarini, J.; Bartoschek, S.; Brönstrup, M.; et al. The lantibiotic peptide labyrinthopeptin A1 demonstrates broad anti-HIV and anti-HSV activity with potential for microbicidal applications. *PLoS ONE* **2008**, *8*, e64010. [CrossRef] [PubMed]

12. Martinez, J.P.; Sasse, F.; Brönstrup, M.; Diez, J.; Meyerhans, A. Antiviral drug discovery: Broad-spectrum drugs from nature. *Nat. Prod. Rep.* **2015**, *32*, 29–48. [CrossRef] [PubMed]

13. Haid, S.; Blockus, S.; Wiechert, S.M.; Wetzke, M.; Prochnow, H.; Dijkman, R.; Wiegmann, B.; Rameix-Welti, M.A.; Eleouet, J.F.; Duprex, P.; et al. Labyrinthopeptin A1 and A2 efficiently inhibit cell entry of hRSV isolates. *Eur. Respir. J.* **2017**, *50*, 4124.

14. Krawczyk, J.M.; Völler, G.H.; Krawczyk, B.; Kretz, J.; Brönstrup, M.; Süssmuth, R.F. Heterologous expression and engineering studies of labyrinthopeptins, class III lantibiotics from *Actinomadura namibiensis*. *Chem. Biol.* **2013**, *20*, 111–122. [CrossRef] [PubMed]

15. Beckmann, A.; Hüttel, S.; Schmitt, V.; Müller, R.; Stadler, M. Optimization of the biotechnological production of a novel class of anti-MRSA antibiotics from *Chitinophaga sancti*. *Microb. Cell Fact.* **2017**, *16*, 143. [CrossRef] [PubMed]

10

Effects of Sterilization Cycles on PEEK for Medical Device Application

Amit Kumar [ID], Wai Teng Yap, Soo Leong Foo and Teck Kheng Lee *

College Central, Institute of Technical Education, 2 Ang Mo Kio Drive, Singapore 567720, Singapore; amittonk@gmail.com (A.K.); yap_wai_teng@ite.edu.sg (W.T.Y.); foo_soo_leong@ite.edu.sg (S.L.F.)
* Correspondence: lee_teck_kheng@ite.edu.sg

Abstract: The effects of the sterilization process have been studied on medical grade thermoplastic polyetheretherketone (PEEK). For a reusable medical device, material reliability is an important parameter to decide its lifetime, as it will be subjected to the continuous steam sterilization process. A spring nature, clip component was selected out of a newly designed medical device (patented) to perform this reliability study. This clip component was sterilized for a predetermined number of cycles (2, 4, 6, 8, 10, 20 ... 100) at 121 °C for 30 min. A significant decrease of ~20% in the compression force of the spring was observed after 30 cycles, and a ~6% decrease in the lateral dimension of the clip was observed after 50 cycles. No further significant change in the compression force or dimension was observed for the subsequent sterilization cycles. Vickers hardness and differential scanning calorimetry (DSC) techniques were used to characterize the effects of sterilization. DSC results exhibited no significant change in the degree of cure and melting behavior of PEEK before and after the sterilization. Hardness measurement exhibited an increase of ~49% in hardness after just 20 cycles. When an unsterilized sample was heated for repetitive cycles without the presence of moisture (121 °C, 10 and 20 cycles), only ~7% of the maximum change in hardness was observed.

Keywords: polyetheretherketone; PEEK; DSC; Vickers hardness; medical grade; medical device

1. Introduction

In the late 1990s, polyetheretherketone (PEEK) was initially developed as a high-performance thermoplastic to replace metal-based orthopedic [1,2] and trauma [3,4] implants. Invibio Ltd. (Thornton Cleveleys, United Kingdom) first offered PEEK commercially as a biomaterial for implant applications in 1998, and since then research on PEEK as a biomaterial has enhanced significantly [5]. For implant applications, many PEEK composite materials have been developed, which have been studied for their behavior under mechanical impact, biotribology, friction, dynamic, damage, and fracture [6–10].

A thermomechanical study performed on PEEK composite [11,12] reported the changes in crystallinity, macroscopic decoloration, large deformation in impact, high strain rate, and heating-induced deformation. As all implants were subjected to body temperature, which is significantly below the glass transition temperature of PEEK, no significant change in elastic properties of PEEK was observed [13]. However, it was reported that the yielding and plastic flow behavior of PEEK was affected at physiological temperatures [12,14].

For a medical device, which will be used in surgery, the steam sterilization process is mandatory to properly disinfect the device before its next use. The influence of the sterilization process on the micromechanical properties of carbon fiber-reinforced PEEK has been studied [15] for bone implant applications. It was reported that after 3 steam sterilization cycles, there was no significant change in elastic modulus, hardness, or coefficient of friction of carbon composite PEEK. The influences of thermal cycling (temperature range between +60 °C and −60 °C, 750 cycles) on the fatigue behavior of carbon/PEEK laminates were also studied [16]. No significant change in tensile properties was observed, but a decrease of 25% in fatigue strength was observed. In another study, the fatigue performance of PEEK was investigated under sterilization and thermal aging cycles and no significant change in fatigue performance of PEEK was found [17]. Most of these studies performed on PEEK are basically for implant applications, low temperature thermal cycling fatigue, or to investigate the behavior after only a few cycles of sterilization. However, PEEK should be investigated for a higher number of sterilization cycles in order to test its reliability in surgical device applications.

In this work, the thermal reliability of medical grade PEEK (natural) was investigated for its use in surgical device applications. Although there are many available medical grade plastic materials, such as PTFE (polytetrafluoroethylene), PEEK, PC (polycarbonate), PS (polysulfone), and PVDF (polyvinylidene fluoride), PEEK presents the best combination of mechanical (stiffness, hardness, and wear) and thermal (high glass transition and melting temperature) properties required for the dimensional and thermal stability of a medical device [18]. An important reason to choose natural PEEK is its radiolucent property [13], which makes it transparent to radiography. If any surgical device is fabricated using natural PEEK, then the real-time monitoring of surgery is possible, as X-ray can clearly see through this material.

2. Materials and Methods

In this work, two different medical grade PEEK materials (SUSTAPEEK MG NAT 1000X φ-12 mm and φ-50 mm from Roechling, Germany and TECAPEEK MT natural rod, φ-50mm from Ensinger, Germany) were used to fabricate samples for thermal reliability studies. Both materials were of medical grade natural PEEK, having nearly similar mechanical and thermal properties, except the water absorption value which was 0.2% for Roechling PEEK and 0.02% for Ensinger PEEK in 24 h at 23 °C [19,20].

2.1. Specimen Preparation

Figure 1a shows a picture of the spring-type clip component part of a medical device [21], which is manufactured from Roechling PEEK using the machining process. Figure 1b shows the image of test samples used for the hardness and differential scanning calorimetry (DSC) characterization, which are also manufactured by the machining process using PEEK materials procured from Roechling and Ensinger. Fifteen samples, each 20 × 10 × 4 mm in size, were fabricated using two different sizes of cylindrical rod, with 12 mm and 50 mm diameters, to investigate the effect of primary processing conditions on the thermal cyclic performance of PEEK during sterilization.

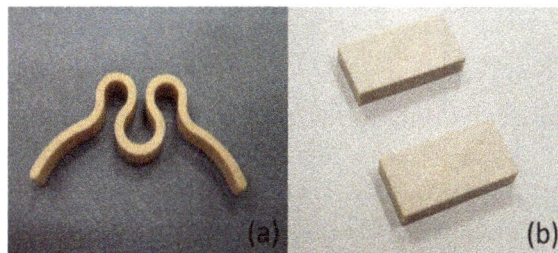

Figure 1. (**a**) Spring-type clip component of a medical device and (**b**) fabricated test specimens for hardness and differential scanning calorimetry (DSC) characterization.

2.2. Autoclave Sterilization

A pressure cooker tester (PC-242HS, HIRAYAMA, Tokyo, Japan) autoclave was used to perform the sterilization test on the clip component and test samples. Samples were heated up to 121 °C for 30 min under 0.1 MPa pressure at a nearly 100% humidity level. Initially, the clip component was subjected to 2, 4, 6, 8, 10, 20 . . . 90, 100 sterilization cycles, and after each set of cycles (for example after 2 cycles, then after 4 cycles) the clip was tested for the compression force and dimensional change until the 100 cycles were finished. Later, 14 test samples of each type (12 and 50 mm Roechling, 50 mm Ensinger) were placed inside the autoclave and after 2, 4, 6, 8, 10, 20 . . . 90, 100 sterilization cycles, one sample of each type was brought out from the autoclave and characterized for hardness and DSC.

2.3. Characterization

2.3.1. Clip Characterization

The clip component, after each set of autoclave cycles (2, 4, 6, 8, 10 . . . 90, 100) and before the compression force measurement, was subjected to 250 compression cycles (force, 1 kgf $\pm$ 20% or 0.8 to 1.2 kgf) using an in-house designed jig to test the complete device assembly [21]. The detailed calculation to determine the number of clip compression cycles after each set of sterilization cycles and based on its actual usage in the device assembly is included in the supplementary information. To compare the effect of the sterilization cycles on the compression force of the clip, another similar clip was tested directly for compression force after each 250 spring compression cycles without sterilization (up to 6600 cycles). The compression force of the clip, in both the condition with and without sterilization, was measured using a force gauge (DigiTech, Osaka, Japan, Model: DTG-10) with a 100 N force capacity. At the same time, the lateral dimension of the clip was also measured using a digital Vernier caliper. This whole process was repeated until the 100 sterilization cycles were completed.

2.3.2. Test Samples Characterization

The hardness of the test samples was measured using a Vickers hardness tester VMT-7, an automatic digital hardness tester (MATSUZAWA, Akita Japan) with a square-based pyramid shaped indenter with a phase angle of 136°. An indenting load of 5 Kgf was used for a dwell time of 15 s. The testing was performed as per ASTM E384-11e1. Each sample was indented for 10 indentations and the average of these values is considered as the hardness of the sample.

The DSC test was performed using professional Q20 equipment from TA Instruments, New castle, DE, USA. Small chips from the test samples were sliced using wire cutting pliers. A small quantity (< 20 mg) of chipped PEEK material was placed inside a hermetic Al pan, which was sealed with an Al pan cover. DSC characterization was performed between the temperature range from 40 °C to 400 °C with a ramp rate of 20 °C/min, as per ISO 11357 standard. A continuous flow of argon gas at 50 mL/min was maintained throughout the DSC testing.

3. Results

3.1. Compression Force and Dimension

Table 1 consists the results of the compression force and clip dimensions after each set of autoclave cycles (2, 4, 6 . . . 90, 100). It was observed that the sterilization has no significant effect on the compression force of the clip up to 4 cycles; this is in good agreement with the work of Godara et al. [15] on sterilization effects on carbon composite PEEK. From 6 to 20 cycles there was a decrease of ~10% in the compression force, which further decreased up to ~20% for 30 cycles. After 30 cycles, the change in compression force became insignificant. A similar trend was observed in the clip dimension results; first, the clip size decreased up to 40 cycles and then the size became stable. Figure 2 shows the plot between the compression force and the number of clip compression cycles with and without sterilization cycles. From the results, it can be seen that the sterilized sample exhibited a sudden decrease in the compression force of the clip after every few sterilization cycles, whereas the results for the unsterilized sample showed a gradual decrease in the compression force of the clip. This difference in behavior of the compression force for both clip samples is due to the absorption of moisture during the sterilization process, which induces stresses due to the expansion and results in the degradation of the structural properties [22]. In general, moisture absorption in polymer occurs through diffusion and capillary processes [23–25], which induce plastic deformation either by plasticization or by differential strain due to the swelling while stretching the polymeric chains [26]. These effects can significantly alter the physical, chemical, or mechanical characteristics of materials at different scales [27]. In the current situation, the clip was subjected to moisture at a high temperature (121 °C). As the PEEK (Roechling) used to manufacture the clip has a moisture absorption capacity of 0.2% in 24 h at 23 °C [20], at the high temperature of sterilization a significant amount of moisture absorption can be expected due to the higher diffusion rate of the moisture. Thus, the change in the compression force behavior of the clip is mainly due to the effects of heat and moisture.

Table 1. Compression force and dimension change results of the clip component in comparison with number of autoclave cycles and test sample's hardness measurement.

Autoclave Cycles	Compression Force (Kgf)	Dimension Change (mm)	Hardness Roechling (φ-12 mm)
	Value (% Change)	Value (% Change)	Value (% Change)
0	1.250	38.65	29 ± 0.8
2	1.249 ($\downarrow$0.08)	38.63 ($\downarrow$0.05)	35 ± 1.6 ($\uparrow$20.69)
4	1.262 ($\downarrow$0.96)	36.79 ($\downarrow$4.81)	35 ± 1.4 ($\uparrow$20.69)
6	1.138 ($\downarrow$8.96)	36.61 ($\downarrow$5.28)	38 ± 2.0 ($\uparrow$31.03)
8	1.113 ($\downarrow$10.96)	36.43 ($\downarrow$5.74)	39 ± 1.8 ($\uparrow$34.48)
10	1.117 ($\downarrow$10.64)	36.44 ($\downarrow$5.72)	39 ± 3.7 ($\uparrow$34.48)
20	1.120 ($\downarrow$10.4)	36.37 ($\downarrow$5.90)	43 ± 1.6 ($\uparrow$48.28)
30	0.983 ($\downarrow$21.36)	36.19 ($\downarrow$6.36)	41 ± 2.3 ($\uparrow$41.38)
40	0.981 ($\downarrow$21.52)	36.03 ($\downarrow$6.78)	34 ± 1.7 ($\uparrow$17.24)
50	0.983 ($\downarrow$21.36)	35.71 ($\downarrow$7.61)	33 ± 1.6 ($\uparrow$13.79)
60	0.999 ($\downarrow$20.08)	35.75 ($\downarrow$7.50)	32 ± 2.0 ($\uparrow$10.34)
70	0.951 ($\downarrow$23.92)	35.64 ($\downarrow$7.79)	34 ± 2.0 ($\uparrow$17.24)
80	0.947 ($\downarrow$24.24)	35.65 ($\downarrow$7.76)	34 ± 1.2 ($\uparrow$17.24)
90	0.954 ($\downarrow$23.68)	35.57 ($\downarrow$7.97)	34 ± 1.4 ($\uparrow$17.24)
100	0.974 ($\downarrow$22.08)	35.54 ($\downarrow$8.05)	33 ± 1.3 ($\uparrow$13.79)

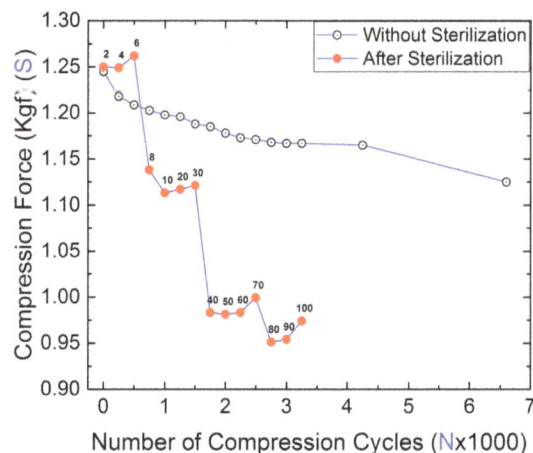

Figure 2. Plot between the compression force and the number of compression cycles of the clip with and without sterilization cycles.

Figure 3 represents the change in the clip lateral dimension before and after 100 cycles of sterilization. The size of the clip initially decreased faster up to 40 cycles and then slowed down for subsequent cycles, with a maximum of 8% change after 100 sterilization cycles. The spring size is directly related to the spring nature of the clip, as this will bring the clip back to its original size after compression. Therefore, both decreased at the same time due to permanent plastic deformation in the clip material.

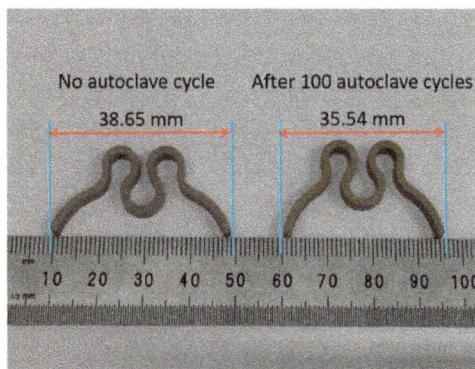

Figure 3. Clip component lateral dimension before and after 100 sterilization cycles.

To understand the permanent plastic deformation of the clip material due to the effects of moisture and heat, the rectangular shape (Figure 1b) test samples were characterized for hardness and DSC analysis after a similar set of sterilization cycles.

3.2. Hardness

Table 1 comprises the hardness characterization results for the test samples made from Roechling PEEK with a 12 mm diameter rod. Figure 4 shows the graphical representation of the change in compression force due to hardness variation of the clip with respect to the number of autoclave cycles. The inset image in Figure 4 shows the indentation cavity formed during the hardness measurement. The hardness results show that the hardness value increased up to ~48% after 20 sterilization cycles and there was a decrease until ~17% after 40 cycles. After 40 cycles, the change in hardness value became stable between 10.34% to 17.24%. These results confirm that there is a change in the material's mechanical property in terms of hardness scale due to the effects of heat and moisture during the sterilization process. Once the moisture absorption capacity of PEEK reached to its saturation value (at nearly 40 cycles), there were no further changes in the hardness value of the test samples (or in the compression force of the clip).

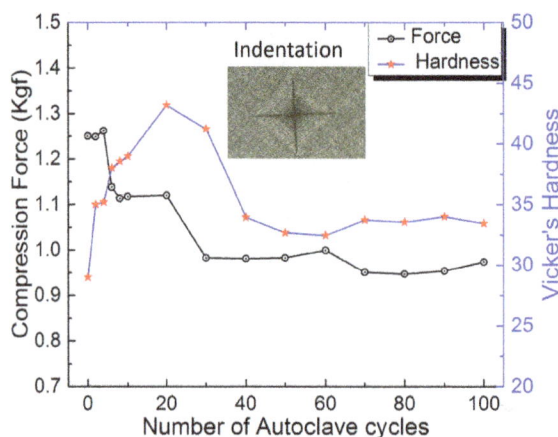

Figure 4. Change in the compression force with the hardness variation of the clip versus the number of autoclave cycles.

Figure 5 shows the plot between the changes in clip dimension with the hardness variation of the clip with respect to the number of autoclave cycles. A similar type of conclusion can be drawn from these results. The clip size decreased faster for the initial 40 cycles of sterilization and became stable for the subsequent number of autoclave cycles. An increase in the hardness represents a decrease in the elastic nature of the material, and therefore a decrease in the compression force and dimension of the spring due to permanent plastic deformation.

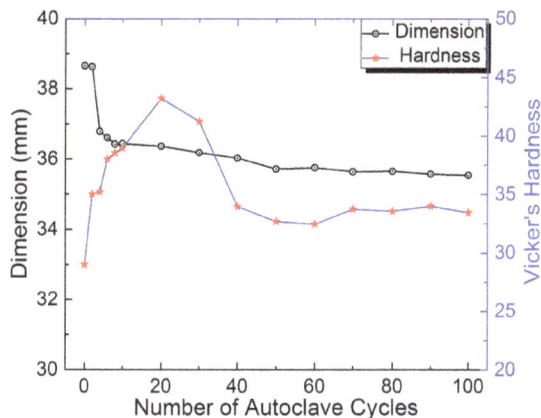

Figure 5. Change in the clip dimension with the hardness variation of the clip with respect to the number of autoclave cycles.

Table 2 contains the hardness comparison results of the samples made from 12 mm and 50 mm diameter PEEK rods procured from Roechling and Ensinger. It was observed that the sample fabricated from the 12 mm rod was more prone to hardening compared to the sample fabricated from the 50 mm rod. Results show that the sample made from the Roechling 50 mm rod exhibited an increase of ~32% and the sample made from the Ensinger 50 mm rod exhibited an increase of ~25% in hardness compared to the ~48% increase in hardness of the sample made from the Roechling 12 mm rod. This indicates that the initial condition of the raw material, which is being used for the fabrication of device components, is also very important, as it may exhibit different types of material degradation behavior under cyclic sterilization conditions. These changes in the degradation behavior could be due to the different cooling rate for 12 mm and 50 mm rods while they are extruded. The 12 mm rod would have cooled faster as the surface area to volume ratio is higher compared to the 50 mm rod. Due to the faster cooling rate, the 12 mm rod might induce more points and line defects in the material, which cause a deeper and faster diffusion of moisture in the PEEK material, resulting in a faster rate of hardening.

Table 2. Comparison of the hardness values of test samples manufactured from 12 mm and 50 mm rods and from two different PEEK manufacturers.

Autoclave Cycles	Hardness Roechling (φ-12 mm)	Hardness Roechling (φ-50 mm)	Hardness Ensinger (φ-50 mm)
	Value (% Change)	Value (% Change)	Value (% Change)
0	29 ± 0.8	28 ± 1.2	28 ± 0.6
2	35 ± 1.6 (↑20.69)	30 ± 1.2 (↑7.14)	30 ± 1.0 (↑7.14)
4	35 ± 1.4 (↑20.69)	31 ± 1.0 (↑10.71)	29 ± 1.1 (↑3.57)
6	38 ± 2.0 (↑31.03)	31 ± 1.0 (↑10.71)	30 ± 1.7 (↑7.14)
8	39 ± 1.8 (↑34.48)	33 ± 0.7 (↑17.86)	31 ± 1.7 (↑10.71)
10	39 ± 3.7 (↑34.48)	34 ± 0.9 (↑21.43)	33 ± 2.2 (↑17.86)
20	43 ± 1.6 (↑48.28)	33 ± 2.2 (↑17.86)	32 ± 1.6 (↑14.29)
30	41 ± 2.3 (↑41.38)	32 ± 1.4 (↑14.29)	30 ± 1.5 (↑7.14)
40	34 ± 1.7 (↑17.24)	33 ± 1.2 (↑17.86)	31 ± 1.8 (↑10.71)
50	33 ± 1.6 (↑13.79)	37 ± 1.4 (↑32.14)	34 ± 1.9 (↑21.43)
60	32 ± 2.0 (↑10.34)	37 ± 2.5 (↑32.14)	35 ± 2.0 (↑25)
70	34 ± 2.0 (↑17.24)	35 ± 1.0 (↑25)	34 ± 1.2 (↑21.43)
80	34 ± 1.2 (↑17.24)	36 ± 1.5 (↑28.57)	35 ± 1.8 (↑25)
90	34 ± 1.4 (↑17.24)	36 ± 0.9 (↑28.57)	35 ± 1.3 (↑25)
100	33 ± 1.3 (↑13.79)	35 ± 1.4 (↑25)	35 ± 1.7 (↑25)

Figure 6 shows the graphic of the hardness variation of different test samples with respect to the number of autoclave cycles. For the first 50 cycles, the rate of hardening was different in the samples fabricated from 12 mm and 50 mm rods. For more than 50 cycles, the variation in hardness value became insignificant, irrespective of the different processing conditions of the raw material (12 mm or 50 mm rods). The slight difference in the hardness value of the samples made from 50 mm Ensinger and Roechling rods was due to their different moisture absorption capacities (0.02% and 0.2% in 24 h at 23 °C) [19,20].

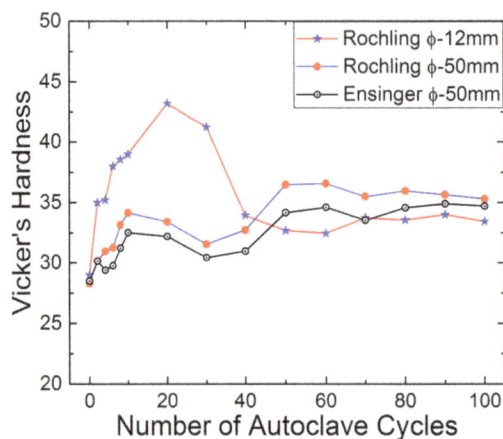

Figure 6. Effect of rod size on the hardness variation with respect to autoclave cycles.

3.3. Differential Scanning Calorimetry

Figure 7 shows the DSC testing results after 0, 6, and 10 autoclave cycles for the test sample fabricated from 12 mm Roechling PEEK. Results show that there was no significant change observed in the DSC behavior of PEEK even after 10 sterilization cycles. This observation indicates that the changes in PEEK properties are not due to a change in its bulk material properties, but rather a surface-driven phenomenon due to the diffusion of moisture at high temperatures. Therefore, the thickness of a component plays an important role in deciding the material's properties and degradation rate to thus decide the component's life.

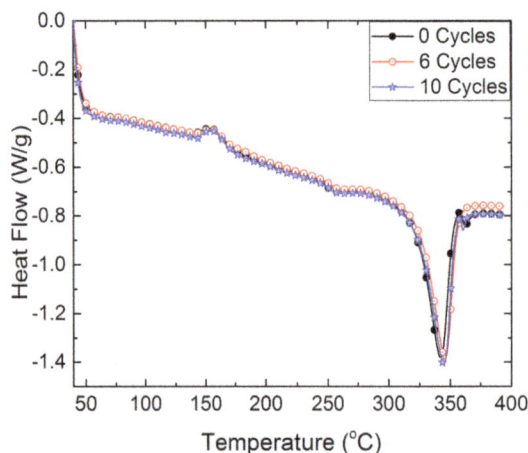

Figure 7. DSC curves between the temperature and heat flow of Roechling PEEK (12 mm) after 0, 6, and 10 autoclave cycles.

In order to further verify that the degradation of PEEK material properties were due to the diffusion of moisture at a high temperature (121 °C), an untreated test sample (50 mm, Roechling) was repetitively heated (at 121 °C temperature for 30 min) for 10 and 20 cycles and cooled to room temperature, in an electric oven without the presence of moisture. Table 3 lists the results of the hardness measurement after repetitive heating and cooling cycles without the presence of moisture and the hardness of samples that had already undergone a similar number of sterilized cycles. The results show that repetitive heating without moisture exhibited only a 7.14% increase in hardness, irrespective of the number of cycles. Conversely, the hardness change due to repetitive heating in the presence of moisture (sterilization) was measured as 21.43% and 17.86% for the same number of cycles. Another experiment was performed on the sterilized samples, which had already undergone 20 and 30 autoclave cycles. Both samples were heated for a long time (20 h, 100 °C) to remove the moisture content and then the hardness of both the samples was measured again. It was observed that the hardness of the samples after a long time heating was reduced tremendously to a level of 6.89% and 3.45% for 20 and 30 cycles, respectively, compared to 48.28% and 42.38% just after the sterilization cycles (see Table 3). Both of these experiments performed on the test sample revealed the role of moisture content in the hardness increment of PEEK after the sterilization cycles. Therefore, the moisture at higher temperatures during the sterilization process caused a very detrimental effect on the mechanical properties of PEEK polymer due to the high diffusion rate of moisture at high temperatures.

Table 3. Effect of heating without moisture on the hardness of Roechling PEEK test samples.

	Repetitive Heating Cycles without Moisture		
No. of cycles	Hardness Roechling (50 mm) MPa	Hardness Sterilized sample MPa (%)	Hardness heated sample MPa (%)
10	28 ± 1.2	34 ± 0.9 (21.43↑)	30 ± 1.3 (7.14↑)
20		33 ± 2.2 (17.86↑)	30 ± 1.0 (7.14↑)
	Long Heating Period (20 h, 100 °C) of Sterilized Sample		
No. of cycles	Hardness Roechling (12 mm) MPa	Hardness Sterilized sample MPa (%)	Hardness heated sample MPa (%)
20	29 ± 0.8	43 ± 1.6 (48.28↑)	31 ± 1.1 (6.89↑)
30		41 ± 2.3 (42.38↑)	30 ± 0.8 (3.45↑)

4. Conclusions

The effects of the sterilization process on the material properties (reliability) of medical grade thermoplastic polyetheretherketone (PEEK) have been studied. Test results on the clip component of the medical device concluded that there is a decrease of ~20% in the compression force of the clip after 30 autoclave cycles and a decrease of ~6% in the lateral dimension of the clip after 50 autoclave cycles. The change in both the compression force and dimension become stable for the subsequent cycles of sterilization. DSC results on the test sample concluded that there are no significant changes in bulk material properties and the degradation behavior of PEEK properties is a surface-driven phenomenon. Hardness testing results further confirmed that the changes in the clip properties for the initial sterilization (30–50) cycles are due to changes in the hardness or due to the absorption of moisture on the surface of the polymer. When an untreated sample of PEEK was subjected to repetitive heating and cooling cycles in the absence of moisture, it was observed that the hardness of the PEEK material increased by only 7.14% after 10 and 20 cycles, respectively, compared to 21.43% and 17.86% in the presence of both heat and moisture during the sterilization process. When the sterilized samples were heated for a long time (20 h) to remove the moisture content absorbed during sterilization, a tremendous decrease in the hardness value of the samples were observed compare to just after the sterilization hardness. Therefore, it can be concluded that the degradation of PEEK is mainly due to the effect of moisture at elevated temperatures. Thus, this thermal reliability study on PEEK suggests that if a reusable medical device is fabricated using PEEK, which will be subjected to repeated sterilization processes, the change in mechanical properties of PEEK needs to be accounted for in the proposed design.

Acknowledgments: This research work was conducted under the grant MOE2014-TIF-1-G-043 awarded by The Ministry of Education of Singapore.

Author Contributions: Amit Kumar, Soo Leong Foo, and Teck Kheng Lee conceived and designed the experiments; Amit Kumar and Wai Teng Yap performed the experiments; Amit Kumar analyzed the results and wrote the paper; Soo Leong Foo and Teck Kheng Lee contributed comments on the manuscript.

Conflicts of Interest: The authors declare no conflict of interest.

References

1. Liao, K. Performance characterization and modeling of a composite hip prosthesis. *Exp. Tech.* **1994**, *18*, 33–41. [CrossRef]

2. Maharaj, G.R.; Jamison, R.D. Intraoperative impact: Characterization and laboratory simulation on composite hip prostheses. In *Composite Materials for Implant Applications in the Human Body: Characterization and Testing*; Jamison, R.D., Gilbertson, L.N., Eds.; ASTM: Philadelphia, PA, USA, 1993; pp. 98–108.

3. Kelsey, D.J.; Springer, G.S.; Goodman, S.B. Composite implant for bone replacement. *J. Compos. Mater.* **1997**, *31*, 1593–1632. [CrossRef]

4. Corvelli, A.A.; Biermann, P.J.; Roberts, C. Design, analysis, and fabrication of a composite segmental bone replacement implant. *J. Adv. Mater.* **1997**, *28*, 2–8.

5. Williams, D. New horizons for thermoplastic polymers. *Med. Device Technol.* **2001**, *12*, 8–9. [PubMed]

6. Wang, A.; Lin, R.; Stark, C.; Dumbleton, J.H. Suitability and limitations of carbon fiber reinforced PEEK composites as bearing surfaces for total joint replacements. *Wear* **1999**, *225*, 724–727. [CrossRef]

7. Jones, E.; Wang, A.; Streicher, R. Validating the limits for a PEEK composite as an acetabular wear surface. In Proceedings of the 27th annual meeting of Society of Biomaterials, St Paul, MN, USA, 24–29 April 2001; p. 27.

8. Joyce, T.J.; Rieker, C.; Unsworth, A. Comparative in vitro wear testing of PEEK and UHMWPE capped metacarpophalangeal prostheses. *Biomed. Mater. Eng.* **2006**, *16*, 1–10. [PubMed]

9. Manley, M.; Ong, K.; Kurtz, S.M.; Rushton, N.; Field, R.E. Biomechanics of a PEEK horseshoe-shaped cup: Comparisons with a predicate deformable cup. In Proceedings of the Transactions of the 53rd Orthopedic Research Society, San Diego, CA, USA, 11–14 February 2007; Volume 32, p. 1717.

10. Arias, A.; Rodríguez-Martínez, J.A.; Rusinek, A. Numerical simulations of impact behaviour of thin steel plates subjected to cylindrical, conical and hemispherical non-deformable projectiles. *Eng. Fract. Mech.* **2007**, *75*, 1635–1656. [CrossRef]

11. Sobieraj, M.; Rimnac, C. Fracture, fatigue and noch behavior of PEEK. In *PEEK Biomaterials Handbook*; Kurtz, S., Ed.; William Andrew Books: Norwich, NY, USA, 2012; pp. 61–73.

12. El-Qoubaa, Z.; Othman, R. Characterization and modeling of the strain rate sensitivity of polyetheretherketone's compressive yield stress. *Mater. Des.* **2015**, *66*, 336–345. [CrossRef]

13. Kurtz, S.; Devine, J. PEEK biomaterials in trauma, orthopedic, and spinal implants. *Biomaterials* **2007**, *28*, 4845–4869. [CrossRef] [PubMed]

14. Rae, P.; Brown, E.; Orler, E. The mechanical properties of poly(ether-ether-ketone) (PEEK) with emphasis on the large compressive strain response. *Polymer* **2007**, *48*, 598–615. [CrossRef]

15. Godara, A.; Raabe, D.; Green, S. The influence of sterilization processes on the micromechanical properties of carbon fiber-reinforced PEEK composites for bone implant applications. *Acta Biomater.* **2007**, *3*, 209–220. [CrossRef] [PubMed]

16. Tai, N.H.; Yip, M.C.; Tseng, C.M. Influences of thermal cycling and low-energy impact on the fatigue behavior of carbon/PEEK laminates. *Compos. Part B Eng.* **1999**, *30*, 849–865. [CrossRef]

17. Xin, H.; Shephered, D.E.T.; Dearn, K.D. Strength of poly-ether-ether-ketone: Effects of sterilisation and thermal ageing. *Polym. Test.* **2013**, *32*, 1001–1005. [CrossRef]

18. Ensinger Engineering plastics: Manual. Available online: https://www.ensingerplastics.com/downloads (accessed on 14 December 2017).

19. Ensinger TECAPEEK. Available online: https://www.ensingerplastics.com/en/shapes/biocompatible-medical-grade/peek (accessed on 14 December 2017).

20. Roechling SUSTAPEEK. Available online: https://www.roechling.com/sg/industrial/materials/thermoplastics/detail/sustapeek-mg-natural-203/ (accessed on 14 December 2017).

21. Leong, F.S.; Wei, N.K.; Quan, G.J.; Hua, T.P.; Kheng, L.T.; Hui, W.Q.; Edmund, C.; Kesavan, E. System and apparatus for guiding an instrument. U.S. Patent US20160206383 A1, 21 July 2016.

22. Ray, B.C. Temperature effect during humid ageing on interfaces of glass and carbon fibers reinforced epoxy composites. *J. Colloid Interface Sci.* **2006**, *298*, 111–117. [CrossRef] [PubMed]

23. Kaelble, D.H.; Dynes, P.J.; Maus, L. Hydrothermal aging of compositematerials. 1. Aspects. *J. Adhes.* **1976**, *8*, 121–144. [CrossRef]

24. Marom, G.; Broutman, L.J. Moisture in epoxy resin composites. *J. Adhes.* **1981**, *2*, 153–164. [CrossRef]

25. Mijovic, M.; Lin, K.F. The effects of hygrothermal fatigue on physical mechanical properties and morphology on neat epoxy-resin and graphite composite. *J. Appl. Poly. Sci.* **1985**, *30*, 2527–2549. [CrossRef]

26. Barraza, H.J.; Aktas, L.; Hamidi, Y.K.; Long, J., Jr.; O'Rear, E.A.; Altan, M.C. Moisture absorption and wet-adhesion properties of resin transfer molded (RTM) composites containing elastomer-coated glass fibers. *J. Adhes. Sci. Technol.* **2003**, *17*, 217–242. [CrossRef]

27. Zheng, Q.; Morgan, R.J. Synergistic thermal—Moisture damage mechanisms of epoxy and their carbon-fiber composites. *J. Compos. Mater.* **1993**, *27*, 1465–1478. [CrossRef]

11

Statistical Modelling of Temperature and Moisture Uptake of Biochars Exposed to Selected Relative Humidity of Air

Luciane Bastistella [1], Patrick Rousset [2,3,*], Antonio Aviz [1], Armando Caldeira-Pires [4],
Gilles Humbert [5] and Manoel Nogueira [6]

[1] Faculty of Mining and Environmental Engineering—Femma, Federal University of South and Southwest of Pará, Marabá 68507-590, PA, Brazil; batistella.luciane@gmail.com (L.B.); antonioaaviz@gmail.com (A.A.)
[2] Joint Graduate School of Energy and Environment, Center of Excellence on Energy Technology and Environment-KMUTT, Bangkok 10140, Thailand
[3] CIRAD, UPR BioWooEB, F-34398 Montpellier, France
[4] Department of Mechanical Engineering, University of Brasilia—UnB, Campus Universitário Darcy Ribeiro, S/N, Asa Norte, Brasília 70910-900, DF, Brazil; armandcp@unb.br
[5] FerroPem, R & D Department, F-73025 Chambéry, France; gilles.humbert@ferroglobe.com
[6] Faculty of Mechanical Engineering—Fem, Federal University of Pará, Belém 66075-900, PA, Brazil; mfmn@ufpa.br
* Correspondence: patrick.rousset@cirad.fr

Abstract: New experimental techniques, as well as modern variants on known methods, have recently been employed to investigate the fundamental reactions underlying the oxidation of biochar. The purpose of this paper was to experimentally and statistically study how the relative humidity of air, mass, and particle size of four biochars influenced the adsorption of water and the increase in temperature. A random factorial design was employed using the intuitive statistical software Xlstat. A simple linear regression model and an analysis of variance with a pairwise comparison were performed. The experimental study was carried out on the wood of *Quercus pubescens*, *Cyclobalanopsis glauca*, *Trigonostemon huangmosun*, and *Bambusa vulgaris*, and involved five relative humidity conditions (22, 43, 75, 84, and 90%), two mass samples (0.1 and 1 g), and two particle sizes (powder and piece). Two response variables including water adsorption and temperature increase were analyzed and discussed. The temperature did not increase linearly with the adsorption of water. Temperature was modeled by nine explanatory variables, while water adsorption was modeled by eight. Five variables, including factors and their interactions, were found to be common to the two models. Sample mass and relative humidity influenced the two qualitative variables, while particle size and biochar type only influenced the temperature.

Keywords: biochars; moisture uptake; statistical modelling

1. Introduction

Spontaneous combustion has long been recognized as a fire hazard in stored coal and fires usually beginning as "hot spots" deep within the stockpile. Understanding the mechanisms by which carbon-based products get heated to the critical temperature is very important to suppress self-ignition, and ensure secure storage, transport, and handling. On the other hand, self-ignition may also be useful in combustion processes if it occurs under controlled conditions. Therefore, self-ignition can be categorized as a favorable or an unfavorable process which can be controlled by managing the desired parameters.

New experimental techniques, as well as modern variants on venerable methods, have recently been employed to investigate the fundamental reactions underlying auto-ignition in great detail [1]. A requirement for self-ignition to occur is that the material is sufficiently porous and reactive so that adequate fuel and oxygen are available throughout the whole self-heating process. According to Miura [2], the initial conditions for coal self-heating include many factors which can be divided into two main types: the properties of coal (intrinsic factors) and the environment/storage conditions (extrinsic factors). Heating results from some chemical and/or physical processes occurring within the material and this phenomenon is mainly attributed to exothermic processes such as low temperature oxidation, microbial metabolism, adsorption-desorption of water, and air oxidation with a production of undissipated energy [3,4]. The first advances in self–heating investigations related to air relative humidity were attributed to Davis, who studied the effect of moisture content on the spontaneous combustion of coal using an adiabatic calorimeter. He compared the heat produced by coal in contact with dry and saturated oxygen and showed that the spontaneous combustion started at 70 °C [5]. In the 1960s, Stott confirmed these results and proposed differential equations describing the high-temperature oxidation of coal [6]. Other recent studies have been carried out to clarify the mechanism of low-temperature oxidation of coal and showed that this process is in general very slow compared to air moisture uptake [7,8]. A literature review has been made on present theories and methods for the prediction of spontaneous ignition and has mainly focused on engineering models and small-scale methods [9].

The tendency to self-heating is also dependent on material size, so no quantification of a material's self-ignition hazard is possible without incorporating system size and ambient temperature [10]. Studying charcoal briquettes self-ignition, these authors concluded that a temperature of at least 121 °C is required for self-ignition to occur in the largest commercially available bag size, 9 kg. No information and no correlation have been provided for small size samples (g). The effect of particle size with a diameter in the range 2–50 mm has been studied in a large-scale apparatus and has shown that the spontaneous heating of coal leads to flaming combustion below a certain critical range [11]. The liability of spontaneous combustion of lignite increases with decreasing particle size, increasing moisture content of the coal, and decreasing humidity of the air [12]. The ignition delay of a biomass packed-bed has also been studied and showed an increase with fuel properties such as moisture content and particle size, while it decreased with process conditions such as gas velocity and temperature [13]. The information with a direct temperature increase measurement is expected to be an index to estimate the propensity to spontaneous combustion. The adsorption of water vapor on the sample has been shown to play a crucial role in raising the sample temperature over the critical self-ignition temperature [14].

The purpose of this work was to measure the adsorption rate of water vapor and temperature change of woody and non-woody biochars under various relative humidity conditions naturally occurring in arid, semi-arid, and humid climates. The role of sample size on spontaneous heating was also investigated based on experiments and statistical analyses.

2. Material and Method

2.1. Material

The four types of biochar used in this study represent major feedstocks and local common bio-reducers in Yunnan Province in China. These are *Quercus pubescens* (Qp), *Cyclobalanopsis glauca* (Cg), *Trigonostemon huangmosun* (Th), and *Bambusa vulgar* (B), which are mainly hardwood except for bamboo which is a non woody biomass. All selected biomass materials were pyrolysed at 500 °C and held for 60 min in order to be roughly similar to the industrial operating conditions of the factory [15]. The proximate analyses followed the standard procedure of the American Society (ASTM D5142). The elemental composition of C, H, and N content was determined using a Thermo FlashEA 1112 Elemental Analyzer (Thermo Fisher Scientific Inc., Waltham, MA, USA) according to the

European standard XP CEN/TS 15104 and ASTM D5373 for solid biofuels and charcoals, respectively. The higher heating value (HHV) was experimentally determined with a calorimeter LECO AC350 (LECO Corporation, Saint Joseph, MI, USA). The BET method was applied to provide a precise specific surface area with Belsorp-max Bel Japan equipment (MicrotracBEL Corp., Osaka, Japan). This gear is designed for a wide range adsorption isotherm for surface area and pore size distribution analysis. It can measure adsorption isotherms from relative pressure as low as 1×10^{-8} (N2 at 77 K, Ar at 87 K), using a 13.3 Pa pressure transducer. The nitrogen adsorption-desorption of the samples was measured at $-196\,°C$. Prior to the measurements, the samples were degassed at $150\,°C$ for 1 h. The properties of the experimental samples are shown in Table 1.

Table 1. Physicochemical characteristics of biochars.

Properties	Biochar			
	Qp	Cg	Tr	B
Proximate analysis (wt %, dry basis)				
Ash	0.5	2.4	1.5	6.5
Volatile matter	12.7	17.4	15.1	15.6
Volatile matter/Ash	25.4	7.3	10.1	2.4
Fixed carbon	86.8	80.2	83.4	77.8
Ultimate analysis (wt %, dry basis and ash free)				
C	89.6	86.3	89.3	82.2
H	2.3	2.3	2.3	1.5
N	0.5	0.3	0.3	0.5
O (by difference)	7.6	11.1	8.1	15.8
H/C	0.02	0.03	0.02	0.02
High heating value (Mj·Kg^{-1})	33.9	32.7	33.8	30.4
BET Surface area (m^2·g^{-1})	292	62	86	40

2.2. Temperature Measurement

The adsorption rate of water vapor followed the adapted procedure for coal [2]. Experiments were performed by simulating dry and humid climate conditions where dried samples at room temperature were exposed to the following saturated salt solutions: Potassium acetate (22.6%), potassium carbonate (43.2%), sodium chloride (75.3%), potassium chloride (84.3%), and barium chloride (90.2%) [16]. The solutions were prepared in 1.0 L wide-mouthed glass jars using distilled water and were closed with rubber-join screw caps and stored at 25 °C. In addition, each chamber was equilibrated for one day at 25 °C and immersed in a water bath to guaranty the desired relative humidity. The carbonized biomass samples were processed in the form of pieces and fine particles (250 μm), each with 0.1 g and 1.0 g. The particle samples were placed in a mesh basket (37 μm opening) as a support for all water adsorption experiments. One K-type thermocouple with a 0.5 mm diameter was inserted in the biomass and another close to the basket to measure the changes in temperature during the water adsorption process, as shown in Figure 1. For clarity and accuracy purposes, the temperature records of only *Bamboo vulgaris* and *Quercus pubescens* biochars are displayed in Figure 2. For the other two biochars (Cg and Th), the trend was the same, showing a significant increase in temperature with an increase in relative humidity. Before exposure to saturated salt solutions, samples were dried at 105 °C in a nitrogen stream used as a purging gas at a flow rate of 0.5 L/min. The mass gain due to the adsorption of water vapor was also calculated by drying and weighting the samples after each experiment.

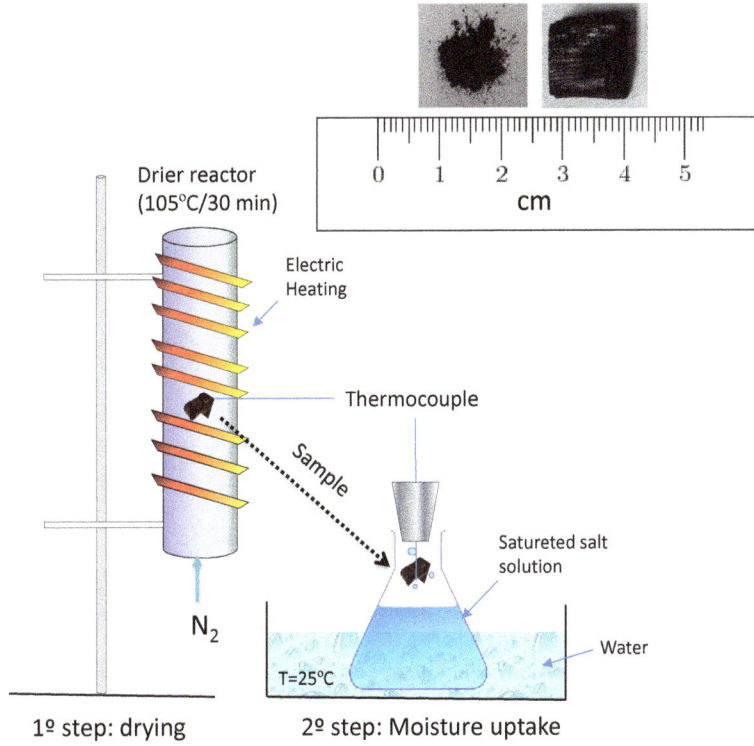

Figure 1. Schematic of experimental setup used for the measurement of temperature change of biomass on exposure to stationary atmosphere (adapted from [2]).

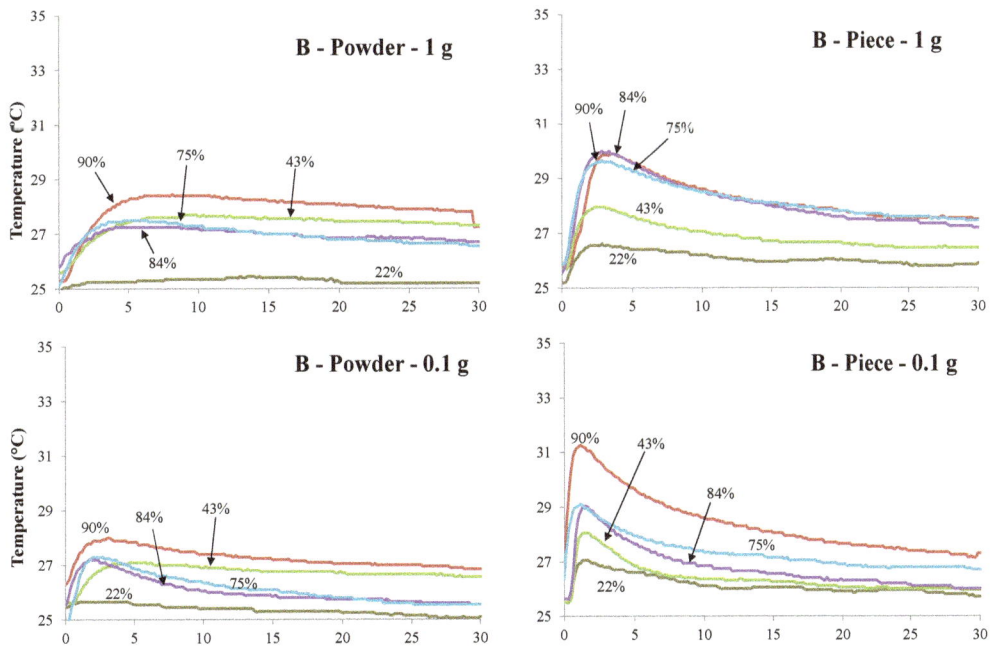

Figure 2. *Cont.*

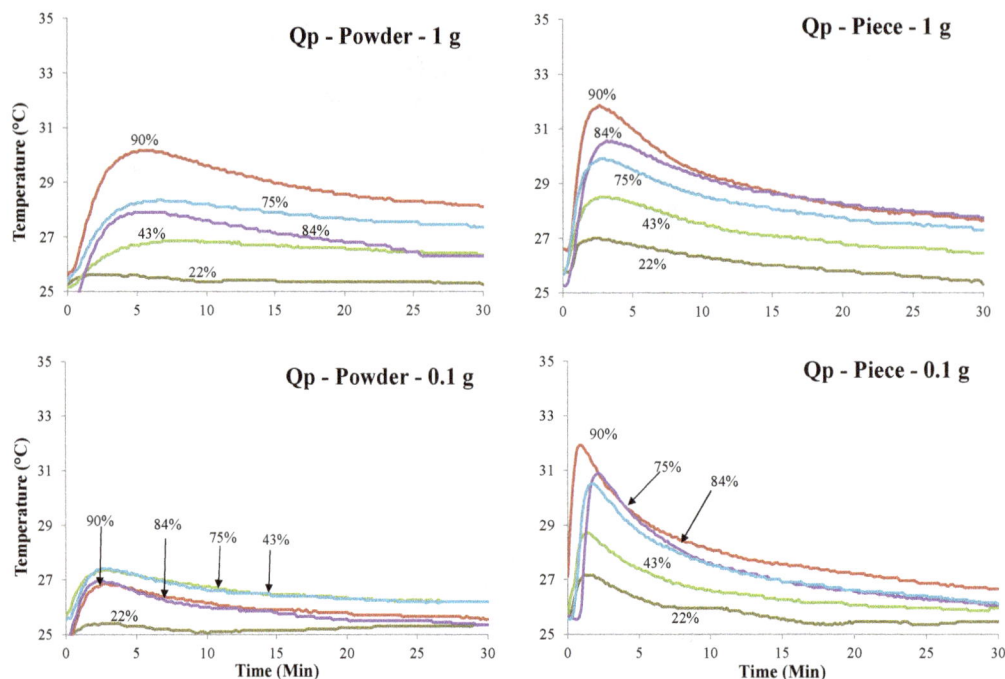

Figure 2. Temperatures (°C) measurement vs. time (min) of *Bamboo* (B) and *Quercus pubescens* (Qp) biochars for powder and piece and 0.1 and 1 g, respectively, exposed to the five selected air humidity (%) conditions.

2.3. Experimental Protocol

One hundred and sixty (160) assays were conducted, corresponding to eighty (80) treatments and two replicates. The XLSTAT software (Addinsoft company, Paris, France) was used to analyze and reformat data within Excel for statistical analysis. Multiple linear regressions, Analysis of Variance (ANOVA) with stepwise model interactions, and Tukey multiple comparisons were used to relate the two dependent variables that are temperature (T) and water adsorption (W) with the four independent variables that are biochar species (B), relative humidity of air (RH), particle size (S), and mass of sample (M). The two variables, temperature and water adsorption, in response to the experiments, were analyzed and discussed following a (5×4^2) random factorial design. The parameters of the experimental design can be found in Table 2.

Table 2. Values of the parameters selected for the experimental design.

Parameters Level	Relative Humidity (%)	Biochars	Particle Size	Mass (g)
1	22.6	*Quercus pubescens* (Qp)	Piece	1
2	43.2	*Cyclobalanopsis glauca* (Cg)	Powder	0.1
3	75.3	*Trigonostemon huangmosun* (Th)	-	-
4	84.3	*Bambusa vulgar* (B)	-	-
5	90.2	-	-	-

The experimental error had a degree of freedom of 33 and 43 for temperature and water sorption, respectively, as dependent variables. The general model for variance analysis (ANOVA) can be described by the following equation, where each independent variable and their interactions are presented:

$$Y_{ijklr} = \mu + [B_i + RH_j + S_k + M_l + (B \times RH)_{ij} + (B \times S)_{ik} + (B \times M)_{il} + (RH \times S)_{jk} + (RH \times M)_{jl}$$

$$+ (S \times M)_{kl} + (B \times RH \times S)_{ijk} + (B \times RH \times M)_{ijl} + (B \times S \times M)_{ikl} + (RH \times S \times M)_{jkl} + (B \times RH \times S \times M)_{ijkl}] + \varepsilon_{ijkl}$$

3. Results and Discussion

3.1. Overall Results

Tables 3 and 4 provide the average values for water adsorption (W) and the temperature (T) of biochar pieces and powdered biochar reached when exposed to five different conditions of relative humidity. All samples show a significant increase in temperature. The higher the air humidity value is, the higher the increase in temperature is. This tendency is observed for almost all samples. The highest and fastest increase in temperature is observed for the piece of biochar characterized by the lowest weight (0.1 g) (Figure 2). The temperature profiles of the powdered biochar samples are significantly different from the biochar piece samples and this is independent of the weight of the samples. While the biochar piece samples required around 2 min to reach the peak temperature, the powdered biochar samples required 5 to 8 min under the highest relative humidity conditions. Cg biochar and piece samples reached the highest temperature (6 °C) for a relative humidity of 90.2%.

Table 3. Averaged values for water adsorption (%) from pieces and powdered biochars exposed to different relative humidities of air based on a random factorial design, considering two replicates per test. (d.b. = dry basis).

Sample	Size	Mass (g)	Relative Humidity (%)				
			22.6	43.2	75.3	84.3	90.2
			Moisture Uptake (%) d.b.				
Qp	Piece	0.1	0.14	2.97	5.30	3.95	1.96
		1	0.43	0.28	2.22	3.23	2.20
Cg	Piece	0.1	0.00	3.25	5.43	6.96	4.87
		1	0.13	1.32	3.55	3.49	1.72
Th	Piece	0.1	0.37	2.53	4.46	4.82	2.72
		1	0.35	1.66	1.92	2.28	2.19
B	Piece	0.1	1.20	2.14	4.29	3.77	4.11
		1	0.61	1.72	2.12	3.06	1.42
Qp	Powder	0.1	2.35	4.97	4.02	4.25	5.21
		1	0.65	1.35	1.78	1.26	1.76
Cg	Powder	0.1	2.73	2.83	5.58	6.15	6.50
		1	0.88	1.07	3.23	1.77	3.60
Th	Powder	0.1	0.31	2.36	2.33	5.10	3.77
		1	0.70	1.10	1.00	2.12	2.05
B	Powder	0.1	1.72	2.96	2.37	6.24	3.52
		1	0.83	1.51	1.74	1.81	2.07

Concerning the water vapor adsorption for all conditions, all the samples with a mass of 1.0 g show a lower water vapor adsorption capacity than samples with a mass of 0.1 g. The values obtained are more dispersed. However, the global trend shows that more mass of water vapor was adsorbed when the samples were exposed to higher levels of relative humidity. This difference is mainly due to the difference in mass transfer in the samples [2]. The highest amount of water adsorbed is observed for Cg (piece 0.1 g/84.3%), Cg (powder 0.1 g/90.2%), and B (powder 0.1 g/84.3%) with 7%, 6.5%, and 6.2%, respectively. For the powdered biochar samples, the better adsorption capacity can be explained by the larger surface area exposed to outside conditions compared to the piece. The surface area is also an important physical property for self-ignition. A direct correlation between oxygen chemisorption and active surface area has been reported by Zhao [17]. However, this phenomenon does not seem to be correlated to the BET results (Table 1), where (Qp) and (B) biochars with 292 and 40 $m^2 \cdot g^{-1}$, respectively, did not show the strongest and lowest potential for spontaneous combustion, respectively. To confirm this, we performed a one-way balanced analysis of variance. As shown in Figure 3, the Tukey's HSD (Honestly Significantly Different) test was applied to all pairwise differences between means. As all the combinations shared the same letter, it can be concluded that the BET does not significantly affect water absorption (W) and temperature (T).

Table 4. Averaged values for temperature ($°C$) increasing from pieces and powdered biochars exposed to different relative humidities of air based on a random factorial design, considering two replicates per test.

Sample	Size	Mass (g)	Relative Humidity (%)				
			22.6	43.2	75.3	84.3	90.2
			ΔT ($°C$)				
Qp	Piece	0.1	1.60	2.95	5.05	5.25	4.80
		1.0	1.20	2.70	4.25	5.30	5.25
Cg	Piece	0.1	1.40	3.05	4.95	5.25	5.70
		1.0	1.55	2.45	4.75	4.61	6.00
Th	Piece	0.1	1.15	2.45	3.85	4.50	5.30
		1.0	1.15	2.50	3.35	4.20	4.40
B	Piece	0.1	1.10	2.55	2.75	3.40	3.50
		1.0	1.40	2.25	3.8	4.45	4.10
Qp	Powder	0.1	0.55	1.70	1.85	2.10	2.25
		1.0	0.35	1.70	2.75	3.10	4.55
Cg	Powder	0.1	0.65	1.40	2.35	2.00	2.55
		1.0	0.65	1.80	5.35	2.00	4.95
Th	Powder	0.1	0.35	1.40	0.90	2.00	2.25
		1.0	0.60	1.70	1.60	2.90	2.85
B	Powder	0.1	0.20	0.95	2.70	1.65	1.65
		1.0	0.35	2.10	2.50	1.45	3.15

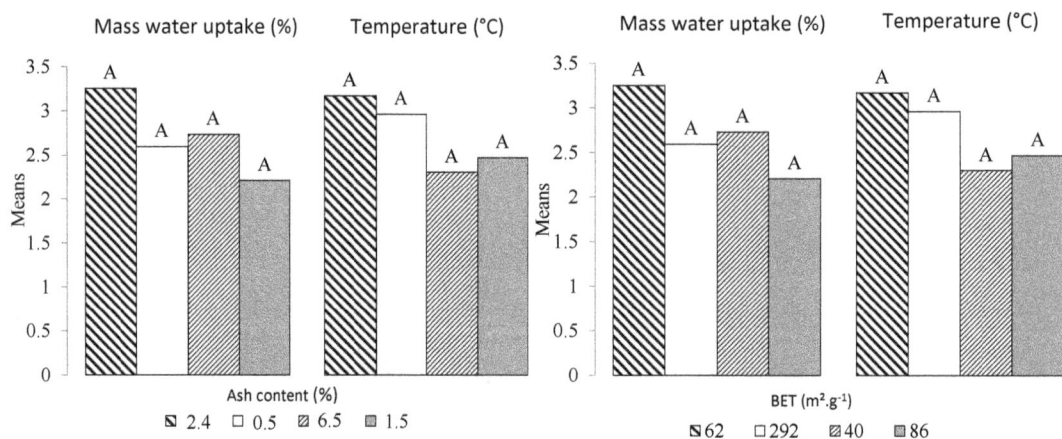

Figure 3. Classification by Tukey's test for water sorption (W) and temperature (T) averages versus BET and ash content. The means with the same letter were not significantly different at 5% ($\alpha = 0.05$).

The above discussion shows that the adsorption of water vapor under different relative humidity conditions (Table 3) has the potential to raise the temperature of the samples. However, other biochar-related physical and chemical properties can also affect water sorption and consequently the increase in temperature (Table 4). For example, the mineral content acts as a heat sink [17]. With the increase in mineral content (indicated by increasing ash content), it has been shown that the crossing point of coal temperature increases (CPT), which is used to evaluate the spontaneous combustion of coal. The Crossing-point Temperature (CPT) is the temperature (temperature and corresponding time) at which the increasing coal temperature is equal to the increasing oven temperature within a Temperature-Programmed System (TPS) [18]. These results suggest that Bamboo with 6.5% ash content should be the most subject to spontaneous combustion, while Qp should be the least affected (0.5%). A recent study by [15] showed the opposite. Figure 3 shows no difference related to ash content for the four biochars. In the literature, a lot of models have been developed to predict spontaneous ignition. They were mainly engineering models and small-scale methods requiring producing input data for such models [9]. In the next sections, a statistical analysis based on linear regressions and analysis of variance (ANOVA) is reported, investigating the weight of each independent or explanatory variable according to the model equation presented in the experimental section.

3.2. Linear Regression Model

Applying a simple linear regression model based on Ordinary Least Squares (OLS), the objective was to determine how temperature (T) varies with water adsorption (W) and to verify if a linear model makes sense. The chart from Figure 4 allows us to visualize the data, the regression line (the fitted model), and two confidence intervals at 95%. It can be clearly seen that there is a linear trend, but also a high variability around the line. This high dispersion of results is corroborated by a low R^2 value (0.126), indicating that only 13% of the variability of the temperature can be explained by water adsorption. The model equation in this case is given by:

$$\Delta T \ (^\circ C) = 1.9 + 0.3 \times \text{Mass water adsorbed (\%)}.$$

Several linear regressions were performed to verify if any linear models limited to selected data from each independent variable (RH, B, M, and S) could better explain the results obtained. Statistics are summarized in Table 5 and enabled us to determine whether or not the explanatory independent variables bring significant information to the model. Despite generally low R^2 values for all explanatory variables, the information brought by size (piece), mass (0.1 and 1 g), and type of biochar (Th) is observed to be more significant than the other variables. These variables explain 60, 40, 30, and 17% of the relation between temperature and water adsorption, respectively. Their probabilities corresponding to the F value were found to be lower than 0.0001. If we can partially conclude with confidence that these four independent variables brought a significant amount of information, the linear regression model still shows limitations; using a simple linear regression is not acceptable for the prediction of temperature increase as a function of water adsorption.

Table 5. Summary statistics for the linear regression model of T ($^\circ C$) vs. W (%) for each qualitative variable. P. = piece; Pow. = powder.

	Biochars				Relative Humidity (%)					Mass (g)		Size	
	Qp	Cg	Th	B	22.6	43.2	75.3	84.3	90.2	0.1	1.0	P.	Pow.
Min	0.31	0.40	0.00	0.10	0.00	0.50	0.92	1.27	1.23	0.02	0.20	0.83	0.04
Max	6.00	6.53	6.10	4.70	1.87	3.50	6.02	5.51	6.53	6.54	6.45	6.52	5.73
Average	3.03	3.20	2.51	2.30	0.90	2.10	3.31	3.43	4.07	2.62	2.92	3.55	1.93
Std.dev.	1.70	1.80	1.50	1.24	0.50	0.70	1.43	1.48	1.52	1.61	1.65	1.51	1.24
R^2	0.05	0.15	0.29	0.06	0.20	0.01	0.05	0.01	0.11	0.17	0.59	0.42	0.07
Pr > F	0.1590	0.015	<0.0001	0.1240	0.0100	0.6430	0.2010	0.8910	0.070	<0.0001	<0.0001	<0.0001	0.0150

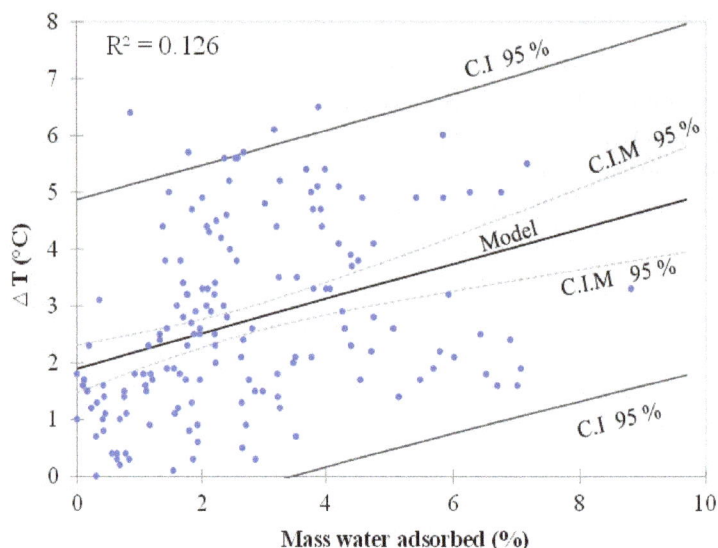

Figure 4. Regression line of temperature (°C) vs. water adsorption (%) with two confidence intervals (C.I) at 95%. The confidence interval on mean (C.I.M) of the prediction for a given value of T is the one closer to the line.

3.3. Analysis of Variance (ANOVA)

The ANOVA function was used to find out if the results would differ according to the formula described in Section 2.3 and, if so, which formula is the most effective. A pairwise comparison was performed to be able to run a Tukey's test, which is generally used in conjunction with an ANOVA to determine which means significantly differ from each other. The test compares the mean of each treatment to the mean of every other treatment. A stepwise method was selected and the statistics corresponding to the different steps were displayed. Finally, the best models for each number or variables with the corresponding statistics and for the criterion chosen were calculated. Table 6 displays the goodness of fit coefficients for the 160 observations, including the R^2 (coefficient of determination). The two dependent variables display a very low coefficient of variation (<1), indicating a good control over the operating conditions. For both water adsorption and temperature, around 89% of the variability is explained. The remaining 11 percent are hidden in other variables including biochar physical and chemical characteristics, which the model classifies as "random effects". Given that the probability (Pr) corresponding to the Fisher's F is lower than 0.0001 for both W and T, we can conclude that the explanatory variables and their interactions have a significant effect.

Table 6. Summary statistics for the experimental factorial design performed considering a mean of two replicates.

Variable	Minimum	Maximum	Mean	Std. Dev.	R^2	F	Pr > F
Mass water uptake (%)	0.00	8.81	2.70	1.83	0.88	19.31	<0.0001
ΔT (°C)	0.00	6.50	2.72	1.60	0.89	30.72	<0.0001

To elaborate the two models for each dependent variable (W and T), the selection process started adding the variable with the largest contribution to the model. If a second variable is such that the probability associated with its "t" is less than the "Probability for entry", it is added to the model. The procedure continues until no more variables can be added. This analysis allowed us to retain eight and nine explanatory variables (Table 7) to predict W and T, respectively. The cumulative coefficient of determination R^2 gives a fair idea of how much of the variability of W and T can be explained by these four qualitative variables and their interactions.

It is observed that the three interactions Size × Biochars, Size × Biochars × RH and Size × Mass × RH do not affect water sorption and temperature, while the variable RH influences them. The same observation is observed with the mass variable (M). This means that when explanatory variables are taken independently, they influence experimental results, but if associated, their effects are limited. The two independent variables Size and Biochars, and the second order interaction Biochars × RH, are found to only influence the temperature variable (T), while Biochars × Mass, Biochars × Mass × RH and Size × Biochars × Mass × RH, (second, third, and fourth order interactions, respectively) only influence the water sorption variable (W). Although the biomasses selected had different morphologic properties, it is noted that these do not affect water adsorption. This observation confirms that the BET area is not correlated to this quantitative variable (W). Finally, all the other explanatory variables and their interactions are observed to significantly influence W and T. For T, around 50% of the variations can be explained by the relative humidity variable, while W variations can be explained by the interaction "Mass × relative humidity" ($R^2 = 0.48$).

Figure 5 allows a comparison of the predictions to the experimental values. The confidence limits permit us to identify outliers, as with the regression plot displayed above. The two models bring significant information to explain the experimental results for W and T. The quite low deviation observed for all points, which remained close to the first bisector line, allow us to conclude that these two models did fit with the experimental results quite well.

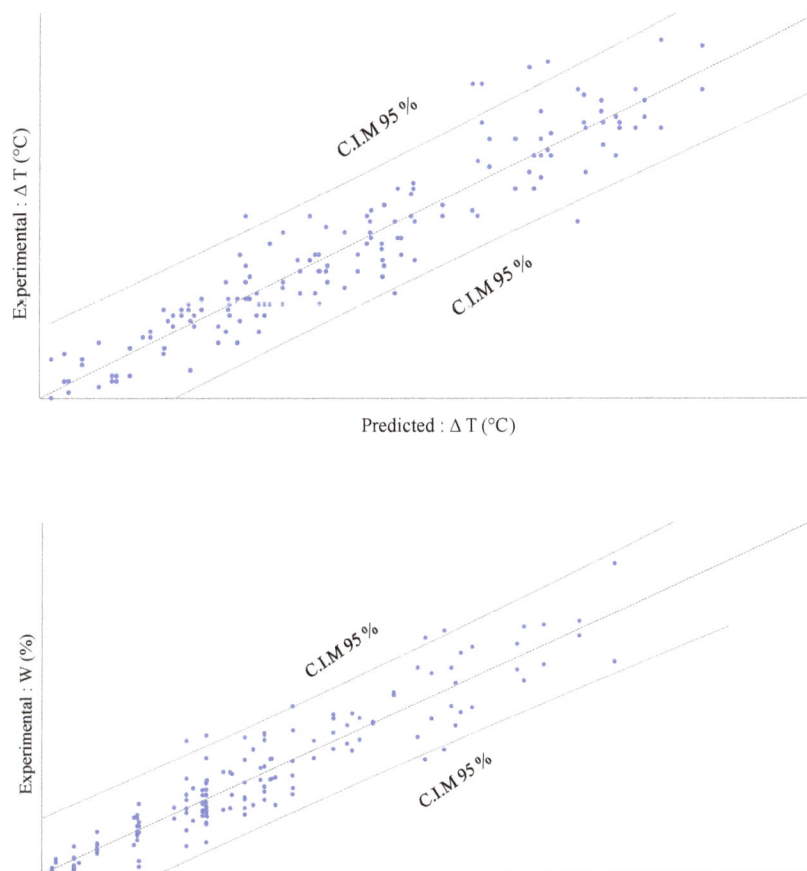

Figure 5. Model predictions vs. experimental results for T and W with two confidence intervals on mean (C.I.M) of the prediction at 95%.

Table 7. Statistics synthesis for explanatory variables and their interactions. R^2 values are cumulated. p-values < 0.0001 = significant; ns = not significant.

		Model: W (%)	Model: ΔT (°C)
Size	R^2		0.72
	F		274.83
	Pr > F	ns	<0.0001
Biochars	R^2		0.77
	F		18.85
	Pr > F	ns	<0.0001
Mass (g)	R^2	0.61	0.85
	F	259.86	13.95
	Pr > F	<0.0001	0.0000
RH (%)	R^2	0.68	0.48
	F	85.11	136.22
	Pr > F	<0.0001	<0.0001
Size × Biochars		ns	ns
Size × Mass (g)	R^2		0.87
	F		17.76
	Pr > F	ns	<0.0001
Size × RH (%)	R^2	0.88	0.80
	F	3.19	9.63
	Pr > F	0.0160	<0.0001
Biochars × Mass (g)	R^2	0.73	
	F	14.26	
	Pr > F	<0.0001	ns
Biochars × RH (%)	R^2		0.84
	F		3.67
	Pr > F	ns	<0.0001
Mass (g) × RH (%)	R^2	0.47	0.89
	F	11.13	2.89
	Pr > F	<0.0001	0.0250
Size × Biochars × Mass (g)	R^2	0.77	0.88
	F	9.74	3.31
	Pr > F	<0.0001	0.0220
Size × Biochars × RH (%)		ns	ns
Size × Mass (g) × RH (%)		ns	ns
Biochars × Mass (g) × RH (%)	R^2	0.81	
	F	3.37	
	Pr > F	0.0000	ns
Size × Biochars × Mass (g) × RH (%)	R^2	0.87	
	F	4.12	
	Pr > F	<0.0001	ns

The previous conclusions drawn from the means are statistically supported by the pairwise multiple comparisons. Significant information arising from Table 7 was summarized. All the combinations between the levels of the four factors and their interactions were associated to letters after applying the Tukey's test. This section focuses on the interpretation of all pairwise differences between means. Two level combinations sharing the same letter translate into not being significantly different. Two combinations with no letter in common translate into being significantly different. Attention is paid to factors and interactions that are the most significant according to the models.

The following two factors and three interactions were identified: Mass, RH, Size × RH, Mass × RH, and Size × Biochars × Mass.

As a reminder, the variable "Mass" is characterized by two values: 0.1 and 1 g. Figure 6 shows two distinct groups (A and B) for both W and T. Although the difference between the water sorption averages is evident (1.9%), this analysis of variance shows a significant difference for T, despite relatively similar average values (2.55 °C and 2.90 °C). Concerning the explanatory variable RH, the four pairs of categories are found to be different. The two RH pairs 84.3 × 75.3 and 90.2 × 75.3 do not show any significant differences, while the means between 84.2 and 90.2 are significantly different (3.1 and 3.9%). Except for 75.2%, all the other air relative humidity conditions show significant differences with regard to the observed adsorption values. The highest and the lowest values are 3.87 and 0.84% for RH values of 84.4 and 22.6, respectively. The maximum temperature is obtained with the highest relative humidity (3.95 °C) and the minimum temperature with the lowest RH (0.84 °C), as corroborated by the literature [14]. We can conclude that these two factors (Mass and RH) played a significant role in both the two models.

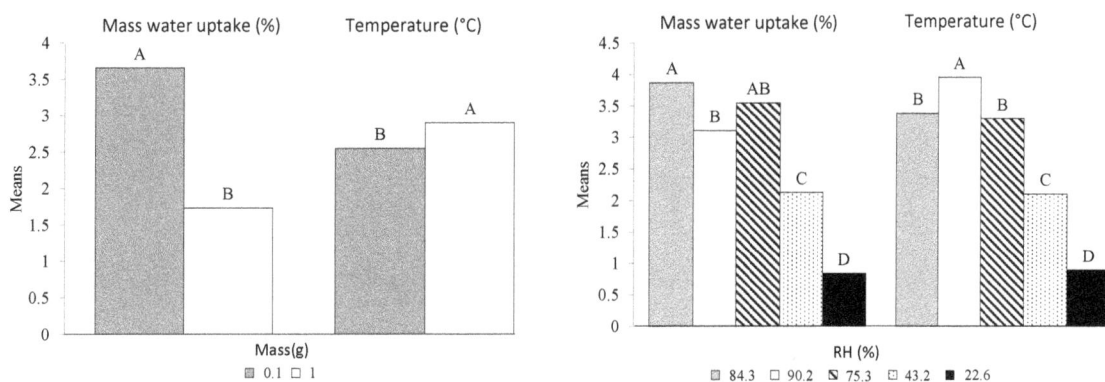

Figure 6. Mass and Relative humidity multiple comparisons according to the Tukey test. The means with the same letter were not significantly different at 5% ($\alpha = 0.05$).

Figure 7 shows the pairwise multiple comparisons for the second order interactions Size × RH and Mass × RH. For each interaction, all the combinations of levels between the two factors are compared to one another. The number of combinations possible between interactions and variables is 10. The variable Size is characterized by grinded biochar (powder) or biochar piece. The interaction Size × RH shows five and seven groups for water adsorption and temperature, respectively. Water sorption shows five combinations with means that are not significantly different. It concerns mainly high relative humidity values regardless of the sample size. Indeed, the difference appears mainly with the lowest humidity values (22 and 43%). The temperature shows two combinations (Piece × 90.2 and powder × 22.6) with no letter in common. Biochar pieces are more sensitive to relative humidity than powdered biochar, showing a higher average temperature (4.8 °C). This can mainly be attributed to the difference in the mass transfer rates and heat generation related to the sample size [2]. Indeed, it has been demonstrated that the effect of both the temperature overshoot and the pressure is dependent on sample thickness [19].

The combinations associated with the interaction Mass × RH show five and four groups for W and T. The variable T shows lesser variability compared to the Size × RH interaction. Five combinations concerning mainly high relative humidities have the same letter (B). Water sorption is mostly influenced by small mass combined with RH, except when RH is equal to 22.6%. When combined with humidity, the explanatory variable Size mostly influences the temperature, while the variable M mostly influences the adsorption of water.

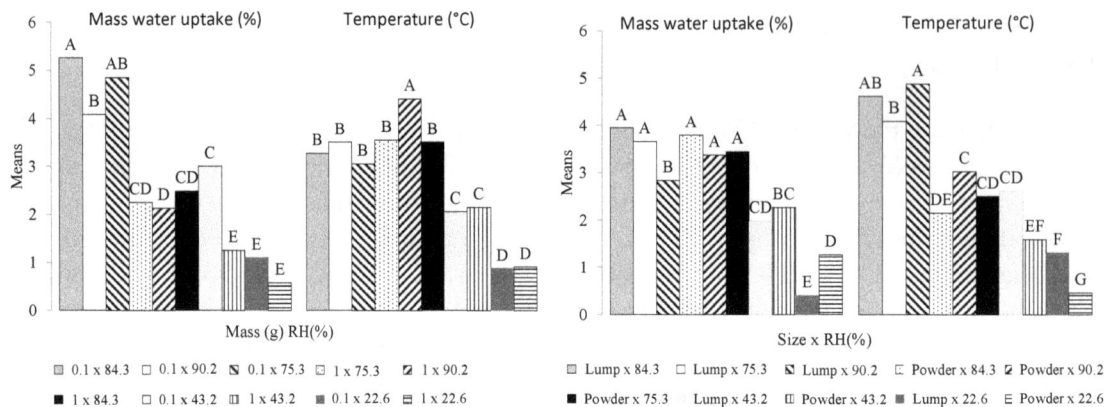

Figure 7. Second order interactions multiple comparisons according to the Tukey test. The means with the same letter were not significantly different at 5% (α = 0.05). Lump = piece.

The last common parameter to W and T models is the third order interaction Size × Biochars × Mass. Table 8 gives the average value classified by the Tukey test. The variable T presents seven groups and W only four groups. The temperature variable is more sensible to this interaction than the water adsorption variable and can be explained by the factors "biochars" and "Size" only present in the T model (Table 7). The overall results confirm that water adsorption is not correlated to the increase in temperature. Therefore, others elements, not yet identified, are likely involved in the increase in temperature.

Table 8. Classification by Tukey's test in decreasing order for T. For each group, the means with the same letter were not significantly different at 5% (α = 0.05).

Parameters	Mass Water Sorbed (%)	ΔT (°C)
Piece × Cg × 0.1	4.10 [ab]	4.07 [a]
Piece × QP × 0.1	2.86 [c]	3.93 [a]
Piece × Cg × 1	1.86 [d]	3.89 [a]
Piece × QP × 1	1.86 [d]	3.66 [ab]
Piece × Th × 0.1	3.00 [c]	3.45 [abc]
Piece × B × 1	1.86 [d]	3.20 [bcd]
Piece × Th × 1	1.86 [d]	3.16 [bcd]
Powder × Cg × 1	1.60 [d]	2.73 [cd]
Piece × B × 0.1	3.10 [bc]	2.66 [cde]
Powder × QP × 1	1.60 [d]	2.50 [de]
Powder × B × 1	1.60 [d]	2.04 [ef]
Powder × Th × 1	1.60 [d]	2.00 [ef]
Powder × Cg × 0.1	4.75 [a]	1.98 [ef]
Powder × QP × 0.1	4.31 [a]	1.75 [fg]
Powder × B × 0.1	4.37 [a]	1.29 [g]
Powder × Th × 0.1	2.74 [c]	1.25 [g]

4. Conclusions

It is well known that the adsorption of water vapour from ambient atmosphere plays a crucial role in raising the temperature of a biochar sample over the critical self-ignition temperature. This study was carried out on the wood of *Quercus pubescens*, *Cyclobalanopsis glauca*, *Trigonostemon huangmosun*, and *Bambusa vulgaris*, and involved five air relative humidity conditions (22, 43, 75, 84, and 90%), two mass samples (0.1 and 1.0 g), and two particle sizes (powder and piece). All experimental results showed a significant increase in temperature with the relative humidity. The highest and fastest increases in temperatures were observed for biochar pieces coupled to the lowest weight (0.1 g). Biochar pieces needed around 2 min to reach the temperature peak; powdered samples needed 5 to 8 min. The global trend showed that a larger mass of water vapor was adsorbed when exposed to a higher relative humidity. All samples with a mass of 1.0 g showed a lower water vapor adsorption compared to samples at 0.1 g. A linear regression model based on the temperature and the water adsorption (W) showed a high dispersion of the results corroborated by a low R^2 value (0.13). Two models were elaborated for each dependent variable (W and T) to simulate water adsorption and temperature. Eight and nine qualitative variables and their interactions were selected for W and T, respectively. Sample mass and relative humidity influenced both W and T, while particle size and type of biochar mainly influenced the temperature. Thus, these findings are very important not only for all scientific aspects, but also in practical applications. They will allow the creation of tabulations giving recommendations for charcoal cooling and storage considering the season (RH) and the critical size of the samples, and consequently to anticipate cool flame phenomena.

Acknowledgments: The authors thank Gilles Humbert for his helpful collaboration providing all biomasses tested in this study.

Author Contributions: Luciane Bastistella conceived and designed the experiments; Luciane Bastistella and Antonio Aviz performed the experiments; Patrick Rousset and Luciane Bastistella analyzed the data; Manoel Nogueira and Armando Caldeira-Pires contributed reagents/materials/analysis tools; Patrick Rousset wrote the paper.

Conflicts of Interest: The authors declare no conflicts of interest.

References

1. Candelier, K.; Dibdiakova, J.; Volle, G.; Rousset, P. Study on chemical oxidation of heat treated lignocellulosic biomass under oxygen exposure by sta-dsc-ftir analysis. *Thermochim. Acta* **2016**, *644*, 33–42. [CrossRef]

2. Miura, K. Adsorption of water vapor from ambient atmosphere onto coal fines leading to spontaneous heating of coal stockpile. *Energy Fuels* **2016**, *30*, 219–229. [CrossRef]

3. Joseph, C. *Combustibles Solides. Charbon: De L'Extraction à La Combustion*; Techniques de l'ingénieur: Paris, France, 2007.

4. Fierro, V.; Miranda, J.L.; Romero, C.; Andrés, J.M.; Arriaga, A.; Schmal, D.; Visser, G.H. Prevention of spontaneous combustion in coal stockpiles: Experimental results in coal storage yard. *Fuel Process. Technol.* **1999**, *59*, 23–34. [CrossRef]

5. Davis, J.D.; Byne, J.F. Influence of moisture on the spontaneous heating of coal. *Ind. Eng. Chem.* **1926**, *18*, 233–236. [CrossRef]

6. Stott, J.B. Influence of moisture on the spontaneous heating of coal. *Nature* **1960**, *188*, 54. [CrossRef]

7. Fujitsuka, H.; Ashida, R.; Kawase, M.; Miura, K. Examination of low-temperature oxidation of low-rank coals, aiming at understanding their self-ignition tendency. *Energy Fuels* **2014**, *28*, 2402–2407. [CrossRef]

8. Zádor, J.; Taatjes, C.A.; Fernandes, R.X. Kinetics of elementary reactions in low-temperature autoignition chemistry. *Prog. Energy Combust. Sci.* **2011**, *37*, 371–421. [CrossRef]

9. Blomqvist, P.; Persson, B. *Spontaneous Ignition of Biofuels–A Literature Survey of Theoretical and Experimental Methods*; Spar: Amsterdam, The Netherlands, 2003.

10. Wolters, F.C.; Pagni, P.J.; Frost, T.R.; Cuzzillo, B.R. Size constraints on self ignition of charcoal briquets. *Fire Saf. Sci.* **2003**, *7*, 593–604. [CrossRef]

11. Akgün, F.; Arisoy, A. Effect of particle size on the spontaneous heating of a coal stockpile. *Combust. Flame* **1994**, *99*, 137–146. [CrossRef]

12. Küçük, A.; Kadıoğlu, Y.; Gülaboğlu, M.Ş. A study of spontaneous combustion characteristics of a turkish lignite: Particle size, moisture of coal, humidity of air. *Combust. Flame* **2003**, *133*, 255–261. [CrossRef]

13. Mahmoudi, A.H.; Hoffmann, F.; Markovic, M.; Peters, B.; Brem, G. Numerical modeling of self-heating and self-ignition in a packed-bed of biomass using xdem. *Combust. Flame* **2016**, *163*, 358–369. [CrossRef]

14. Miura, K.; Muangthong, T.; Wanapeera, J.; Ohgaki, H. Examination of rates of oxidation reaction and water vapor adsorption at low temperatures for understanding spontaneous combustion of coal and biomass. In Proceedings of the Conference on Energy & Climate Change, Innovative for a Sustainable Future, SEE, Bangkok, Thailand, 28–30 November 2016.

15. Rousset, P.; Mondher, B.; Candellier, K.; Volle, G.; Dibdiakova, J.; Humbert, G. Comparing four bio-reducers self-ignition propensity by applying heat-based methods derived from coal. *Thermochim. Acta* **2017**, *655*, 13–20. [CrossRef]

16. Messineo, A.; Ciulla, G.; Messineo, S.; Volpe, M.; Volpe, R. Evaluation of equilibrium moisture content in ligno-cellulosic residues of olive culture. *ARPN J. Eng. Appl. Sci.* **2014**, *9*, 5–11.

17. Zhao, M.Y.; Enders, A.; Lehmann, J. Short- and long-term flammability of biochars. *Biomass Bioenerg.* **2014**, *69*, 183–191. [CrossRef]

18. Xu, Y.-l.; Wang, L.-Y.; Tian, N.; Zhang, J.-P.; Yu, M.-G.; Delichatsios, M.A. Spontaneous combustion coal parameters for the crossing-point temperature (CPT) method in a temperature–programmed system (TPS). *Fire Saf. J.* **2017**, *91*, 147–154. [CrossRef]

19. Turner, I.; Rousset, P.; Rémond, R.; Perré, P. An experimental and theoretical investigation of the thermal treatment of wood (*Fagus sylvatica* L.) in the range 200–260 °C. *Int. J. Heat Mass Transf.* **2010**, *53*, 715–725. [CrossRef]

12

Deep Artificial Neural Networks for the Diagnostic of Caries using Socioeconomic and Nutritional Features as Determinants: Data from NHANES 2013–2014

Laura A. Zanella-Calzada [1,†] **⬤**, **Carlos E. Galván-Tejada** [1,*,†] **⬤**, **Nubia M. Chávez-Lamas** [2],
Jesús Rivas-Gutierrez [2], **Rafael Magallanes-Quintanar** [1], **Jose M. Celaya-Padilla** [3] **⬤**,
Jorge I. Galván-Tejada [1] **and Hamurabi Gamboa-Rosales** [1] **⬤**

[1]   Unidad Académica de Ingeniería Eléctrica, Universidad Autónoma de Zacatecas, Jardín Juarez 147, Centro,
      Zacatecas 98000, Zac, México; lzanellac@uaz.edu.mx (L.A.Z.-C.); tiquis@uaz.edu.mx (R.M.-Q.);
      gatejo@uaz.edu.mx (J.I.G.-T.); hamurabigr@uaz.edu.mx (H.G.-R.)
[2]   Unidad Académica de Odontología, Universidad Autónoma de Zacatecas, Jardín Juarez 147, Centro,
      Zacatecas 98000, Zac, México; nubiachavez@uaz.edu.mx (N.M.C.-L.); rigj002959@uaz.edu.mx (J.R.-G.)
[3]   CONACYT—Universidad Autónoma de Zacatecas—Jardín Juarez 147, Centro, Zacatecas 98000, Zac, Mexico;
      jose.celaya@uaz.edu.mx
*   Correspondence: ericgalvan@uaz.edu.mx
†   These authors contributed equally to this work.

Abstract: Oral health represents an essential component in the quality of life of people, being a determinant factor in general health since it may affect the risk of suffering other conditions, such as chronic diseases. Oral diseases have become one of the main public health problems, where dental caries is the condition that most affects oral health worldwide, occurring in about 90% of the global population. This condition has been considered a challenge because of its high prevalence, besides being a chronic but preventable disease which can be caused depending on the consumption of certain nutritional elements interacting simultaneously with different factors, such as socioeconomic factors. Based on this problem, an analysis of a set of 189 dietary and demographic determinants is performed in this work, in order to find the relationship between these factors and the oral situation of a set of subjects. The oral situation refers to the presence and absence/restorations of caries. The methodology is performed constructing a dense artificial neural network (ANN), as a computer-aided diagnosis tool, looking for a generalized model that allows for classifying subjects. As validation, the classification model was evaluated through a statistical analysis based on a cross validation, calculating the accuracy, loss function, receiving operating characteristic (ROC) curve and area under the curve (AUC) parameters. The results obtained were statistically significant, obtaining an accuracy $\simeq 0.69$ and AUC values of 0.69 and 0.75. Based on these results, it is possible to conclude that the classification model developed through the deep ANN is able to classify subjects with absence of caries from subjects with presence or restorations with high accuracy, according to their demographic and dietary factors.

Keywords: NHANES; oral health; dental caries; classification multivariate models; computer-aided diagnosis; artificial neural networks; deep learning; statistical analysis

1. Introduction

Chronic diseases are the main problems in public health worldwide. The pattern of disease has been transformed and oral diseases are considered one of the main public health problems due to its high incidence and prevalence in all regions of the world, and, as in all diseases, the greatest burden is on populations disadvantaged and socially marginalized; treatment of the conditions is extremely expensive and is not feasible in most low and middle income countries. This characteristic represents an important problem, since, according to the World Health Organization (WHO), oral diseases are the fourth most expensive cause to treat in the most industrialized countries [1].

Oral health is an essential component for quality of life of people due to its influence as a determinant factor in general health of individuals and communities becoming a relevant point in health care. Therefore, the serious repercussions in terms of pain and suffering, impairment of function and effect on quality of life should also be considered. An important aspect to consider is that oral diseases increase the risk of chronic diseases, such as: cardiovascular and cerebrovascular, diabetes mellitus and respiratory. On the other hand, epidemiological surveillance of oral diseases becomes important insofar as it provides useful elements for the planning, programming, organization, integration, control and direction of the oral health program, at the same time that it guides the attention to the population [2].

Caries is the most frequent condition and according to the WHO; it affects between 60% and 90% of children of school age between 5 to 17 years old. The determinants and conditions of oral health status are multicausal, multisectoral and interdisciplinary categories that encompass a series of situations related to the historical and political process that each country experiences. In addition, the report of the Organización Panamericana de la Salud (OPS) describes how the development of this condition depends on the frequency of carbohydrate consumption, the characteristics of the food, the time exposure, plaque removal and susceptibility of the guest, adding the few preventive measures in oral health and the difficulty of making use of specialized dental medical services. These factors interact simultaneously, and their variables correspond to different orders from biological processes, to complex historical-cultural structures and social relationships, socioeconomic level, educational level, among others, making health phenomena complex [3,4].

There are too many risk factors and important determinants related to the presence of caries, and it is clear that the greater the degree of risk exposure, the greater the probability of contracting or developing it. Due to the difficulty of representing the control of the incidence of caries due to its high prevalence and the large number of factors that can influence this state, at present, some studies have implemented algorithms and performed analyses based on computer-aided diagnosis (CAD) where prediction models are used when it is necessary to know in the future the behavior of highly related complex data; these have utilities in clinical and research dentistry [5].

An example of these algorithms are the artificial neural networks (ANNs), a method used for the prediction of diseases, such is the case of oral diseases, as well as being mathematic models based on a principle of learning that is based on the concepts of artificial intelligence and the biological response of the human brain. Likewise, they are involved in systems' construction processes that allow classifying, modeling and predicting information. ANNs are semiparametric nonlinear models, which allow the integration of variables and easily handle large amounts of data compared to linear analyses. They have processing elements or neurons, which are the units of the system that can be adjusted or trained through a process of learning and generalization. The information of each of these neurons is grouped and processed [6].

According to the literature, dental caries incidence has been a problem that has been studied by a series of researchers, trying to identify caries risk determinants. Lam et al. [7] found that dental visits, brushing frequency, lower parental perceived importance of baby teeth, and weaning onto solids are determinants associated with plaque accumulation, validating these results specifically in children. In the work of Fernandes et al. [8], a cross-sectional study of 274 children and their mothers based on demographic/socio-economic status is presented; through a Poisson regression approach, an analysis was performed, obtaining that dental caries can be mainly found in mothers of children aged 1, demonstrating the relationship between demographic/socio-economic status and caries.

As mentioned in the text, there are different types of risk factors that are associated with genetic and nutritional data. Lips et al. [9] presented a work where the association between genetic polymorphisms and risk of dental caries is demonstrated for most of the salivary proteins.

In addition, diabetic and hypertensive patients have dietary risk factors that can be caused by oral health problems. This affirmation was demonstrated in the work of Asif Ahmed et al. [10] through a statistical analysis using clinical data, in order to identify that patients with oro-dental problems were hemodynamically stressed.

In the present study, it is intended to analyze the determinants that affect this oral health situation based on dietary and demographic factors. Based on this general description of the oral health problem, specifically caries, the main contribution of this work is to determine that, given the socioeconomic and dietary features, using an ANN can determine if the patient presents an optimal state of health or with the presence of repairs or caries, obtaining a classification system that allows for knowing the differences in the features that affect the population, as well as looking for the future prediction of oral health.

This method is an important tool for human resources and dental services because it reflects information collected in the population groups studied, which could unveil that oral health problems are not simple conditions, but they are obtained from different processes that have their trigger, which can enhance or mitigate them, impacting positively or negatively on the state of oral and general health.

Related Work

The condition of dental caries has been described in the scientific literature under different terms, as a multifactorial disease that is characterized by localized and progressive demineralization of the inorganic portions of the tooth and the subsequent deterioration of its organic part, with a high degree of morbidity and high prevalence [11–13].

A related work that suggests CADx to improve oral health prediction is developed by Zhang et al. [14], where an autoregressive integrated moving average model is performed and a grey predictive model for the estimation of the national prevalence of early childhood caries from 2014 to 2018, finding that the highest annual prevalence would be of 55.8%, being attributed to the socioeconomic developments and the public health service. In the work of Asif et al. [15], a caries preventive tool based on the influence of genetic pattern using the frequency of occurrence of fingerprint patterns among children, using as a validation measure the p-value, finding that dermatoglyphics could be an appropriate method to explore the possibility of a noninvasive and early predictor for dental caries. Chapple et al. [16] performed a systematic appraisal to identify potential risk factors for caries and periodontal diseases using genetic, role of diet and nutrition risk factors, obtained as results that the genetic contribution to these conditions present an attributable risk up to 50%, while controlled diabetes and obesity are common acquired factors. Hayes et al. [17] had as an objective to determine the risk indicators associated with root caries conducting a prospective longitudinal study using a regression analysis with data related to hygiene habits, diet, smoking habits and education level and as the outcome the root caries experience, suggesting a correlation between root caries and the variables poor plaque control, xerostomia, coronal decay and exposed root surfaces, obtaining a prevention tool for the root caries condition. In the work of Sudhir et al. [18], evaluating Caries Management by Risk Assessment (CAMBRA) as a tool for caries risk prediction was proposed, using as validation parameter OR value and ROC curve, obtaining that CAMBRA was found to be 47.62% with a specificity of 80% and AUC was found to be 0.638, which means that it is valid and highly predictive in determining the caries risk. Finally, Twetman [19] proposed to summarise the findings of recent systematic reviews covering caries risk assessment, finding that caries risk assessment should be carried out at the subject's first dental visit and reassessments should be done during childhood; multivariate models display a better accuracy than the use of single predictors; there is no clearly superior method to predict future caries and no evidence to support the use of one model, program, or technology before the other; and the risk category should be linked to appropriate preventive care with recall intervals based on the individual need.

This work is organized as follows: Section 2 briefly describes the materials and methods used for the development of the data analysis, including data preprocessing, data classification and data evaluation. Section 3 exposes the results obtained. Finally, results are discussed in Section 4 and concluded in Section 5.

2. Materials and Methods

In this section, the process that was carried on for the classification between subjects with presence of caries and subjects with absence was presented, based on the information obtained from a series of socioeconomic and nutritional features. The databases that were used in this work were obtained from the National Health and Nutrition Examination Survey (NHANES, 2013–2014). In addition, the description of the subjects and the methods used for their classification are presented.

In Figure 1, the flowchart of the steps followed for the experimentation of this work is presented. Section (A) presents the data acquisition of the public databases "Demographic" and "Dietary" from NHANES 2013–2014. Section (B) represents the data preprocessing, where a feature reduction is performed due to the high amount of missing data and singular values that are presented, in addition to separating the data in two sets, one for training and one for testing. Section (C) refers to the data classification of subjects according to their oral health status. Finally, in section (D), the evaluation of the ANN performance is presented, in order to know the accuracy with which the model classifies the subjects.

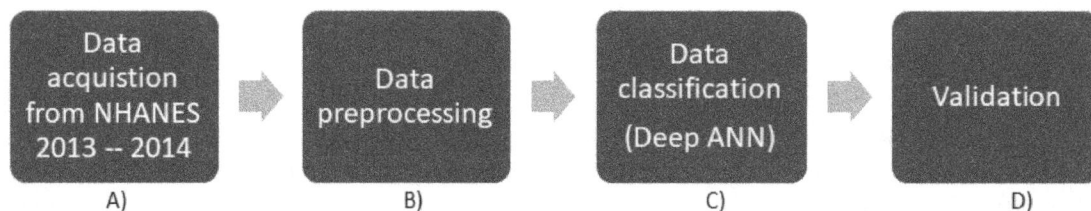

Figure 1. Flowchart of the methodology followed.

2.1. Features Description

NHANES is a program of studies designed to evaluate the health and nutritional status of children and adults in United States, and it was founded by Centers of Disease Control and Prevention (CDC) and National Center for Health Statistics (NCHS). The surveys conducted by this program are unique, since it combines interviews and physical exams [20].

NHANES collects information from different types of data, and, in turn, this information is included in six main contexts; demographic, dietetic, examination, laboratory, questionnaire and limited access. These contexts are contained by the information described as follows:

- Demographic: it provides individual, family and household level information in different topics (income of households and families, size of households and families, pregnancy status, among others).
- Dietetic: it provides detailed information on dietary intake, in order to estimate the types and amount of food and beverages consumed, in addition to estimating the intake of energy, nutrients, and other food components.
- Examination: it provides information of the health status, indicators of disease risk, and access to preventive and treatment services, from different aspects, including oral health.
- Laboratory: it provides information of the results obtained from laboratory analysis of different components (components of urine, proteins, triglycerides, plasma, among others).
- Questionnaire: it provides information of the data obtained from the interviews conducted through a system of computer-aided personal interview, of different topics (alcohol use, cardiovascular health, dermatology, among others).

- Limited access: it provides similar information to that found in the questionnaire; however, it isn't publicly available.

There were 189 features analyzed for this work, of which 188 belonged to demographic data, described in Appendix A, and dietary data, described in Appendix B. These features were used as input variables for the classification of subjects, while the remaining feature belonged to examination data, describing the oral health status of the subjects based on the presence/restoration or absence of caries, for which it was used as an output feature.

2.2. Subjects Description

The subjects of these databases were submitted to a series of different questionnaires, related to the different features. These subjects belong to different counties in the USA and they were randomly selected with a computer algorithm by NHANES.

Figure 2 presents a flowchart of the randomly selection process followed by NHANES. (A) presents the first stage that corresponds to the sampling of counties. In this step, all the counties are divided into 15 groups according to their characteristics. Then, from each group, one county is selected, forming the 15 counties in the NHANES surveys for each year. (B) presents the second stage, corresponding to the sampling of segments, where from each county smaller groups are formed (with a large number of households in each group), selecting between 20 and 24 of these groups. (C) presented the third stage, which is referred to the sampling of households, where from all of the houses that correspond to the groups that were selected in the past stage, a sample of about 30 households are selected within each group. Finally, (D) presents the fourth stage, corresponding to the sampling of persons, where the NHANES interviewers go to the selected households and ask for the information of the surveys (age, race, and gender). The members of the households that responded to the surveys were randomly selected through a computer algorithm; they can be selected some, all, or none of the household members.

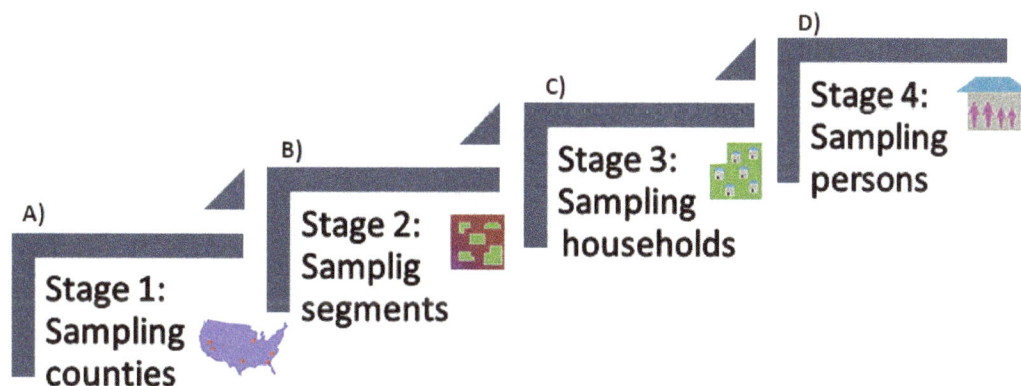

Figure 2. Flowchart of the stages followed by NHANES 2013–2014 for the sampling of participants.

The main target population for NHANES is the non-institutionalized civilian resident population of the USA. The design of the population selection is based on the sampling of a larger number of specific subgroups that present particular public health characteristics, in order to increase the reliability and precision of health status indicators. NHANES started these design changes in 2011, including in its population the oversampled subgroups survey cycle:

- Hispanic persons;
- Non-Hispanic black persons;
- Non-Hispanic Asian persons;
- Non-Hispanic white and other* persons at or below 130% of the poverty level; and
- Non-Hispanic white and other* persons aged 80 years and older.

This random selection of subjects avoid presenting any bias problem, favoring the important variety of demography characteristics of the subjects reducing the probability that the methodology followed could be influenced by the data used.

Figure 3 presents the total of subjects contained in the database used. From 9812 subjects, 3690 were control subjects and 6122 were case subjects. The subjects were in an age range of 0 to 80 years old; 4830 belonged to the masculine gender and 4982 to the feminine gender.

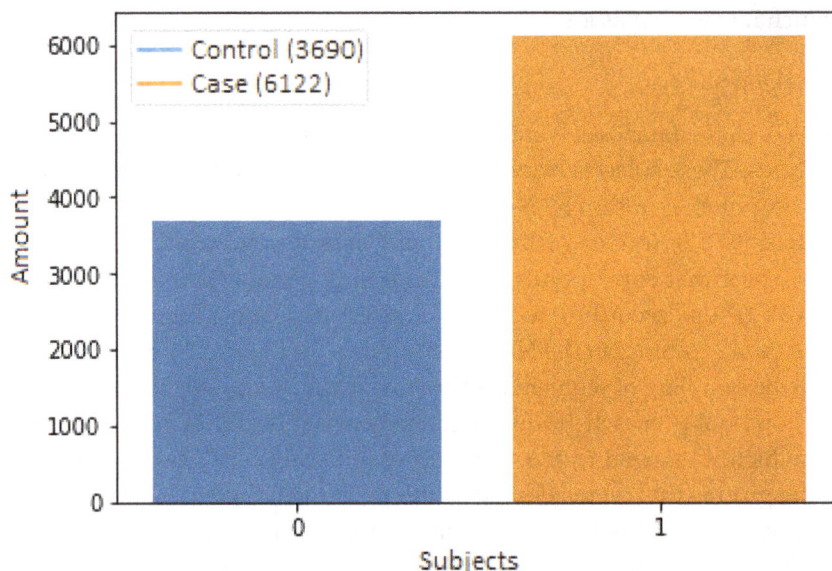

Figure 3. Graph of the total number of subjects used, classified according to their oral health status. The subjects contained in the bar with '0' presented absence of caries, while the subjects contained in the bar with '1' presented presence/restoration of caries.

It is important to mention that this specific database was chosen in order to show the consistency of the data and the results obtained in the relationship to its background work. In this work [21] three different models obtained through a fast backward selection (FBS) of features are presented, each model corresponding to a different age group, then a posterior evaluation using a net reclassification improvement (NRI) technique is performed, besides the AUC parameter, obtaining a maximum true positive-true negative rate of 0.787. On the other hand, this work [22] presents a multivariate model obtained through a FBS method based on the p-value, in order to classify between three different classes, "caries", "restorations" and "control"; this model was evaluated using a statistical analysis, obtaining a maximum AUC value of 0.664. Finally, this work [23] presents a univariate analysis using a linear regression approach in order to classify between subjects with the presence of caries and with absence; then, from the most significant univariate models, a multivariate model contained by three features was developed, which obtained an AUC value of 0.572.

2.3. Data Analysis

The data analysis of this work was performed through a multivariate approach, subjecting the demographic and dietary features to a deep ANN for the classification of subjects according to their health and then a statistical validation was carried out in order to evaluate the results obtained from the ANN developed for these specific data.

2.3.1. Data Preprocessing

The data preprocessing consisted initially in the manual elimination of the features that presented a high percentage of missing values ($>30\%$), represented as "Not a Number" (NaN). Then, of the remaining features, those that presented a low percentage of missing data ($\leq30\%$) were imputed with the "rfimput" function, of the "randomForest" package (version 4.6-12, 6 October 2015) [24], for R, which consists of replacing all NaN with the average of the values that present the column where the missing data is located.

The columns containing singular values were also removed because they didn't provide significant information. Two values are singular if they have multiple values between them or if they present the same value in the whole column.

Finally, the database was separated into two sets, randomly and balanced selecting a set for the training stage, contained with 70% of the data and a set for the testing stage, contained with the remaining 30% of data.

2.3.2. Data Classification

The classification of subjects was performed according to the presence or absence of caries through a dense ANN that was designed based on the data of the subjects, using the package "Keras". "Keras" is a high-level ANN application programming interface written in Python. It was developed with the approach of allowing fast experimentation, facilitating the creation of prototypes in a simple way [25].

ANNs seek a solution to a specific task, based on the correlation between features, through learning or training that resembles the behavior of biological neural networks. Using different layers composed of nodes or neurons, the ANNs look for a model of the relationship between the input features and the output feature. The number of nodes that exist for each layer is configurable, as is the number of layers and, depending on the data, there is usually a greater capacity of the network, the greater the number of layers and neurons; however, if there isn't the right amount of these elements, problems of overfitting can be caused .

ANNs present three main elements: (1) a set of synapses or connections, characterized by a "weight", where the input signal is connected to a neuron through its product with the weight of that connection; (2) an adder, which adds the contributions of a signal weighted by all the weights; and (3) an activation function, which is equivalent to a transfer function, affecting the neurons, allowing for limiting the amplitude of the network output, providing a permissible range for the output signal in terms of finite values. Among the most common activation functions are the lineal, quadratic, geometric, logistic, and rectified linear unit function, among others.

On the other hand, an ANN is dense when each node of a layer is connected with every node of the next layer; while the recurrence term refers to the network presenting at least one feedback loop, which will provide a deeper impact on the learning capacity of the network, as well as on its performance, since it allows for optimizing its behavior through a parameter that gives knowledge of the behavior of the data and allows for guiding the adjustment of the configuration of the network to improve its accuracy.

The ANN designed was composed by a series of dense and dropout hidden layers, as shown in the diagram of Figure 4. The input layer (A) was assigned 104 neurons, making reference to the 104 features of the dataset. The first dense hidden layer (B) was composed of 100 neurons, as was the first dropout hidden layer (C), which had a loss percentage of 50%. The second dense hidden layer (D) was assigned 1000 neurons, as was the second dropout hidden layer (E), which had a loss percentage of 25%. The third dense hidden layer (F) was composed of 100 neurons, as was the third dropout hidden layer (G), which had a loss of 50%. Finally, the fourth dense hidden layer was the output layer (H), which was characterized by two neurons, making reference to the two outputs or possible classifications.

The input layer and the dense hidden layers used as activation function, the Rectified Linear Unit (ReLU) function, which consists of assigning 0 to the values of the neurons that are <0, and to respect the value of the neurons when their values are ≥ 0, as shown in Equation (1) [26]:

$$\text{ReLU}(x) = \begin{cases} x, & x \geq 0, \\ 0, & x < 0. \end{cases} \tag{1}$$

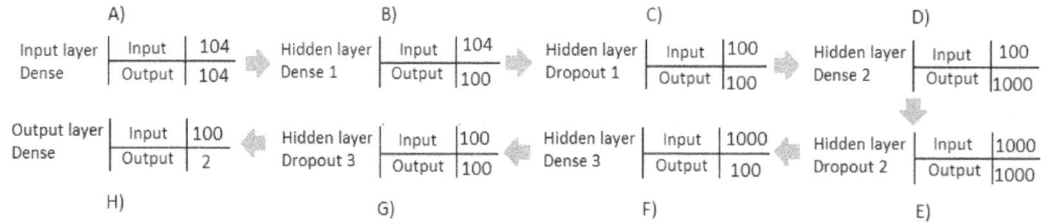

Figure 4. Diagram of the ANN designed.

The output layer used the Normalized Exponential function as an activation function, also known as "Softmax", which represents a general form of the logistic function and it is used to compress a vector of arbitrary values into a vector of real values in a range of $[0, 1]$. This function is shown in Equation (2), where $\sigma(z)$ represents the K-dimensional vector, z, of binary values [27]:

$$\sigma(z)_j = \frac{e^{z_j}}{\sum_{k=1}^{K} e^{z_k}}, \; j = 1, ..., K. \tag{2}$$

Finally, the dropout hidden layers were included in order to avoid overfitting problems, as mentioned before, and its performance is based on assigning the value of 0 to a percentage of the data and thus they are not taken into account for the classification of subjects in that layer. The data subset selected by these layers is changing with each iteration, causing that in each epoch the classification will be done omitting a different percentage of data. On the other hand, due to the recurrence of the ANN, it was possible to optimize its performance. The optimization algorithm, "Adam", was selected, which bases its operation on stochastic gradient descent algorithms, making use of the average of the first and second moments of the gradients to adapt the rate of the learning parameter. Specifically, "Adam" calculates the exponential moving average of the gradient and the square gradient, and controls the decay rates of that moving average [28]. Some of the benefits that present "Adam" are that it is a method that is straightforward to implement, is computationally efficient, has little memory requirements, is invariant to diagonal rescaling of the gradients, and one of the most important points is that is suitable for problems that are large in terms of data or features. In addition, the parameters have intuitive interpretations and typically require little tuning [28].

It is important to mention that the number of iterations or epochs that the ANN will have for the optimization of its behavior, based on the recurrence, is also configurable and will depend on the type of data. To know what is the appropriate number of epochs, there are some parameters that allow for evaluating the behavior through each iteration, such as the loss function and the accuracy [29]. The number of the epochs that were selected through a series of tests with different numbers were 100 epochs, since they demonstrated having the best performance in terms of the accuracy of the ANN.

Based on the above, "keras" has the advantage that it allows for designing the architecture of the ANN according to the type of data, configuring the type of ANN, number of nodes, validation, loss function, among others [30,31].

2.3.3. Evaluation

Finally, in order to validate the results of classification obtained by the ANN, three parameters were evaluated; loss function, accuracy and ROC curve. The loss function and accuracy were calculated on each epoch, allowing to know if the performance of the ANN was improving, while the ROC curve was calculated based on the average of the general performance of the ANN.

ANNs are mostly trained using gradient methods through an iterative process of decreasing the loss function. A loss is designed to have the main property that the lower its value, the better the model that fits its data and, in addition, it will be differentiable, which will optimize the network directly, giving information of the capacity of the system. Therefore, the loss function is based on the search of the global minimum, which corresponds to the minimum error, based on a learning factor. Some of the most common techniques to calculate the loss function are the mean square error, mean absolute error, binary cross-entropy, and Poisson, among others [31].

The loss function that was used is preset in "keras" as "binary-crossentropy" and calculates the cross-entropy value for binary classification problems. This method uses the Kullback–Leibler distance, which is a measure between two density functions g and h, known as the cross-entropy between g and h, as shown in Equation (3). Its operation is based on iterations, generating a random set of values estimating the value that wants to be obtained and then actualizing the parameters in the next iteration to generate "better" values or more approximate, in terms of the Kullback–Leibler distance [32]:

$$D(g,h) = \int g(x) ln \frac{g(x)}{h(x)} \mu(dx) = \int g(x) ln g(x) \mu(dx) - \int g(x) ln h(x) \mu(dx). \tag{3}$$

On the other side, the accuracy allows for measuring the performance of the ANN through a non-differentiable function. This metric doesn't allow to optimize the network, but it allows to select the model that shows the most suitable performance in the training of the network, and it is based on the calculation of the average of the differences that exist between the classification calculated by the ANN model and the true classification of the data, as shown in Equation (4), where the accuracy is reported as 1-*error*. V_{pred} refers to the classification value that was calculated by the ANN, while V_{actual} refers to true classification value [33]:

$$error = V_{pred} - V_{true}. \tag{4}$$

In this work, the accuracy was calculated with the "binary-accuracy" function from "keras", which calculates the average accuracy rate across all predictions for binary classification problems.

The ROC curve is a standard method that allows for evaluating the precision with which the model classifies, based on the relationship between sensitivity and specificity. Sensitivity is defined as the proportion of subjects with a condition that were classified as positive, which means the positive predictive values (PPV); this value is calculated with Equation (5), where TP represents the number of true positives and FP represents the number of false positives [34,35]:

$$PPV = \frac{TP}{TP + FP}. \tag{5}$$

Specificity is defined as the proportion of subjects without a condition that was classified as negative; this means the negative predictive values (NPV); this value is calculated with Equation (6), where TN represents the number of true negatives and FN represents the number of false negatives [34,35]:

$$NPV = \frac{TN}{TN + FN}. \tag{6}$$

The ROC curves were calculated to obtain the true positive and true negative rate for each class, for the macro-average precision and for the micro-average precision.

The macro-average precision is obtained through the sum of the true positives, false positives and false negatives of the system, for different sets, as shown in Equation (7), where TP_1 refers to the true positives of the set one, TP_2 refers to the true positives of the set two, FP_1 refers to the false positives of the set one and FP_2 refers to the false positives of the set two [36,37]:

$$\text{Micro-average} = \frac{TP_1 + TP_2}{TP_1 + TP_2 + FP_1 + FP_2}. \tag{7}$$

On the other hand, the macro-average precision is a direct method; it takes the average of the accuracy of the system in different sets. It is calculated with Equation (8), where A_1 refers to the average of the set one and A_2 refers to the average of the set two [36,37]:

$$\text{Micro-average} = \frac{A_1 + A_2}{2}. \qquad (8)$$

The analysis of this work was performed in Python (version 3.6), using the packages, "Keras" (version 2.1.5) [25], "Scipy" (version 1.0.0) [38], "Pandas" (version 0.22.0) [39] and "Sklearn" (version 0.19.1) [36].

3. Results

The two databases that were used in this work were contained by a total of 188 demographic and dietary features, besides a feature being used that was contained by the oral health status as an output feature for the classification of subjects.

After the data preprocessing, only 105 features were conserved, of which 25 belonged to demographic features and 79 to dietary features, the remaining feature refers to the output feature.

Of the total set of 9812 subjects, 3690 belonged to controls and 6122 belonged to cases. This dataset was divided into two subsets, one for training which was contained by 70% of the data (2596 controls/4272 cases), and one for testing which was contained with the remaining 30% of the data (1094 controls/1850 cases). These data are graphically shown in Figure 5.

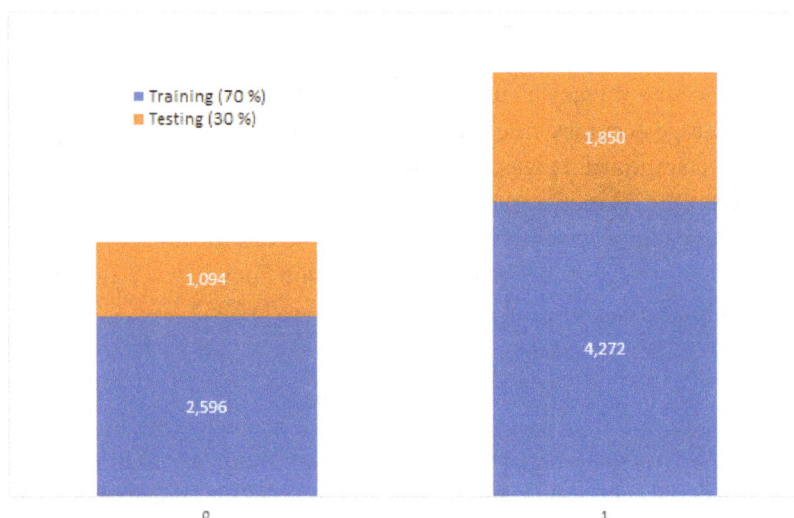

Figure 5. Graph of the number of subjects that belongs to each dataset, training and testing. The content in the bar with "0" presented a status of absence of caries, while the contest in the bar with "1" presented a status of presence/restoration of caries.

The dataset that was designed for the training stage was evaluated at each of the 100 epochs through the accuracy and loss function values. This number of epochs was selected through a comparison between the values obtained using different number of epochs; Table 1 presents the values obtained for ten different epochs.

On the other hand, in Table 2, a comparison of results using different number of layers and neurons is shown. The number of epochs selected was 100, establishing this number according to the result obtained from the previous table. It is possible to observe that the most statically significant values of accuracy and loss function were obtained using an ANN designed with seven layers, four dense layers and three dropout layers. The dense layers were contained by 104, 1000, 100 and two neurons, in descendant order, while the dropout layers represented 0.50, 0.25 and 0.50 percentages, in descendant order too.

Table 1. Comparison of the accuracy and loss function values obtained using different number of epochs.

Epochs	Accuracy	Loss Function	Processing Time (s)
10	0.67	0.60	10.44
30	0.68	0.60	32.02
50	0.68	0.59	55.46
80	0.69	0.59	93.57
100	0.69	0.58	124.55
150	0.68	0.60	188.21
200	0.68	0.61	246.74
300	0.69	0.62	369.06
500	0.70	0.62	646.86
1000	0.70	0.66	4152.96

Table 2. Comparison of the accuracy and loss function values obtained using different number of layers and neurons.

Layers Dense/Dropout	Neurons	Accuracy	Loss Function	Processing Time (s)
2/1	104>0.50>2	0.68	0.60	36.84
3/1	104>0.50>1000>2	0.68	0.59	71.10
3/2	104>0.25>1000>0.50>2	0.69	0.61	93.86
4/1	104>1000>0.50>100>2	0.68	0.73	117.63
4/2	104>0.50>1000>0.50>100>2	0.68	0.59	119.77
4/3	104>0.50>1000>0.25>100>0.50/2	0.69	0.58	124.55
5/1	104>100>1000>0.50>100>2	0.68	0.91	129.53
5/2	104>100>0.50>1000>0.25>100>2	0.68	0.65	131.66
5/3	104>100>0.50>1000>0.25>100>0.50>2	0.69	0.64	139.37
5/4	104>0.25>100>0.50>1000>0.25>100>0.50>2	0.68	0.59	138.38

The graph of the performance of the accuracy is shown in Figure 6A, where the blue line represents the behavior of the training data, obtaining a final accuracy of 0.69, while the orange line represents the behavior of the testing data, obtaining a final accuracy of 0.68.

The graph of the performance of the loss function is shown in Figure 6B, where the blue line represents the behavior of the training data, obtaining a final value of 0.58, while the orange line represents the behavior of the testing data, obtaining a final value of 0.60.

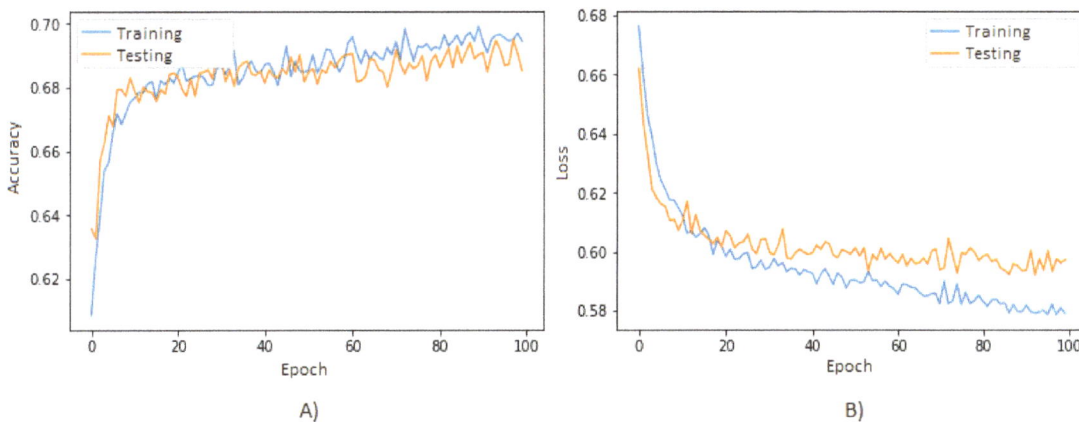

A)   B)

Figure 6. Graphs of the performance of the accuracy (**A**) and the loss function (**B**) of the ANN in each epoch.

Both accuracy and loss function values were obtained in order to know the performance of the classification of subjects on each iteration; nevertheless, the accuracy is a parameter that may show an optimistic response if the data presents any bias. Therefore, there were also measured the specificity and sensitivity parameters in order to ensure that the classification of subjects was statistically significant.

The graph of the Figure 7 shows the ROC curves of the mean performance of the ANN, where the pink line refers to the proportion of sensitivity and specificity for the class "0", which are the subjects that presented absence of caries, obtaining an AUC value of 0.69. The light blue line refers to the proportion of sensitivity and specificity for the class "1", which are the subjects that presented presence/restoration of caries, obtaining an AUC value of 0.69. The dotted orange line refers to the curve calculated with the micro-average of the proportion of sensitivity and specificity for the classification of subjects, obtaining an AUC value of 0.75; the dotted dark blue line refers to the curve calculated with the macro-average, obtaining an AUC value of 0.69.

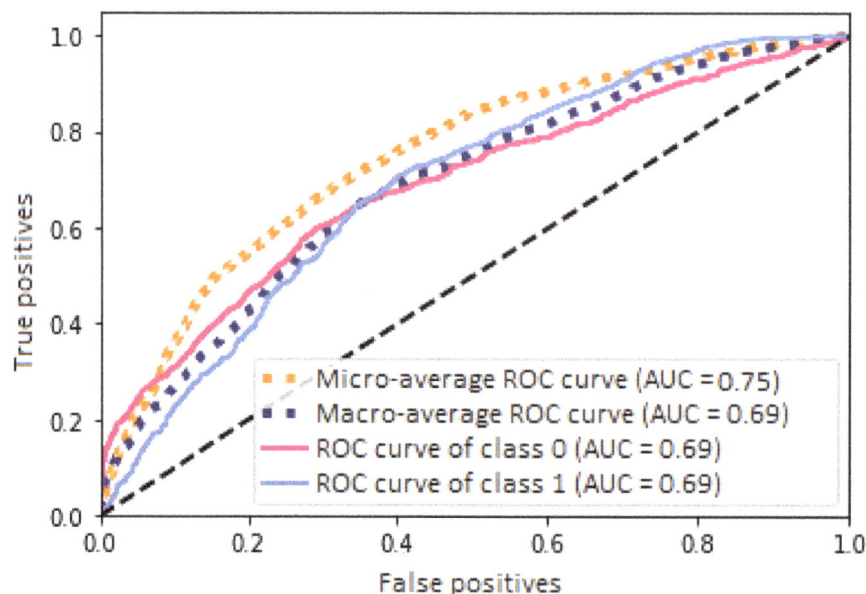

Figure 7. Graph of the ROC curves generated based on the average performance of the ANN in the classification of subjects.

4. Discussion

Of the two datasets that were obtained through the preprocessing step, it is possible to observe that the set of cases contains more data than the set of controls as it is observed in Figure 5; this was due to the fact that the cases included the subjects with presence of caries and with restorations. Nevertheless, the number of subjects that were contained on each dataset was enough to train the ANN, obtaining statistically significant results.

The data were separated in two sets based on an aleatory and balanced selection; one of the sets was contained with 70% of the data and the other dataset was contained with 30%, as shown in Figure 5. The dataset that was contained with the largest amount of data had the purpose of training the ANN. The dense ANN was trained during 100 epochs and was optimized through the optimization algorithm, "Adam", which had the purpose of improving the performance of the ANN with feedback.

Finally, for the validation of the ANN modeling, the accuracy and the loss function in each of the epochs with the dataset that was contained with the least amount of data were obtained. The loss function was calculated using cross-entropy in order to know how the performance of the ANN was modified. The validation was carried out in both datasets, training and testing, in order to ensure that there was no adjustment in the classification of the subjects; that is, if the performance of these parameters is similar for both datasets, it means that the ANN is classifying based on a generalized model, since it manages to classify with an accuracy and a loss function similar in the training dataset and in the testing dataset.

In the graph of Figure 6A, it is observed that the performance of the accuracy is similar for both datasets, being slightly higher for the training data, which is normal since they are the data on which the ANN modeling is being based. As the performance is similar for both datasets, it is possible to know that the ANN model became generalized through the learning by being able to classify unknown data with a similar accuracy to that obtained with the training data. The final value obtained for the accuracy was 0.69, which is statistically significant since it means that at least 69% of the data was correctly classified.

On the other hand, in the graph of Figure 6B, it is observed that the performance of the loss function is also similar for both datasets. It is notable that, for the training dataset, this parameter is smaller compared to that obtained with the testing dataset; however, as in the case of the accuracy parameter, this performance is normal because the testing dataset was unknown for the model. In addition, having a similar performance allows for knowing that the model doesn't have overfitting problems and that it manages to classify unknown data in a similar way as it classifies known data. The final value obtained for the accuracy had an average of 0.59. This value isn't close to the ideal, which is zero; however, it is possible to observe that the data remains tending toward zero for both datasets, showing that the testing dataset decreases slower than the training dataset. This problem can be solved increasing the amount of data, removing features that are not significant or that are redundant for the model and increasing the size if the ANN or the number of epochs.

Although the graph of the loss function doesn't present very favorable results, it allows for knowing that the modeling of the ANN is robust among the two datasets, in addition to continuing to decrease the loss function and reducing the error through the epochs, which is one of the main purposes of the training stage.

Finally, Figure 7 shows that all curves presented statistically significant values, obtaining AUC values ≥ 0.69. These curves are generated based on the proportion of true positives and true negatives, where for class "0", class "1" and the macro-average, an AUC value of 0.69 was obtained, which implies that 69% of the subjects were classified appropriately for each of the classes and for their general average, while. for the curve generated for the micro-average, an AUC value of 0.75 was obtained, which implies that 75% of the subjects were classified appropriately when the measure was based in subsets of data. The ROC curves of class "0" and class "1" presented a very similar performance, besides obtaining the same AUC value, which means that the classification performance is equitable and the generalization of the model allows for classifying subjects of both classes.

On the other hand, the AUC value is greater for the micro-average than for the macro-average; this may be because the calculation of the macro-average is based in obtaining the true positives/true negatives proportion through the whole amount of data, while, for the micro-average, the calculation of the ROC curve is based in subsets of data, where some subsets can present better AUC values than obtained with the whole set, thus achieving a better average of classification.

Is it important to remark that this generalized model presents an impartial performance, since it was trained with data that cover a reasonable spectrum of the possible outcomes; in addition, the moment of time and geographical location are parameters that were taken into account in the initial data acquisition; regardless of whether the data was acquired in the USA, most of the subjects were from other regions, of which a percentage belongs to Mexican subjects, providing robustness in the results and avoiding any bias problem related to the structure, the moment of time and the geographical location of the data.

Finally, the results obtained allow for supporting the main motivation of this work, which was to develop a tool for human resources and dental services, since this model is able to give a statistically significant response of the possible development of caries based on information that doesn't require any extra equipment for its acquisition, only the willingness of people to answer the demographic and diet surveys, providing support to the dental specialists in the work of reducing the high incidence of dental caries

Based on these discussions, it can be proposed as future work to add a stage in the methodology that consists of a step for the feature selection using a genetic algorithm approach. This stage will help to remove redundant information and correlated features, keeping only those features that provide the most significant information for the classification of subjects, besides reducing the computational cost. Additionally, for the validation of the feature selection, a forward selection and backward elimination may be used in order to test the accuracy of the set of features obtained through the selection process, ensuring the certainty of the model behavior.

5. Conclusions

According to the results obtained, it is possible to conclude that the amount of data that was used to model the classification of the subjects was adequate for the preliminary generalized learning of the ANN, basing this on the performance that the classification of cases had, being similar to the classification of controls. This performance was observed in the accuracy and loss function graphs, which allows for recognizing that the number of epochs that were assigned to the ANN was sufficient to maintain stable performance of the parameters in both datasets.

The accuracy achieved with this modeling was around 0.70, a value that is statistically significant since it implies that 70% of the time will be correctly classified to the subjects with presence or absence/restoration of caries.

The loss function decreased by approximately 10% since the beginning of the training, showing minor changes throughout the epochs, implying an approximation to the global minimum sought.

As a final evaluation, all the ROC curves obtained an AUC value statistically significant, implying that, around 70% of the time, subjects were correctly classified, according to the true positives and true negatives proportion, a value that corroborates the accuracy obtained. Nevertheless, a methodology was proposed based in convolutional neural networks in order to identify features in a spectral or spatial domain to obtain time-independent abstract features, looking for the improvement in the modeling for the classification of subjects.

Based on this, it is possible to classify subjects with the absence of caries from subjects with presence/restorations, through demographic and dietary data, with a statistically significant accuracy, demonstrating that the socioeconomic and nutritional status are important determinants in the development of caries.

Then, according to the results obtained, it was demonstrated that the demographic situation can significantly affect the prevalence of dental caries. Based on this, the analysis of oral health data from exclusively Mexican subjects is proposed, comparing those results with the ones obtained in this work, with the purpose of proving how demography can influence the oral health status.

On the other hand, even though the results do not show an ideal behavior, they give preliminary knowledge of the benefit that its implementation would have in a real environment, since the requirements for its use are minimal, no in-depth knowledge of the techniques used is required, results are presented quickly and the computational cost is very low, besides the accuracy obtained being statistically significant. In addition, an important point to take into account is that the performed experimental tests were based in a real production environment, since all the subjects that were part for the development of this work were real control and cases, and the demographic and dietary databases were obtained from real information.

The hardware tool that is required for the implementation of this work is a computer, while the free software tool that is required is Python, which is a programming language that allows for working quickly and integrating systems more effectively. As extra information, it is mentioned that the computer that was used for the development of this work was a laptop Acer Aspire F5-573-70LX 15.6" (Acer America Corporation, 333 West San Carlos Street, Suite 1500, San Jose, CA 95110), Intel Core i7-7500U 2.70 GHz (Plot 6, Bayan Lepas Technoplex Medan Bayan Lepas 11900 Bayan Lepas Penang,

Malaysia, Georgetown, Pulau Pinang, Malasia), 16 GB, 1 TB + 128 GB Solid State Drive, Windows 10 Home (15010 NE 36th Street, Microsoft Campus, Building 92, Redmond, WA 98052), 64-bit; and the version of Python that was used is 3.6 [40].

Finally, based on the last point, it is evident that the implementation of the model proposed in this work could provide an easy, free and fast tool that helps the specialists in the preventive diagnosis of dental caries, besides offering an option that collaborates in the decrement of the incidence of this public health problem.

Author Contributions: C.E.G.-T. and L.A.Z.-C., performed the study. C.E.G.-T., L.A.Z.-C., and J.I.G.-T. performed the study design and data analysis. N.M.C.-L. and J.R.-G. contributed to materials and methods (selection of patients from database) used in this study. C.E.G.-T., J.I.G.-T., and J.M.C.-P. performed statistical analysis and statistical validation with critical feedback to authors. J.M.C.-P., R.M.-Q. and L.A.Z.-C. contributed with the neural network implementation used in this study. L.A.Z.-C. and H.G.-R. provide feedback from results. All authors interpreted findings from the analysis and drafted the manuscript.

Funding: This research received no external funding.

Conflicts of Interest: The authors declare no conflict of interest.

Appendix A

Table A1. Description of demographic features.

Feature	Description
AIALANGA	Language of the MEC ACASI Interview Instrument.
DMDBORN4	In what country was Sample Person (SP) born?
DMDCITZN	Is SP a citizen of the United States?
DMDHHSIZ	Total number of people in the Household.
DMDHHSZE	Number of adults aged 60 years or older in the household.
DMDEDUC2	Highest grade or level of school completed or the highest degree received.
DMDHHSZB	Number of children aged 6–17 years old in the household (HH).
DMDHSEDU	HH reference person's spouse's education level.
DMDMARTL	Marital status.
DMDFMSIZ	Total number of people in the Family.
DMDHHSZA	Number of children aged 5 years or younger in the household.
DMDHRGND	HH reference person's gender.
DMDHREDU	HH reference person's education level.
DMDEDUC3	Highest grade or level of school completed or the highest degree received.
DMDHRAGE	HH reference person's age in years.
DMDHRBR4	HH reference person's country of birth.
DMDHREDU	HH reference person's education level.
DMDHRMAR	HH reference person's marital status.
DMQADFC	Did SP ever serve in a foreign country during a time of armed conflict or on a humanitarian or peace-keeping mission?
DMDYRSUS	Length of time the participant has been in the US.
DMQMILIZ	Has SP ever served on active duty in the U.S. Armed Forces, military Reserves, or National Guard?
FIAINTRP	Was an interpreter used to conduct the Family interview?
FIALANG	Language of the Family Interview Instrument.
FIAPROXY	Was a Proxy respondent used in conducting the Family Interview?
INDFMPIR	A ratio of family income to poverty guidelines.
INDHHIN2	Total household income (reported as a range value in dollars).
INDFMIN2	Total family income (reported as a range value in dollars).
LATVPSU	Variance unit: PSU variable for variance estimation.
LATVSTRA	Variance unit: stratum variable for variance estimation.
MIAINTRP	Was an interpreter used to conduct the MEC CAPI interview?
MIALANG	Language of the MEC CAPI Interview Instrument.
MIAPROXY	Was a Proxy respondent used in conducting the MEC CAPI Interview?
RIAGENDR	Gender of the participant.
RIDAGEEX	Age in years of the participant at the time of examination. Individuals aged 959 months and older are topcoded at 959 months.

Table A1. *Cont.*

Feature	Description
RIDAGEMN	ge in months of the participant at the time of screening. Reported for persons aged 24 months or younger at the time of exam.
RIDRETH3	Recode of reported race and Hispanic origin information, with Non-Hispanic Asian Category.
RIDEXMON	Six month time period when the examination was performed.
RIDEXPRG	Pregnancy status for females between 20 and 44 years of age at the time of MEC exam.
RIDAGEYR	Age in years of the participant at the time of screening.
RIDRETH1	Recode of reported race and Hispanic origin information.
RIDEXAGM	Age in months of the participant at the time of examination.
RIDSTATR	Interview and examination status of the participant.
SDMVPSU	Masked variance unit pseudo-PSU variable for variance estimation.
SDDSRVYR	Data release cycle.
SDMVSTRA	Masked variance unit pseudo-stratum variable for variance estimation.
WTLAF8YR	Subsample 8-year fasting weight for participants aged 12 years and older who were examined in the morning sessions in Los Angeles, CA, USA.
WTLAI8YR	Full sample 8-year interview weight for participants in Los Angeles, CA, USA.
WTLAM8YR	Full sample 8-year MEC exam weight for participants in Los Angeles, CA, USA.

Appendix B

Table A2. Description of dietary features.

Feature	Description
WTDRD1	Dietary day one sample weight.
DBQ095Z	Type of salt usually add to food at the table.
DR1TKCAL	Energy (kcal).
DR1TSUGR	Total sugars (g).
DR1TFIBE	Dietary fiber (g).
DR1TSFAT	Total saturated fatty acids (g).
DR1TMFAT	Total monounsaturated fatty acids (g).
DR1TPFAT	Total polyunsaturated fatty acids (g).
DR1TLYCO	Lycopene (μg).
DR1TFA	Folic acid (μg).
DR1TB12A	Added vitamin B12 (μg).
DR1_300	Was the amount of food that you ate yesterday much more than usual, usual, or much less than usual?
DR1_320Z	Total plain water drank yesterday—including plain tap water, water from a drinking fountain, water from a water cooler, bottled water, and spring water.
DR1_330Z	Total tap water drank yesterday—including filtered tap water and water from a drinking fountain.
DR1BWATZ	Total bottled water drank yesterday (g).
DR1DAY	Intake day of the week.
DR1DBIH	Number of days between intake day and the day of family questionnaire administered in the household.
DR1SKY	What type of salt was it? (Was it ordinary or seasoned salt, lite salt, or a salt substitute?).
DR1STY	Did SP add any salt to her/his food at the table yesterday?
DR1TACAR	Alpha-carotene (μg).
DR1TBCAR	Beta-carotene (μg).
DR1TCAFF	Caffeine (mg).
DR1TCALC	Calcium (mg).
DR1TCOPP	Copper (mg).
DR1TCRYP	Beta-cryptoxanthin (μg).
DR1TFDFE	Folate as dietary folate equivalents (μg).
DR1TFF	Food folate (μg).
DR1TFOLA	Total folate (μg).
DR1TLZ	Lutein + zeaxanthin (μg).
DR1TM161	MFA 16:1 (Hexadecenoic) (g).
DR1TM181	MFA 18:1 (Octadecenoic) (g).
DR1TM201	MFA 20:1 (Eicosenoic) (g).
DR1TM221	MFA 22:1 (Docosenoic) (g).
DR1TMAGN	Magnesium (mg).
DR1TMOIS	Moisture (g).
DR1TNIAC	Niacin (mg).

Table A2. *Cont.*

Feature	Description
DR1TNUMF	Total number of foods/beverages reported in the individual foods file.
DR1TP182	PFA 18:2 (Octadecadienoic) (g).
DR1TP183	PFA 18:3 (Octadecatrienoic) (g).
DR1TP184	PFA 18:4 (Octadecatetraenoic) (g).
DR1TP204	PFA 20:4 (Eicosatetraenoic) (g).
DR1TP205	PFA 20:5 (Eicosapentaenoic) (g).
DR1TP225	PFA 22:5 (Docosapentaenoic) (g).
DR1TP226	PFA 22:6 (Docosahexaenoic) (g).
DR1TPOTA	Potassium (mg).
DR1TRET	Retinol (μg).
DR1TS040	SFA 4:0 (Butanoic) (g).
DR1TS060	SFA 6:0 (Hexanoic) (g).
DR1TS100	SFA 10:0 (Decanoic) (g).
DR1TS120	SFA 12:0 (Dodecanoic) (g).
DR1TS140	SFA 14:0 (Tetradecanoic) (g).
DR1TS180	SFA 18:0 (Octadecanoic) (g).
DR1TSELE	Selenium (μg).
DR1TSODI	Sodium (mg).
DR1TTHEO	Theobromine (mg).
DR1TVARA	Vitamin A as retinol activity equivalents (μg).
DR1TVB1	Thiamin (Vitamin B1) (mg).
DR1TVB12	Vitamin B12 (μg).
DR1TVB2	Riboflavin (Vitamin B2) (mg).
DR1TVB6	Vitamin B6 (mg).
DR1TVC	Vitamin C (mg).
DR1TVK	Vitamin K (μg).

Table A3. Description of dietary features.

Feature	Description
DR1TWS	When you drink tap water, what is the main source of the tap water? Is the city water supply; a well or rain cistern; a spring; or something else?
DR1TZINC	Zinc (mg).
DRABF	Indicates whether the sample person was an infant who was breast-fed on either of the two recall days.
DRD340	During the past 30 days did you eat any types of shellfish listed on this card?
DRD350A	Clams eaten during the past 30 days.
DRD350AQ	Number of times clams were eaten in the past 30 days.
DRD350B	Crabs eaten during the past 30 days.
DRD350BQ	Number of times crab was eaten in the past 30 days.
DRD350C	Crayfish eaten during the past 30 days.
DRD350CQ	Number of times crayfish was eaten in the past 30 days.
DRD350D	Lobsters eaten during the past 30 days.
DRD350DQ	Number of times lobster was eaten in the past 30 days.
DRD350E	Mussels eaten during the past 30 days.
DRD350EQ	Number of times mussels were eaten in the past 30 days.
DRD350F	Oysters eaten during the past 30 days.
DRD350FQ	Number of times oysters were eaten in the past 30 days.
DRD350G	Scallops eaten during the past 30 days.
DRD350GQ	Number of times scallops were eaten in the past 30 days.
DRD350H	Shrimp eaten during the past 30 days.
DRD350HQ	Number of times shrimp was eaten in the last 30 days.
DRD350I	Other shellfish (ex. octopus, squid) eaten during the past 30 days.
DRD350IQ	Number of times other shellfish (ex. octopus, squid) was eaten in the past 30 days.
DRD350J	Other unknown shellfish eaten during the past 30 days.
DRD350JQ	Number of times other unknown shellfish was eaten in the past 30 days.
DRD350K	Refused to give detailed information on shellfish eaten during the past 30 days.
DRD360	During the past 30 days did you eat any types of fish listed on this card?
DRD370A	Breaded fish products eaten during the past 30 days.
DRD370AQ	Number of times breaded fish products were eaten in the past 30 days.
DRD370B	Tuna eaten during the past 30 days.

Table A3. *Cont.*

Feature	Description
DRD370BQ	Number of times tuna was eaten in the past 30 days.
DRD370C	Bass eaten during the past 30 days.
DRD370CQ	Number of times bass was eaten in the past 30 days.
DRD370D	Catfish eaten during the past 30 days.
DRD370DQ	Number of times catfish was eaten in the past 30 days.
DRD370E	Cod eaten during the past 30 days.
DRD370EQ	Number of times cod was eaten in the past 30 days.
DRD370F	Flatfish eaten during the past 30 days.
DRD370FQ	Number of times flatfish was eaten in the past 30 days.
DRD370G	Haddock eaten during the past 30 days.
DRD370GQ	Number of times haddock was eaten in the past 30 days.
DRD370H	Mackerel eaten during the past 30 days.
DRD370HQ	Number of times mackerel was eaten in the past 30 days.
DRD370I	Perch eaten during the past 30 days.
DRD370IQ	Number of times perch was eaten in the past 30 days.
DRD370J	Pike eaten during the past 30 days.
DRD370JQ	Number of times pike was eaten in the past 30 days.
DRD370K	Pollock eaten during the past 30 days.
DRD370KQ	Number of times pollock was eaten in the past 30 days.
DRD370TQ	Number of times other type of fish was eaten in the past 30 days.
DRD370V	Refused to give detailed information on fish eaten during the past 30 days.
DRQSDIET	Are you currently on any kind of diet, either to lose weight or for some other health-related reason?
DRQSDT1	What kind of diet are you on? (Is it a weight loss or a low calorie diet: low fat or cholesterol diet; low salt or sodium diet; sugar free or low sugar diet; low fiber diet; high fiber diet; diabetic diet; or another type of diet?).
DRD370L	Porgy eaten during the past 30 days.
DRD370LQ	Number of times porgy was eaten in the past 30 days.
DRD370M	Salmon eaten during the past 30 days.
DRD370MQ	Number of times salmon was eaten in the past 30 days.
DRD370N	Sardines eaten during the past 30 days.
DRD370NQ	Number of times sardines were eaten in the past 30 days.
DRD370O	Sea bass eaten during the past 30 days.
DRD370OQ	Number of times sea bass was eaten in the past 30 days.
DRD370U	Other unknown type eaten during the past 30 days.

Table A4. Description of dietary features.

Feature	Description
WTDR2D	Dietary two-day sample weight.
DR1DRSTZ	Dietary recall status.
DR1TPROT	Protein (g).
DR1TCARB	Carbohydrate (g).
DR1TTFAT	Total fat (g).
DR1TCHOL	Cholesterol (mg).
DR1TFA	Folic acid (µg).
DR1TCHL	Total choline (mg).
DR1TIRON	Iron (mg).
DBD100	Frequency with which ordinary salt is added to the food on the table.
DRQSPREP	Frequency with which ordinary salt or seasoned salt is added in cooking or preparing foods in the household.
DR1TALCO	Alcohol (g).
DR1TS080	SFA 8:0 (Octanoic) (g).
DR1TATOC	Vitamin E as alpha-tocopherol (mg).
DR1TATOA	Added alpha-tocopherol (Vitamin E) (mg).
DR1TVD	Vitamin D (D2 + D3) (µg).
DR1TPHOS	Phosphorus (mg).
DR1TS160	SFA 16:0 (Hexadecanoic) (g).
DRD370UQ	Number of times other unknown type of fish was eaten in the past 30 days.
DRD370V	Refused to give detailed information on fish eaten during past 30 days.
DRDINT	Indicates whether the sample person has intake data for one or two days.
DRQSDIET	Are you currently on any kind of diet, either to lose weight or for some other health-related reason?

References

1. Oral Health. Available online: http://www.who.int/oral_health/disease_burden/global/en/ (accessed on 5 June 2017).

2. Ridao Marín, D. *Desarrollo de un Sistema de Ayuda a la Decisión para Tratamientos Odontológicos con Imágenes Digitales*; Universidad de Málaga: Málaga, Spain, 2017; pp. 10–12.

3. Espinoza Solano, M.; León-Manco, R.A. Prevalencia y experiencia de caries dental en estudiantes según facultades de una universidad particular peruana. *Rev. Estomatol. Hered.* **2015**, *25*, *3*, 187–193. [CrossRef]

4. Acuña Aguilar, L.D.; Porras Cerón, D.; Ríos Rueda, L.D. *Prevalencia de Lesiones Cariosas y Factores Asociados Presentes en Pacientes con SíNdrome de Down en las Fundaciones Fundown y san Luis Guanella de Bucaramanga*; Universidad Santo Tomás: Bucaramanga, Colombia, 2017; pp. 12–16.

5. Gispert Abreu, E.D.L.Á.; Castell-Florit Serrate, P.; Herrera Nordet, M. Salud bucal poblacional y su producción intersectorial. *Rev. Cubana Estomatol.* **2015**, *52*, 62–67.

6. Niño, T.C.; Guevara, S.V.; González, F.A.; Jaque, R.A.; Infante, C. Uso de redes neuronales articiales en predicción de morfología mandibular a través de variables craneomaxilares en una vista posteroanterior. *Univ. Odontol.* **2016**, *35*, 1–28.

7. Lam, C.U.; Khin, L.; Kalhan, A.; Yee, R.; Lee, Y.; Chong, M.F.; Kwek, K.; Saw, S.; Godfrey, K.; Chong, Y.; et al. Identification of Caries Risk Determinants in Toddlers: Results of the GUSTO Birth Cohort Study. *Caries Res.* **2017**, *51*, 271–282. [CrossRef] [PubMed]

8. Fernandes, I.; Sá-Pinto, A.; Marques, L.S.; Ramos-Jorge, J.; Ramos-Jorge, M. Maternal identification of dental caries lesions in their children aged 1–3 years. *Eur. Arch. Paediatr. Dent.* **2017**, *18*, 197–202. [CrossRef] [PubMed]

9. Lips, A.; Antunes, L.S.; Antunes, L.A.; Pintor, A.V.B.; Santos, D.A.B.D.; Bachinski, R.; Küchler, E.C.; Alves, G.G. Salivary protein polymorphisms and risk of dental caries: A systematic review. *Braz. Oral Res.* **2017**, *31*. [CrossRef] [PubMed]

10. Ahmed, A.; Ikram, K.; Masood, H.; Urooj, M. Identification of relationship between oral disorders & hemodynamic parameters. *Pak. Oral Dent. J.* **2017**, *37*, 202–204.

11. Sarmiento, R.V.; Barrionuevo, F.P.; Huamán, Y.S.; Loyola, M.C. Prevalencia de caries de infancia temprana en niños menores de 6 años de edad, residentes en poblados urbano marginales de Lima Norte. *Rev. Estomatol. Herediana* **2011**, *21*, 79–86. [CrossRef]

12. Oropeza-Oropeza, C.D.; Molina-Frechero, N.; Castañeda-Castaneira, E.; Zaragoza-Rosado, C.D.; Cruz Leyva, C.D. Caries dental en primeros molares permanentes de escolares de la delegación Tláhuac. *Rev. ADM* **2012**, *69*, 63–68.

13. Cardozo, B.J.; Gonzalez, M.M.; Pérez, S.R.; Vaculik, P.A.; Sanz, E.G. Epidemiología de la caries dental en niños del Jardín de Infantes "Pinocho" de la ciudad de Corrientes. *Rev. Fac. Odontol.* **2017**, *9*, *1*, 35–41.

14. Zhang, X.; Zhang, L.; Zhang, Y.; Liao, Z.; Song, J. Predicting trend of early childhood caries in mainland China: A combined meta-analytic and mathematical modelling approach based on epidemiological survey. *Sci. Rep.* **2017**, *7*, 6507. [CrossRef] [PubMed]

15. Asif, S.M.; Babu, D.B.; Naheeda, S. Utility of Dermatoglyphic Pattern in Prediction of Caries in Children of Telangana Region, India. *J. Contemp. Dent. Pract.* **2017**, *18*, 490–496. [CrossRef] [PubMed]

16. Chapple, I.L.; Bouchard, P.; Cagetti, M.G.; Campus, G.; Carra, M.C.; Cocco, F.; Nibali, L.; Hujoel, P.; Laine, M.L.; Lingstrom, P.; et al. Interaction of lifestyle, behaviour or systemic diseases with dental caries and periodontal diseases: Consensus report of group 2 of the joint EFP/ORCA workshop on the boundaries between caries and periodontal diseases. *J. Clin. Periodontol.* **2017**, *44*, s39–s51. [CrossRef] [PubMed]

17. Hayes, M.; Da Mata, C.; Cole, M.; McKenna, G.; Burke, F.; Allen, P.F. Risk indicators associated with root caries in independently living older adults. *J. Dent.* **2016**, *51*, 8–14 [CrossRef] [PubMed]

18. Sudhir, K.M.; Kanupuru, K.K.; Fareed, N.; Mahesh, P.; Vandana, K.; Chaitra, N.T. CAMBRA as a Tool for Caries Risk Prediction Among 12-to 13-year-old Institutionalised Children-A Longitudinal Follow-up Study. *Oral Health Prev. Dent.* **2016**, *14*, 355–362. [PubMed]

19. Twetman, S. Caries risk assessment in children: How accurate are we? *Eur. Arch. Paediatr. Dent.* **2016**, *17*, 27–32. [CrossRef] [PubMed]

20. National Health and Nutrition Examination Survey Data. 2013–2014. Available online: http://www.cdc.gov/nchs/nhanes.htm (accessed on 5 November 2017).

21. Zanella-Calzada, L.A.; Galván-Tejada, C.E.; Chávez-Lamas, N.M.; Gracia-Cortés, M.D.C.; Moreno-Báez, A.; Arceo-Olague, J.G.; Celaya-Padilla, J.M.; Galván-Tejada, J.I.; Gamboa-Rosales, H. A Case–Control Study of Socio-Economic and Nutritional Characteristics as Determinants of Dental Caries in Different Age Groups, Considered as Public Health Problem: Data from NHANES 2013–2014. *Int. J. Environ. Res. Public Health* **2018**, *15*, 957. [CrossRef] [PubMed]

22. Zanella-Calzada, L.A.; Galván-Tejada, C.E.; Chávez-Lamas, N.M.; Galván-Tejada, J.I.; Celaya-Padilla, J.M. Multivariate features selection from demographic and dietary descriptors as caries risk determinants in oral health diagnosis: Data from NHANES 2013–2014. In Proceedings of the International Conference on Electronics, Communications and Computers (CONIELECOMP), Cholula, Mexico, 21–23 Feburary 2018; pp. 217–224.

23. Chávez-Lamas, N.M.; Zanella-Calzada, L.A.; Galván-Tejada, C.E. An Analysis of Dietary and Demographic Data in Oral Health, Data from the National Health and Nutrition Examination Survey: A Preliminary Study. *Adv. Comput. Netw. Appl.* **2017**, *142*, 79–88.

24. Liaw, A.; Wiener, M. Classication and regression by randomforest. *R News* **2002**, *2*, 18–22.

25. Francois Chollet. Keras: Deep Learning Library for Theano and Tensorflow. pp. 145–151. Available online: https://keras.io/k (accessed on 5 May 2018).

26. Lomuscio, A.; Maganti, L. An approach to reachability analysis for feed-forward relu neural networks. *arXiv* **2017**. [CrossRef]

27. Carlini, N.; Wagner, D. Towards evaluating the robustness of neural networks. In Proceedings of the IEEE Symposium on Security and Privacy (SP), San Jose, CA, USA, 22–26 May 2017; pp. 39–57.

28. Kingma, D.P.; Ba, J. Adam: A method for stochastic optimization. *arXiv* **2014**. [CrossRef]

29. Haykin, S.S. *Neural Networks and Learning Machines*; Pearson: Upper Saddle River, NJ, USA, 2009; Volume 3, pp. 1–46.

30. Chollet, F. *Deep Learning with Python*; Manning Publications Co.: Greenwich, CT, USA, 2017; Volume 1, pp. 1–93.

31. Antona Cortés, C. Herramientas Modernas en Redes Neuronales: La LibreríA Keras. Bachelor's Thesis, UAM, Departamento de Ingeniería Informática, Madrid, Spain, 2017; pp. 21–38.

32. Helene Bischel, S. El método de la EntropíA Cruzada. Algunas Aplicaciones. Master's Thesis, Universidad de Almería, Almería, Spain, 2015; pp. 4–14.

33. Nye, M.; Saxe, A. *Are Efficient Deep Representations Learnable?* UAM, Departamento de Ingeniería Informática: Madrid, Spain, 2017; pp. 1–4.

34. Lobo, J.M.; Jiménez-Valverde, A.; Real, R. AUC: A misleading measure of the performance of predictive distribution models. *Glob. Ecol. Biogeogr.* **2008**, *17*, 145–151. [CrossRef]

35. Hanley, J.A.; McNeil, B.J. The meaning and use of the area under a receiver operating characteristic (ROC) curve. *Radiology* **1982**, *143*, 29–36. [CrossRef] [PubMed]

36. Pedregosa, F.; Varoquaux, G.; Gramfort, A.; Michel, V.; Thirion, B.; Grisel, O.; Blondel, M.; Prettenhofer, P.; Weiss, R.; Dubourg, V.; et al. Scikit-learn: Machine learning in Python. *J. Mach. Learn. Res.* **2011**, *12*, 2825–2830.

37. Rosa, K.D.; Shah, R.; Lin, B.; Gershman, A.; Frederking, R. Topical clustering of tweets. In Proceedings of the ACM SIGIR: SWSM, Beijing, China, 28 July 2011.

38. Jones, E.; Oliphant, T.; Peterson, P. SciPy: Open Source Scientific Tools for Python. Available online: https://docs.scipy.org/doc/scipy (accessed on 5 May 2018).

39. Wes McKinney. Data Structures for Statistical Computing in Python. In Proceedings of the 9th Python in Science Conference, Austin, TX, USA, 28 June–3 July 2010; pp. 51–56.

40. Python Software Foundation. Python Language Reference, Version 3.6. Available online: http://www.python.org (accessed on 5 June 2018).

PERMISSIONS

LIST OF CONTRIBUTORS

Federica Poli, Stefano Selleri and Annamaria Cucinotta
Department of Engineering and Architecture, University of Parma, Parco Area delle Scienze 181/A, 43124 Parma, Italy

Carlo Fornaini
Department of Engineering and Architecture, University of Parma, Parco Area delle Scienze 181/A, 43124 Parma, Italy
Micoralis Laboratory, Faculty of Dentistry, University of Cote d'Azur, 24 Avenue des Diables Bleus, 06357 Nice, France

Elisabetta Merigo, Nathalie Brulat-Bouchard, Ahmed El Gamal and Jean-Paul Rocca
Micoralis Laboratory, Faculty of Dentistry, University of Cote d'Azur, 24 Avenue des Diables Bleus, 06357 Nice, France

Erica L. Rex and Karen C. T. Arlt
Biomedical Engineering and McCaig Institute for Bone and Joint Health, University of Calgary, Calgary, AB T2N 1N4, Canada

Carolyn Anglin
Biomedical Engineering and McCaig Institute for Bone and Joint Health, University of Calgary, Calgary, AB T2N 1N4, Canada
Department of Civil Engineering, University of Calgary, Calgary, AB T2N 1N4, Canada

Cinzia Gaudelli
Section of Orthopaedic Surgery, Cumming School of Medicine, University of Calgary, Calgary, AB T2N 1N4, Canada

Emmanuel M. Illical
Department of Orthopaedic Surgery, SUNY Downstate Medical Center, Brooklyn, NY 11203, USA

John Person
Box 13 Engineering, Calgary, AB T3L 2P5, Canada

BarryWylant
Q Industrial Design Corporation, Calgary, AB T2T 0E7, Canada

Nádia S. Parachin
Grupo de Engenharia Metabólica Aplicada a Bioprocessos, Instituto de Ciências Biológicas, Universidade de Brasília, CEP 70.790-900 Brasília-DF, Brazil

Nadiele T. M. Melo
Grupo de Engenharia Metabólica Aplicada a Bioprocessos, Instituto de Ciências Biológicas, Universidade de Brasília, CEP 70.790-900 Brasília-DF, Brazil
Pós-Graduação em Ciências Genômicas e Biotecnologia, Universidade Católica de Brasília, CEP 70.790-900 Brasília-DF, Brazil

Lucas S. Carvalho
Grupo de Engenharia Metabólica Aplicada a Bioprocessos, Instituto de Ciências Biológicas, Universidade de Brasília, CEP 70.790-900 Brasília-DF, Brazil
Integra Bioprocessos e Análises, Campus Universitário Darcy Ribeiro, Edifício CDT, Sala AT-36/37, CEP 70.790-900 Brasília-DF, Brazil

Kelly C. L. Mulder, Gisele S. Menino and Eduardo Mulinari
Integra Bioprocessos e Análises, Campus Universitário Darcy Ribeiro, Edifício CDT, Sala AT-36/37, CEP 70.790-900 Brasília-DF, Brazil

André Moraes Nicola
Faculty of Medicine, University of Brasilia, Campus Universitário Darcy Ribeiro, Faculdade de Medicina, Sala BC-103, CEP 70.790-900 Brasília-DF, Brazil

Firouz Abbasian, Ebrahim Ghafar-Zadeh and Sebastian Magierowski
Biologically Inspired Sensors and Actuators Laboratory, Department of EECS, Lassonde School of Engineering, York University, Toronto, ON M3J 1P3, Canada

Sameer Joshi, Atul A. Chaudhari, Vida Dennis and Shree Ram Singh
Center for NanoBiotechnology Research, Alabama State University, Montgomery, AL 36016, USA

Daniel J. Kirby
Aston Pharmacy School, Life and Health Sciences, Aston University, Birmingham B4 7ET, UK

Yvonne Perrie
Strathclyde Institute of Pharmacy and Biomedical Sciences, University of Strathclyde, 161 Cathedral Street, Glasgow G4 0RE, UK

Patrick S. Doyle
Department of Chemical Engineering, Massachusetts Institute of Technology, Cambridge, MA 02139, USA

George E. Kapellos
Department of Chemical Engineering, Massachusetts Institute of Technology, Cambridge, MA 02139, USA
School of Environmental Engineering, Technical University of Crete, 73100 Chania, Greece

Nicolas Kalogerakis
School of Environmental Engineering, Technical University of Crete, 73100 Chania, Greece

Christakis A. Paraskeva
Department of Chemical Engineering, University of Patras, 26504 Rion Achaia, Greece

Claire Mayer-Laigle, Nicolas Blanc, Rova Karine Rajaonarivony and Xavier Rouau
UMR Ingénierie des Agropolymères et des Technologies Emergentes (IATE), University of Montpellier, CIRAD, INRA, Montpellier SupAgro, Montpellier, France

Martin Kornecki and Jochen Strube
Institute for Separation and Process Technology, Clausthal University of Technology, Leibnizstr. 15, 38678 Clausthal-Zellerfeld, Germany

Zeljka Rupcic, Stephan Hüttel, Steffen Bernecker and Marc Stadler
Department Microbial Drugs, Helmholtz Centre for Infection Research GmbH, Inhoffenstraße 7, 38124 Braunschweig, Germany
German Centre for Infection Research (DZIF), partner site Hannover-Braunschweig, 38124 Braunschweig, Germany

Sae Kanaki
Toyama Prefectural University, 5180 Kurokawa Imizu-shi, Toyama 939-0398, Japan

Amit Kumar, Wai Teng Yap, Soo Leong Foo and Teck Kheng Lee
College Central, Institute of Technical Education, 2 Ang Mo Kio Drive, Singapore 567720, Singapore

Luciane Bastistella and Antonio Aviz
Faculty of Mining and Environmental Engineering – Femma, Federal University of South and Southwest of Pará, Marabá 68507-590, PA, Brazil

Patrick Rousset
Joint Graduate School of Energy and Environment, Center of Excellence on Energy Technology and Environment-KMUTT, Bangkok 10140, Thailand
CIRAD, UPR BioWooEB, F-34398 Montpellier, France

Armando Caldeira-Pires
Department of Mechanical Engineering, University of Brasilia – UnB, Campus Universitário Darcy Ribeiro, S/N, Asa Norte, Brasília 70910-900, DF, Brazil

Gilles Humbert
FerroPem, R & D Department, F-73025 Chambéry, France

Manoel Nogueira
Faculty of Mechanical Engineering – Fem, Federal University of Pará, Belém 66075-900, PA, Brazil

Laura A. Zanella-Calzada, Carlos E. Galván-Tejada, Rafael Magallanes-Quintanar, Jorge I. Galván-Tejada and Hamurabi Gamboa-Rosales
Unidad Académica de Ingeniería Eléctrica, Universidad Autónoma de Zacatecas, Jardín Juarez 147, Centro, Zacatecas 98000, Zac, México

Nubia M. Chávez-Lamas and Jesús Rivas-Gutierrez
Unidad Académica de Odontología, Universidad Autónoma de Zacatecas, Jardín Juarez 147, Centro, Zacatecas 98000, Zac, México

Jose M. Celaya-Padilla
CONACYT – Universidad Autónoma de Zacatecas – Jardín Juarez 147, Centro, Zacatecas 98000, Zac, Mexico

Index